METHODS IN MOLECULAR BIOLOGY

Series Editor
John M. Walker
School of Life and Medical Sciences
University of Hertfordshire
Hatfield, Hertfordshire, AL10 9AB, UK

For further volumes:
http://www.springer.com/series/7651

Adipose-Derived Stem Cells

Methods and Protocols

Second Edition

Edited by

Bruce A. Bunnell

Center for Stem Cell Research and Regenerative Medicine, Tulane University School of Medicine, New Orleans, LA, USA

Jeffrey M. Gimble

Center for Stem Cell Research and Regenerative Medicine, Tulane University School of Medicine, New Orleans, LA, USA; LaCell LLC, New Orleans, LA, USA

Humana Press

Editors
Bruce A. Bunnell
Center for Stem Cell Research and Regenerative Medicine
Tulane University School of Medicine
New Orleans, LA, USA

Jeffrey M. Gimble
Center for Stem Cell Research and Regenerative Medicine
Tulane University School of Medicine
New Orleans, LA, USA

LaCell LLC
New Orleans, LA, USA

ISSN 1064-3745 ISSN 1940-6029 (electronic)
Methods in Molecular Biology
ISBN 978-1-4939-7797-0 ISBN 978-1-4939-7799-4 (eBook)
https://doi.org/10.1007/978-1-4939-7799-4

Library of Congress Control Number: 2018938174

Cover caption: Cover image courtesy of Rachel Sabol.

Printed on acid-free paper

This Humana Press imprint is published by the registered company Springer Science+Business Media, LLC part of Springer Nature.
The registered company address is: 233 Spring Street, New York, NY 10013, U.S.A.

Preface

Adipose stromal/stem cells (ASC) remain of intense interest in the field of regenerative medicine for their potent differentiation, immunomodulatory potential, and paracrine actions via their secretome. Over the intervening time since the publication of the first edition of this book, the use of the stroma vascular fraction (SVF) as a potential therapeutic strategy has gained momentum. The SVF is obtained from processed adipose tissue and is composed of a heterogeneous mesenchymal population of cells that includes not only adipose stromal and hematopoietic stem and progenitor cells but also endothelial cells, erythrocytes, fibroblasts, lymphocytes, monocyte/macrophages, and pericytes. The SVF has been demonstrated to possess both anti-inflammatory and differentiation potential. In this second volume of *Methods in Molecular Biology* focused on ASC and SVF, we have again obtained well-defined and thoroughly vetted protocols from leaders in the field. The chapters are focused on protocols centered around the discovery, preclinical, and clinical processes. Emphasis was placed on human ASC again; however, additional small and large animal species with preclinical relevance are included. Many of the protocols are new to this edition and represent scientific advances in the field of ASC biology and application, while others represent refinements and enhancements in previously published protocols.

The editors have many people to thank for their contributions to this work: First and foremost our colleagues who were kind enough to find time in their busy schedules to author the chapters found in this edition. We would also like to thank Professor John Walker, Series Editor, and the editorial staff at Springer whose direction, advice, and occasional prodding were greatly needed and deeply appreciated. Finally, we would also like to thank our families for permitting us the time required on nights and weekends to complete this work.

New Orleans, LA, USA

Bruce A. Bunnell
Jeffrey M. Gimble

Contents

Contributors

JOEL A. ARONOWITZ · *Cedars-Sinai Medical Center, Los Angeles, CA, USA; University Stem Cell Center, Los Angeles, CA, USA*
JOSIE C. BODLE · *Joint Department of Biomedical Engineering, North Carolina State University and University of North Carolina Chapel Hill, Raleigh, NC, USA*
ANNIE C. BOWLES · *Department of Cell and Molecular Biology, Tulane University School of Science and Engineering, New Orleans, LA, USA; Center for Stem Cell Research and Regenerative Medicine, Tulane University School of Medicine, New Orleans, LA, USA*
CODY BROWN · *Department of Anatomy and Cell Biology, Schulich School of Medicine and Dentistry, The University of Western Ontario, London, ON, Canada*
BRUCE A. BUNNELL · *Center for Stem Cell Research and Regenerative Medicine, Tulane University School of Medicine, New Orleans, LA, USA; Department of Pharmacology, Tulane University School of Medicine, New Orleans, LA, USA*
VALERIO CERVELLI · *Department of Plastic and Reconstructive Surgery, University of Rome "Tor Vergata", Rome, Italy*
RODNEY K. CHAN · *United States Army Institute of Surgical Research, JBSA Fort Sam Houston, TX, USA*
NICHOLAS CLARK · *Regenerative Medicine Department, Naval Medical Research Center, Silver Spring, MD, USA*
J. LOWRY CURLEY · *Center for Stem Cell Research and Regenerative Medicine, Tulane University School of Medicine, New Orleans, LA, USA; LaCell LLC, New Orleans, LA, USA*
CHRISTIAN DANI · *Faculté de Médecine, Université Nice Sophia Antipolis, iBV, UMR CNRS/INSERM, Nice Cedex 2, France*
ALEXES C. DAQUINAG · *Brown Foundation Institute of Molecular Medicine for the Prevention of Human Diseases, University of Texas Health Science Center at Houston, Houston, TX, USA*
THOMAS A. DAVIS · *Regenerative Medicine Department, Naval Medical Research Center, Silver Spring, MD, USA; Department of Surgery, Uniformed Services University of the Health Sciences, Bethesda, MD, USA*
RAM DEVIREDDY · *Bioengineering Laboratory, Department of Mechanical Engineering, Louisiana State University, Baton Rouge, LA, USA*
MARILYN DIETRICH · *Pennington Biomedical Research Center, Louisiana State University System, Baton Rouge, LA, USA*
WEI DUAN · *Laboratory for Equine and Comparative Orthopedic Research, School of Veterinary Medicine, Louisiana State University, Baton Rouge, LA, USA*
FRANCESCO EGRO · *University of Pittsburgh, Pittsburgh, PA, USA*
MARGO EMONT · *Life Sciences Institute, University of Michigan, Ann Arbor, MI, USA; Department of Molecular and Integrative Physiology, University of Michigan, Ann Arbor, MI, USA*
CARMEL FARGASON · *Laboratory for Equine and Comparative Orthopedic Research, School of Veterinary Medicine, Louisiana State University, Baton Rouge, LA, USA*

Z. ELIZABETH FLOYD · *Ubiquitin Laboratory, Pennington Biomedical Research Center, Louisiana State University System, Baton Rouge, LA, USA*

LAUREN E. FLYNN · *Department of Anatomy and Cell Biology, Schulich School of Medicine and Dentistry, The University of Western Ontario, London, ON, Canada; Department of Chemical and Biochemical Engineering, Thompson Engineering Building, The University of Western Ontario, London, ON, Canada; Department of Chemical Engineering, Queen's University, Kingston, ON, Canada*

ANOOSHA FORGHANI · *Bioengineering Laboratory, Department of Mechanical Engineering, Louisiana State University, Baton Rouge, LA, USA*

ANTHONY D. FOSTER · *Regenerative Medicine Department, Naval Medical Research Center, Silver Spring, MD, USA; Department of Surgery, Uniformed Services University of the Health Sciences, Bethesda, MD, USA*

JULIE FRADETTE · *Centre de Recherche en Organogénèse Expérimentale de l'Université Laval/ LOEX, CRCHU de Québec-Université Laval, Québec, QC, Canada; Department of Surgery, Faculty of Medicine, Université Laval, Québec, QC, Canada; CMDGT/LOEX, Aile–R, Pavilion Hôpital Enfant-Jésus, CRCHU de Québec-Université Laval, Québec, QC, Canada*

PIETRO GENTILE · *Department of Plastic and Reconstructive Surgery, University of Rome "Tor Vergata", Rome, Italy; Catholic University, Tirane, Albania*

JEFFREY M. GIMBLE · *Center for Stem Cell Research and Regenerative Medicine, Tulane University School of Medicine, New Orleans, LA, USA; LaCell LLC, New Orleans, LA, USA*

WARREN L. GRAYSON · *Department of Biomedical Engineering, Johns Hopkins University School of Medicine, Baltimore, MD, USA; Translational Tissue Engineering Center, Johns Hopkins University School of Medicine, Baltimore, MD, USA; Department of Materials Science and Engineering, Johns Hopkins University Whiting School of Engineering, Baltimore, MD, USA*

ANNE-LAURE HAFNER · *Faculté de Médecine, Université Nice Sophia Antipolis, iBV, UMR CNRS/INSERM, Nice Cedex 2, France*

CLOE S. HAKAKIAN · *University Stem Cell Center, Los Angeles, CA, USA*

YUAN DI C. HALVORSEN · *Center for Computational and Integrative Biology, Harvard Medical School, Boston, MA, USA*

PINAR YILGOR HURI · *Department of Biomedical Engineering, Johns Hopkins University School of Medicine, Baltimore, MD, USA; Translational Tissue Engineering Center, Johns Hopkins University School of Medicine, Baltimore, MD, USA; Department of Biomedical Engineering, Ankara University Faculty of Engineering, Ankara, Turkey*

HEEJIN JUN · *Life Sciences Institute, University of Michigan, Ann Arbor, MI, USA*

GAIL KILROY · *Pennington Biomedical Research Center, Louisiana State University System, Baton Rouge, LA, USA*

MIKHAIL G. KOLONIN · *Brown Foundation Institute of Molecular Medicine for the Prevention of Human Diseases, University of Texas Health Science Center at Houston, Houston, TX, USA*

KAI P. LEUNG · *United States Army Institute of Surgical Research, JBSA Fort Sam Houston, TX, USA*

Hui Li · *National Engineering Laboratory for Oral Regenerative Medicine, West China Hospital of Stomatology, Sichuan University, Chengdu, China; Life Sciences Institute, University of Michigan, Ann Arbor, MI, USA*
Jie Li · *Center for Stem Cell Research and Regenerative Medicine, Tulane University School of Medicine, New Orleans, LA, USA; LaCell LLC, New Orleans, LA, USA; National Engineering Laboratory for Oral Regenerative Medicine, West China Hospital of Stomatology, Sichuan University, Chengdu, China*
Elizabeth G. Loboa · *College of Engineering, University of Missouri, Columbia, MO, USA*
Ryan A. Lockhart · *University Stem Cell Center, Los Angeles, CA, USA*
Mandi J. Lopez · *Laboratory for Equine and Comparative Orthopedic Research, School of Veterinary Medicine, Louisiana State University, Baton Rouge, LA, USA*
Kacey G. Marra · *University of Pittsburgh, Pittsburgh, PA, USA*
Pascal Morissette Martin · *Department of Anatomy and Cell Biology, Schulich School of Medicine and Dentistry, The University of Western Ontario, London, ON, Canada*
Dominique Mayrand · *Centre de Recherche en Organogénèse Expérimentale de l'Université Laval/LOEX, CRCHU de Québec-Université Laval, Québec, QC, Canada; Department of Surgery, Faculty of Medicine, Université Laval, Québec, QC, Canada*
Hiroshi Mizuno · *Department of Plastic and Reconstructive Surgery, Juntendo University School of Medicine, Tokyo, Japan*
Tala Mohsen-Kanson · *Faculté de Médecine, Université Nice Sophia Antipolis, iBV, UMR CNRS/INSERM, Nice Cedex 2, France; Department of Biology, Faculty of Sciences II, Lebanese University, Fanar, Lebanon*
Justin Morrissette-McAlmon · *Department of Biomedical Engineering, Johns Hopkins University School of Medicine, Baltimore, MD, USA; Translational Tissue Engineering Center, Johns Hopkins University School of Medicine, Baltimore, MD, USA*
Rachel C. Nordberg · *Joint Department of Biomedical Engineering, North Carolina State University and University of North Carolina Chapel Hill, Raleigh, NC, USA*
Christina L. Rettinger · *United States Army Institute of Surgical Research, JBSA Fort Sam Houston, TX, USA*
Shahensha Shaik · *Bioengineering Laboratory, Department of Mechanical Engineering, Louisiana State University, Baton Rouge, LA, USA*
Arthi Shridhar · *Department of Chemical and Biochemical Engineering, Thompson Engineering Building, The University of Western Ontario, London, ON, Canada*
Glauco R. Souza · *Nano3D Biosciences, Houston, TX, USA; Oncology Division, Department of Internal Medicine, UT Health Science Center at Houston, Houston, TX, USA*
Satoshi Tajima · *Department of Plastic and Reconstructive Surgery, Juntendo University School of Medicine, Tokyo, Japan; Department of Dentistry and Oral Surgery, AOI Universal Hospital, Kanagawa, Japan*
Weidong Tian · *National Engineering Laboratory for Oral Regenerative Medicine, West China Hospital of Stomatology, Sichuan University, Chengdu, China*
Morikuni Tobita · *Department of Plastic and Reconstructive Surgery, Juntendo University School of Medicine, Tokyo, Japan*
Hubert Tseng · *Nano3D Biosciences, Houston, TX, USA*
Alan Tucker · *Center for Stem Cell Research and Regenerative Medicine, Tulane University School of Medicine, New Orleans, LA, USA*

JUN WU · *Life Sciences Institute, University of Michigan, Ann Arbor, MI, USA; Department of Molecular and Integrative Physiology, University of Michigan, Ann Arbor, MI, USA*
XIYING WU · *Center for Stem Cell Research and Regenerative Medicine, Tulane University School of Medicine, New Orleans, LA, USA; LaCell LLC, New Orleans, LA, USA*
CLAIRE YU · *Department of Chemical Engineering, Queen's University, Kingston, ON, Canada*
HUI YU · *Life Sciences Institute, University of Michigan, Ann Arbor, MI, USA*
NAN ZHANG · *Laboratory for Equine and Comparative Orthopedic Research, School of Veterinary Medicine, Louisiana State University, Baton Rouge, LA, USA*

Chapter 1

Isolation and Flow Cytometric Analysis of the Stromal Vascular Fraction Isolated from Mouse Adipose Tissue

Annie C. Bowles, Alan Tucker, and Bruce A. Bunnell

Abstract

Evidence from preclinical research and clinical trials demonstrates the use of the stromal vascular fraction (SVF) as therapy for numerous indications. These results demonstrate that autologous SVF is not only safe and effective but provides robust anti-inflammatory, immunomodulatory, and reparative effects in vivo. The potency of the SVF is attributed to the cellular composition which includes adipose-derived stem cells (ASCs), adipocytes, endothelial cells, and various immune cells. As the name would suggest, these SVF cells are derived from the stromal compartment of adipose, or fat. Once digested, the cells that constitute adipose are released and collected as the SVF. The cellular frequencies within the SVF can then be assessed using a fluorescent antibody-based technique known as flow cytometry. The following chapter provides a standard operating protocol that describes the procedures from harvesting the fat tissue from experimental mice to isolating and characterizing the SVF.

Key words Stromal vascular fraction, Adipose-derived stem cells, Adipose tissue, Flow cytometry, Immunophenotyping

1 Introduction

Fat or adipose is the largest endocrine organ of the body involved in thermoregulation, insulation, cushioning, and energy storage [1, 2]. It is a highly vascularized tissue that is mainly composed of connective tissue and adipocytes, and is an attractive source of stem cells, called adipose-derived stem cells (ASCs) [3, 4]. When fat is collected, digested and centrifuged, three layers are created. The bottom layer, or pellet, is called the stromal vascular fraction (SVF) [4, 5]. This concentrated cellular pellet contains the adipocytes, ASCs, endothelial cells, and numerous types of immune cells [2, 3]. Clinically, human donors of autologous fat can be removed by a minimally invasive liposuction procedure then rapidly digested, processed, and administered back to the patient as SVF during the same procedure [6]. This abundant source of cells has demonstrated

Bruce A. Bunnell and Jeffrey M. Gimble (eds.), *Adipose-Derived Stem Cells: Methods and Protocols*, Methods in Molecular Biology, vol. 1773, https://doi.org/10.1007/978-1-4939-7799-4_1,

potent anti-inflammatory, immunomodulatory, and reparative effects in vivo [7].

In humans, SVF is commonly used to supplement fat grafting in cosmetic and reconstructive surgery and is believed to attenuate the resorption process [6, 8]. Veterinary medicine has demonstrated the safe and effective treatment of bone, tendon, and joint injuries and osteoarthritis in over thousands of large animals [9, 10]. In a murine model of multiple sclerosis, the SVF has demonstrated attenuation of disease, modulation of the inflammatory disease milieu and improvement of motor function [11, 12]. Collectively, this evidence demonstrates the safe and effective use of SVF for treating many medical indications.

This chapter described a widely used method of SVF isolation from murine adipose that yields consistent outcomes and is easily reproduced. The protocol begins with harvesting the inguinal white adipose tissue from mice, and then describes the methods of digestion, isolation, and characterization of the composition of the SVF. Additionally, a comprehensive panel of cellular markers is provided for flow cytometric analysis to determine the frequency of desired cell types within the SVF.

2 Materials

2.1 Animals

1. C57Bl/6 mice.

2.2 Reagents

1. Dulbecco's modified Eagle medium: Ham's F-12 (DMEM/F-12).
2. Fetal bovine serum (FBS), premium select, hybridoma qualified not heat inactivated.
3. L-glutamine (200 mM) in solution of 0.85% NaCl.
4. Antibiotic–antimycotic (anti-anti; 100×).
5. Phosphate buffered saline (PBS) without Ca^{2+} or Mg^{2+}, 1×.
6. Bovine serum albumin (BSA), lyophilized powder, essentially fatty acid free.
7. Collagenase type-I, 390 U/mg.
8. 0.25% trypsin and 1 mM ethylenediaminetetraacetic acid (EDTA) in Hanks' Balanced Salt Solution (HBSS).
9. 0.4% Trypan Blue in solution of 0.85% NaCl.
10. Distilled, deionized water (DDI-H_2O).
11. 4% paraformaldehyde (PFA) in PBS.
12. Fixation/permeabilization concentrate.
13. Fixation/permeabilization diluent.

2.3 Supplies

1. Flow cytometry antibodies (*see* Table 1).
2. 500 ml vacuum-driven filtration unit with 0.22-μm pore size.

Table 1
Anti-mouse antibodies for determining the cell frequencies in the SVF

Antibody	Clone	Isotype	Affinity
CD3	17A2	IgG	All T cells
CD4	GK1.5	IgG2b	Helper T cells, regulatory T cells
CD8	53–6.7	IgG2b	Cytotoxic T cells
CD14	SA2–8	IgG2a	Monocytes/macrophages
CD19	1D3	IgG2a	B cells
CD25	7D4	IgG2a	Regulatory T cells
CD31	390	IgG2a	Endothelial cells, lymphocytes
CD34	RAM34	IgG2a	HSC, dendritic cells
CD36	72-1	IgG2a	Dendritic cells, monocytes/macrophages
CD45	30-F11	IgG2b	All leukocytes
CD68[a]	FA11	IgG2a	Monocytes/macrophages
CD90	HIS51	IgG2a	HSC, MSC, monocytes/macrophages
F4/80	BM8	IgG2a	Dendritic cells, monocytes
FCeR1	MAR-1	IgG	Mast cells
foxp3[a]	FJK-16 s	IgG2a	Regulatory T cells
Ly6G	RB6-8C5	IgG2	Granulocytes
Sca-1	D7	IgG2a	HSC, MSC

HSC hematopoietic stem cells, *MSC* mesenchymal stem cells
[a]Intracellular markers that require fixation and permeabilization protocol

3. Sterile scissors and forceps.
4. Sterile 50 ml polystyrene collection tubes.
5. 50 ml vacuum-driven filtration unit with 0.22-μm pore size.
6. Sterile microcentrifuge tubes.
7. Micropipettes and pipettes with sterile tips.
8. Parafilm.

2.4 Equipment

1. Scale milligram to gram measurements.
2. Biological safety cabinet, class II with vacuum aspiration source with tubing and waste container.
3. Incubator, water jacketed and humidified with 5% CO_2, maintained at 37 °C
4. Inverted phase-contrast microscope.
5. Large and small centrifuges.

6. Rocker incubator.
7. Flow cytometer, with eight-color-capability.
8. Water bath with autoclaved water, maintained at 37 °C.

2.5 Solutions

1. Complete culture media (CCM): Dulbecco's modified Eagle medium: Ham's F-12.

   10% fetal bovine serum.

   1% 2 mM L-glutamine.

   1% antibiotic–antimycotic.
2. Adipose digestion solution (w/v; grams of adipose to milliliters of PBS):

   0.1% collagenase Type I.

   1% BSA.
3. 1% PFA (optional): 4% PFA in DDI-H_2O (1:3).
4. Working fixation/permeabilization solution: fixation/permeabilization concentrate in diluent (1:3).
5. 1× Working Perm. Buffer.

   10× Perm. Buffer in DDI-H_2O (1:10).

3 Methods

3.1 Isolation of the Inguinal White Adipose Tissue

The following procedure is performed after euthanasia of the experimental animals. It is recommended that each animal is secured dorsum down where the ventral surface is accessible for the isolation of white fat tissue from the inguinal area (Fig. 1).

1. Make a small incision beneath the sternum using forceps and surgical scissors.
2. Insert scissors into incision and blunt dissect between the skin and peritoneum forming a pocket.
3. Make a large, vertical incision from the initial incision down to the genitalia area.
4. Make pockets around each flank by securing the skin with forceps and inserting the scissors between the skin and peritoneum followed by blunt dissections.
5. Make a horizontal incision from the initial incision to the back of one side of the mouse.
6. Expose the inguinal fat pad lying between the skin and the peritoneum.
7. Fold the skin inside-out and securely pin the loose corner.

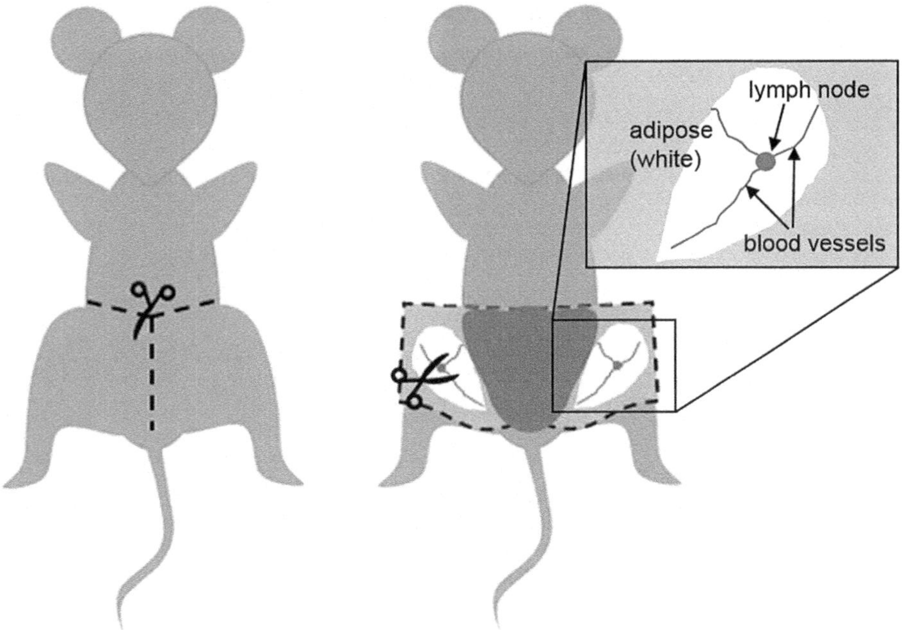

Fig. 1 Isolation of the inguinal white adipose tissue from mice

8. Identify and remove the lymph node located in the middle of the fat pat (*see* **Note 1**).
9. Carefully excise the entire fat pad and collect the fat in a conical tube.
10. Repeat **steps 5–9** for excision of the alternate inguinal fat pad.

3.2 Isolation of the Stromal Vascular Fraction

1. Collect fat tissue in a 50 ml conical tube.
2. Weigh fat.
   Under a biosafety hood:
3. Wash fat by adding ample PBS and vigorously shake the capped conical tube.
4. Transfer fat to a fresh tube.
5. Repeat **steps 3** and **4** until solution is clear.
6. Transfer washed fat to a fresh tube. Put aside conical tube containing fat tissue.
7. In a fresh 50 ml conical tube, make digestion solution.
8. Gently rock conical tube to dissolve solutes; do not vortex.
9. Secure the 50 ml disposable vacuum filtration system with 0.22-μm pore size to conical tube with digestion solution.
10. Attach vacuum tubing to filter unit and vacuum port.

11. Filter digestion solution into attached collection tube.
12. Add filtered digestion solution to tube containing the fat tissue.
13. Start a timer for 1 h.
14. Quickly mince fat in digestion solution with scissors by downward motion cutting in the tube (*see* **Note 2**).
15. Add Parafilm around top of tube to seal.
16. Place tube with fat tissue in digestion solution into an incubator shaker.
17. Incubate fat in digestion solution at 37 °C in incubator for time remaining. Set shaker speed to 100 rpm.

    Proceed under a biological safety cabinet henceforth.

18. Prepare 500 ml CCM.
19. Using the 500 ml vacuum-driven filtration unit with 0.22-μm pore size, filter CCM into the collection reservoir.
20. Warm CCM in water bath at 37 °C (*see* **Note 3**).
21. After 1 h incubation of fat in the digestion solution, remove tube from incubator shaker and return under biosafety hood.
22. Neutralize digestion solution by adding (v/v) CCM to the tube.
23. Cap the tube and invert several times to mix.
24. Centrifuge conical tube for 5 min at 500 ×*g* at room temperature (RT).
25. Remove and shake conical tube to assist in the release of cells from tissue.
26. Centrifuge conical tube again for 5 min at 500 ×*g* at RT.
27. Carefully remove conical tube from centrifuge. Be sure not to disturb layers.
28. Carefully aspirate top liquid layers leaving pellet of stromal vascular fraction (SVF).
29. Add CCM to conical tube to resuspend cells of SVF by pipetting up and down several times.
30. Filter cell suspension through sterile gauze to remove any hair or large debris and collect in a fresh 50 ml conical tube.
31. Count live cells using Trypan Blue exclusion method and determine total live cell count.

Solution now consists of a single-cell suspension of SVF cells. Cells can be prepared in a saline solution for experimental use or analyzed by flow cytometric analysis as described below.

3.3 Staining Cells for Flow Cytometric Analysis

1. Determine the concentration of live SVF cells in known volume of PBS.
2. Aliquot between 2.5×10^5 and 5×10^5 cells in approximately 100 μl per microcentrifuge tube for each protocol (Table 2; *see* **Note 4**).
3. Stain cells in each microcentrifuge tube by pipetting 1–5 μl of each antibody in each panel according to Table 2. Initial titration of antibodies will determine appropriate volume of antibody used per sample. (Additional antibody for live/dead staining is optional).
4. Cap and vortex tubes for 5 s.
5. Incubate tubes for 15 min at RT in the dark.
6. Wash by adding 1.0–1.5 ml of PBS to each tube.
7. Centrifuge tubes at 500 ×*g* for 5 min.
8. Decant by quickly inverting tube once over a waste beaker. This should allow approximately 100 μl to remain over the cell pellet.
9. (Optional) Repeat **steps 6–11** for secondary antibodies.
10. Wash again by repeating **steps 9–11**.

Table 2
Sample panel for analysis of the frequencies of subpopulations within the SVF

Panel	Antibody–fluorochrome
Protocol 1 Requires fixation and permeabilization[a]	CD4–PE CD3–PE610 CDCD8–PeCy5 foxp3–APC[a]
Protocol 2	CD19–PE610 FCεRI–PeCy7 CD45–APC
Protocol 3	CD31–PE CD90–PE610 Sca-1–PeCy5
Protocol 4 Isotype controls	IgG1–FITC IgG2a–PE

†This panel can be used to analyze either GFP^+ cells or GFP^- cells utilizing the FL 1 (488/530 nm) to evaluate the GFP-expressing populations [13]
[a]Intracellular markers that require fixation and permeabilization protocol

Cells are prepared for immediate analysis using a flow cytometer. Alternatively, tagged cells can be preserved by fixation and stored at 4 °C for analysis up 1 week. If this is desired, additional steps to this protocol must then be performed:

1. Fix cells by adding 500 μl of 1% paraformaldehyde to each tube.
2. Incubate cells for 5 min at RT in the dark.
3. Add 1.0 ml of PBS.
4. Centrifuge tubes at 500 ×*g* for 5 min.
5. Decant by quickly inverting tube once over a waste beaker.
6. Add 1.0 ml of PBS to each tube.
7. Centrifuge tubes at 500 ×*g* for 5 min.
8. Repeat **steps 5** and **6**.
9. Add 400 μl of PBS to each tube (*see* **Note 5**).
10. Vortex. Samples can be immediately analyzed or stored at 4 °C (Table 3).

Table 3
Gating strategy for identification of subpopulations of cells in SVF

Cell type	Cell marker expression
Adipocytes	$CD45^+$ $CD34^-$ $CD36^+$
Hematopoietic stem cells	$CD45^+$ $CD34^+$ $CD31^-$ $Sca\text{-}1^+$ $CD90^+$
Mesenchymal stem cells	$CD45^-$ $CD31^-$ $Sca\text{-}1^+$ $CD90^+$
Smooth muscle cells	$CD45^-$ $CD34^{dim}$ $CD31^-$
Endothelial cells	$CD45^-$ $CD34^+$ $CD31^+$ $CD36^+$
Granulocytes	$CD45^+$ $CD34^-$ $Ly6G^+$
B cells	$CD45^+$ $F4/80^-$ $CD19^+$
Natural killer cells	$CD45^+$ $F4/80^-$ $CD19^-$
Dendritic cells	$CD45^+$ $F4/80^+$ $CD34^+$
Monocytes/macrophages	$CD45^+$ $F4/80^+$ $CD14^+CD68^+$
Regulatory T cells	$CD3^+$ $CD4^+$ $CD25^+$ $foxp3^+$
Cytotoxic T cells	$CD3^+$ $CD8^+$
Helper T cells	$CD3^+$ $CD4^+CD25^-$
Mast cells	$CD45^+$ $Fc\varepsilon R1^+$

4 Notes

1. The lymph node can be easily identified at the convergence of blood vessels.
2. The more the tissue is minced, the higher surface area for SVF cells to be released for higher yield.
3. Place CCM in a 37 °C water bath until use.
4. It is important to use microcentrifuge tubes with a pointed bottom in order to properly decant. For the initial setup of flow cytometric analysis, it is necessary to analyze controls such as unstained samples, single-stained samples of each antibody, and isotype control-stained samples.
5. A final volume of 500 μl is recommended for analysis.

References

1. Fruhbeck G (2008) Overview of adipose tissue and its role in obesity and metabolic disorders. Methods Mol Biol 456:1–22
2. Coelho M, Oliveira T, Fernandes R (2013) Biochemistry of adipose tissue: an endocrine organ. Arch Med Sci 9:191–200
3. Boutens L, Stienstra R (2016) Adipose tissue macrophages: going off track during obesity. Diabetologia 59(5):879–894
4. Lin G, Garcia M, Ning H et al (2008) Defining stem and progenitor cells within adipose tissue. Stem Cells Dev 17:1053–1063
5. Gimble JM, Katz AJ, Bunnell BA (2007) Adipose-derived stem cells for regenerative medicine. Circ Res 100:1249–1260
6. Wu AY (2013) Autologous fat transfer with in-situ mediation (AIM): a novel and compliant method of adult mesenchymal stem cell therapy. J Transl Med 11:136
7. Leto Barone AA, Khalifian S, Lee WP et al (2013) Immunomodulatory effects of adipose-derived stem cells: fact or fiction? Biomed Res Int 2013:383685
8. Trojahn Kolle SF, Oliveri RS, Glovinski PV et al (2012) Importance of mesenchymal stem cells in autologous fat grafting: a systematic review of existing studies. J Plast Surg Hand Surg 46:59–68
9. Volk SW, Theoret C (2013) Translating stem cell therapies: the role of companion animals in regenerative medicine. Wound Repair Regen 21:382–394
10. Black LL, Gaynor J, Gahring D et al (2007) Effect of adipose-derived mesenchymal stem and regenerative cells on lameness in dogs with chronic osteoarthritis of the coxofemoral joints: a randomized, double-blinded, multicenter, controlled trial. Vet Ther 8:272–284
11. Semon JA, Zhang X, Pandey AC et al (2013) Administration of murine stromal vascular fraction ameliorates chronic experimental autoimmune encephalomyelitis. Stem Cells Transl Med 2:789–796
12. Bowles AC, Strong AL, Wise RM, Thomas RC, Gerstein BY, Dutreil MF, Hunter RS, Gimble JM, Bunnell BA (2017) Adipose stromal vascular fraction-mediated improvements at late-stage disease in a murine model of multiple sclerosis. STEM CELLS 35(2):532–544
13. Ludwig DB, Trotter JT, Gabrielson JP, Carpenter JF, Randolph TW (2011) Flow cytometry: A promising technique for the study of silicone oil-induced particulate formation in protein formulations. Anal Biochem 410(2):191–199

Chapter 2

A Method for Isolation of Stromal Vascular Fraction Cells in a Clinically Relevant Time Frame

Joel A. Aronowitz, Ryan A. Lockhart, and Cloe S. Hakakian

Abstract

There is increasing interest in the clinical applications of adipose-derived stem cells (ASCs) and the stromal vascular fraction (SVF) based on promising preclinical data. As adipose-derived therapeutics begin to translate into the clinical setting, it is important to maintain patient safety as well as uniformity in technique. Here, we describe a method for isolation of stromal vascular fraction cells in a clinically relevant time frame. Analytical laboratory techniques are mentioned, but respective protocols are not provided here.

Key words Adipose-derived stem cell, Stromal vascular fraction, SVF isolation, SVF, ADSC

1 Introduction

Adipose-derived stem cells (ASCs) have drawn significant attention in the past decade for potential clinical applications. The combined paracrine effects and multipotent differentiation capacity of ASCs [1–7] has been examined for a variety of therapeutic clinical applications including breast reconstruction/augmentation, facilitation of diabetic ulcer healing, facial rejuvenation, various orthopedic conditions, repair of cardiac muscle tissue after myocardial infarction, and many more [8–12]. Since many of these applications have shown promise for clinical application based on findings in the laboratory setting, both in vitro and in animal models, groups around the world are beginning to translate these technologies into the clinical setting.

Therapies based on adipose-derived stem cells begin with the isolation of the stromal vascular fraction (SVF) cells. Stromal vascular fraction cells can be isolated in two different ways: with and without the use of tissue dissociation enzymes. Nonenzymatic methods tend to be relatively inexpensive compared to their enzymatic counterparts, but yield fewer cells per milliliter of lipoaspirate processed. Additionally, the resulting composition of the cell populations is different, with mechanical methods yielding a lower frequency of progenitor cells and a higher frequency of

Bruce A. Bunnell and Jeffrey M. Gimble (eds.), *Adipose-Derived Stem Cells: Methods and Protocols*, Methods in Molecular Biology, vol. 1773, https://doi.org/10.1007/978-1-4939-7799-4_2, © Springer Science+Business Media, LLC, part of Springer Nature 2018

CD45^{+} cells [13–16]. Overall, the proteolytic disruption of the extracellular matrix allows for greater cell recovery. While seemingly the preferred method, enzymatic isolation adds additional factors to consider in regards to patient safety in the clinical setting, including residual enzyme activity and allergic reaction to tissue dissociation enzymes.

After SVF isolation, therapies can use the SVF cells for treatment or cultural expanded cells. Cultural expansion offers the advantage of a more homogeneous population and a greater number of ASCs, but requires a minimum of two procedures to complete treatment: one for liposuction/harvest and a second for treatment after the cells have been culturally expanded. Using SVF has the advantage of being able to be completed in a single procedure, including liposuction, isolation and treatment. Here, we describe a method for isolation of SVF cells in a clinically relevant time frame of 60–90 min. This method is able to isolate an average of 2.0×10^5 nucleated cells per milliliter of lipoaspirate with average viability >90%.

2 Materials

2.1 Supply List for Isolation

1. Package of sterile, 25 count 50 mL disposable flip-top centrifuge tubes.
2. 1000 cc bags of sterile lactated Ringer's solution (130 mmol/L Na^{+}, 109 mmol/L Cl^{-}, 28 mmol/L lactate, 1.5 mmol/L Ca^{2+}, 4 mmol/L K^{+}).
3. 500 mL glass beakers (for waste and buffer).
4. 500 mL separatory funnel with ring stand.
5. Disposable, individually packaged, sterile latex gloves (as needed).
6. Sterile, individually packaged 3 cc syringe with 18 gauge needle.
7. Sterile, individually packaged 1 cc syringe with 18 gauge needle.
8. 12 mL sterile water.
9. Four vials Vitacyte CIzyme AS (35 Wünsch units collagenase activity per vial).
10. 100 μm cell strainer, sterile, individually packaged.

2.2 Enzyme Preparation

1. Using a sterile 3 cc syringe, transfer 3 cc of sterile water at room temperature into each of the four vials of lyophilized tissue dissociation enzymes in order to resuspend (*see* **Note 1**).
2. Once the enzyme has fully dissolved, remove the enzyme solution from vials using the 3 cc syringe and transfer the contents of each vial into 50 mL centrifuge tubes. Raise the volume of each of the four tubes to a final volume of 50 cc using warm lactated Ringer's solution.

3. Place tubes into the heated shaker for a minimum of 30 min at 37 °C and 200 rpm. Be sure to keep warm and shaking until use.

2.3 Tumescent Solution Preparation

Tumescent solution is prepared by a trained surgical technician in the operating room.

1. Fifty cubic centimeters of lidocaine plain is injected into a sterile 1000 cc bag of saline solution.
2. Two milliliters of AMP epinephrine 1:1000 solution is injected into the saline/lidocaine solution.

3 Methods

All preparation and isolation occurs inside of a P&C Multi-Station (Fig. 1). The Multi-Station contains a centrifuge (400 cc capacity) and a heated shaking unit contained within the confines of a laminar flow hood. All procedures and preparations should be conducted inside of the laminar flow hood. All glassware is autoclaved prior to use. The laminar flow hood should be fully sterilized before and after isolation. Sterile gloves should be worn at all times to avoid contamination of samples.

3.1 Lipoaspiration

All equipment/supplies required for liposuction are not included in this protocol as they can vary from facility to facility. Suction-assisted liposuction was conducted using a "super-wet" tumescent solution (*see* Subheading 2, **item 3**) and a 2.7 mm blunt tipped cannula with a vacuum pressure of −25 mmHg.

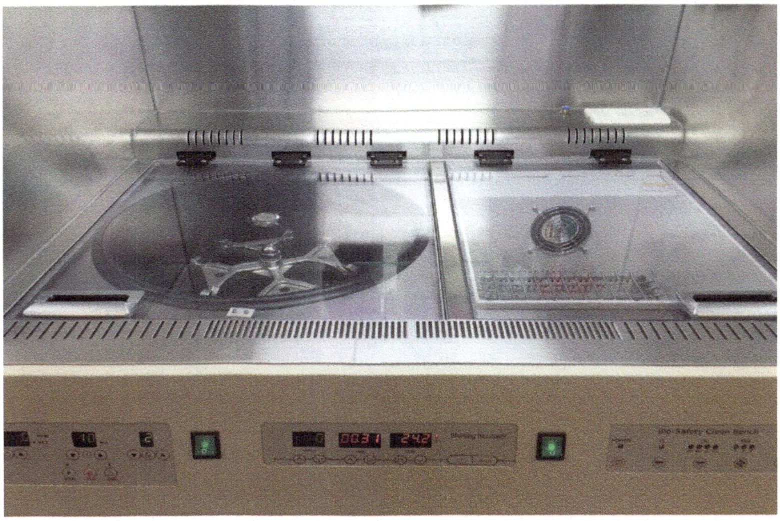

Fig. 1 Inside the P&C Multi-Station. The left side contains a centrifuge with a 400 cc capacity. The right side contains a heated shaking unit

1. Harvest site varies from patient to patient based on donor site availability and aesthetic desires of the patient. Donor site selection is left to the discretion of the patient and surgeon.
2. Prior to harvest, the patient is infiltrated with 750 cc–1000 cc of tumescent solution (*see* Subheading 2, **item 3** for recipe) to limit blood loss, pain and bruising due to harvest. Tumescent is allowed to infiltrate for 10–15 min prior to harvest.
3. Using a 2.7 mm blunt tipped cannula, fat tissue is aspirated into sterile canisters (*see* **Note 2**).
4. Lipoaspirate is then transferred into a sterile 500 mL separatory funnel for isolation. Isolation is carried out by a technician in a sterile laminar flow biohood (*see* **Note 3**).

3.2 Isolation of the Stromal Vascular Fraction

1. Wash lipoaspirate with an equal volume of warm (37 °C) lactated Ringer's (LR) solution. After addition of LR, gently mix and allowed the mixture to separate. The mixture will separate into two distinct portions, a lower aqueous layer (Red/clear) and the upper fatty layer (yellow/pink). The lower, aqueous layer is discarded. Wash three times or until fat layer is yellow (no longer pink) (Fig. 2).

Fig. 2 Lipoaspirate separated in the separatory funnel. The upper, yellow layer is lipoaspirate. The lower pink layer is a mixture of blood and tumescent

2. Aliquot 25 cc washed lipoaspirate into sterile 50 cc centrifuge tubes (*see* **Note 4**).
3. Add 25 cc of warm (37 °C) tissue dissociation enzyme solution to each 25 cc aliquot of washed lipoaspirate. Invert to mix.
4. Incubate lipoaspirate/enzyme solution in a heated shaker at 37 °C, 200 rpm for 20–30 min (*see* **Note 5**).
5. Following incubation, centrifuge the tubes at 2000 rpm (700 × *g*) for 10 min to separate components.
6. After centrifugation, three layers are distinctly visible in the tubes. The uppermost layer will contain fat and oil. The middle layer will be the aqueous layer, which will appear red/clear. At the bottom of the tubes there is a pellet. The pellet contains the stromal vascular fraction cells (Fig. 3).
7. Remove the aqueous and fatty/oil layers. Dispose of all biohazard materials properly in biohazard waste containers.
8. Combine the SVF pellets into two sterile 50 cc tubes (four pellets into each tube). Pellets are combined by suspending them in a small amount of warm LR (*see* **Note 6**).
9. Bring each tube containing pellets to a volume of 50 cc using warm LR (*see* **Note 7**).

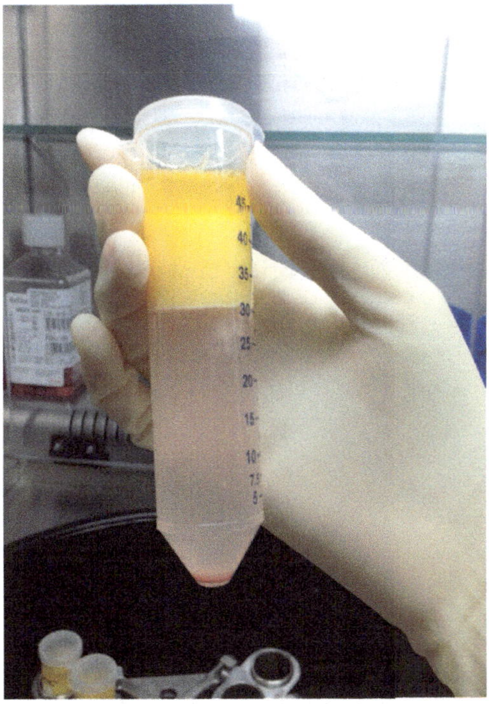

Fig. 3 The result of centrifuging the lipoaspirate after enzymatic digestion

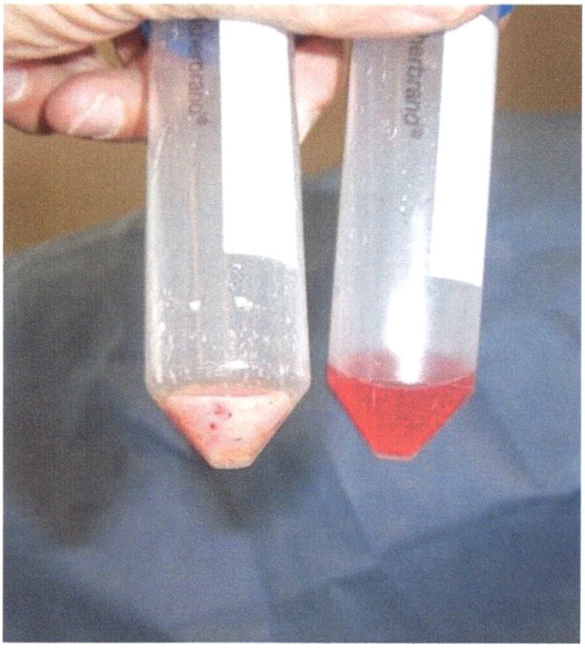

Fig. 4 The left tube contains the detritus and tissue fragments which have been removed via the 100 μm cell strainer. The right tube contains the strained SVF

10. Washing: Centrifuge at 700 × *g* for 5 min. Save the pellet and pour off the aqueous layer. Repeat two more times.
11. After the third and final wash, combine the two remaining pellets into a single tube. Adding a small amount of LR may make the transfer easier.
12. Filter through a 100 μm cell strainer in order to remove tissue fragments which remain from the digestion of the adipose tissue. Use LR to wash the cells through the strainer (Fig. 4) (*see* **Note 8**).
13. Bring final volume to 15 cc using warm LR (*see* **Note 9**).
14. The following volumes are taken for analysis (*see* **Note 10**):
    (a) 0.1–0.2 cc for cell count and viability using Chemometec NC-200 (*see* **Note 11**) (Fig. 5)
    (b) 2.1 cc for sterility testing
    Gram stain: <0.1 cc SVF needed.
    2 week aerobic culture: 1 cc SVF in 9 cc Tryptic Soy Broth (TSB).
    2 week anaerobic culture: 1 cc SVF in 9 cc Thioglycollate broth.
    (c) 2 cc for flow cytometry
    Markers tested: FITC-CD31, APC-CD34, PE-CD45.
15. The remaining volume is returned to the surgeon.

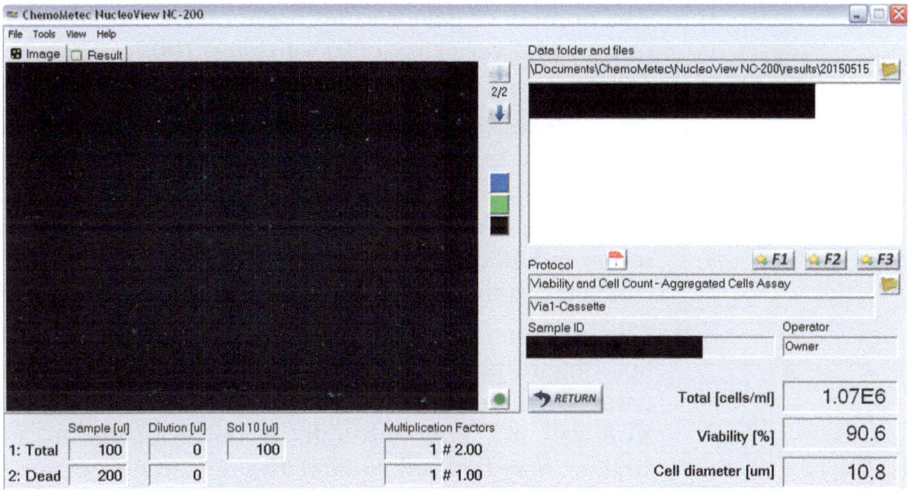

Fig. 5 A representative readout from the nucleocounter NC-200. The sample tested was a 10% dilution of the final SVF output. The final volume of SVF was 10 cc total. 400 cc of lipoaspirate was processed. The total nucleated cell count for the whole SVF sample was 128.4 million nucleated cells. 321,000 nucleated cells were isolated per cc of washed lipoaspirate processed

4 Notes

1. The tissue dissociation enzyme mixture used here contains collagenase and neutral protease (Dispase). The level of collagenase activity is 35 Wünsch units [17] per vial, but the total collagen degrading activity is reported to be 330,000 units [18] per vial. Make sure that enzymes are of GMP grade. Be aware that the level of activity and the units used to express the amount of activity varies between manufacturers and activity may vary from batch to batch from the same manufacturer. The certificate of analysis received with enzymes will give the exact level of activity. Proper enzyme preparation may vary between products and should be done according to the recommendations by the manufacturer.
2. Volume of lipoaspirate is measured as volume after washing. Larger volumes are usually received from the surgeon, but this often contains a significant amount of tumescent solution which is removed during washing. Using this method, we have processed between 50 cc and 800 cc of washed lipoaspirate.
3. A sterile funnel may make the transfer easier for larger volumes of lipoaspirate.
4. Transfer the washed lipoaspirate from the separatory funnel to a 500 mL beaker. It is easier to aliquot washed lipoaspirate from a beaker as opposed to a separatory funnel.
5. Tubes may need to be manually inverted and mixed every 5–10 min, as the lipoaspirate may collect at one end of the tube. Mixing will provide more even digestion of tissue.

6. The number of pellets varies based on the amount of fat tissue processed. Every 25 cc of fat = 1 pellet. 200 cc of lipoaspirate = 8 pellets.
7. In a previous study we were able to demonstrate negligible collagenase levels after three washing steps [19]. If desired, collagenase can be neutralized using autologous serum. This is done by washing in 20% serum which can be made using 10 cc serum and 40 cc LR for one of the wash steps. Serum can be generated by centrifuging heparinized blood at 800 × *g* for 10 min.
8. The amount of detritus varies depending on the volume of lipoaspirate processed. If sample contains a large amount of tissue fragments, the strainer may become clogged and the SVF will not flow through. If this happens an additional strainer may be used to filter the remaining portion. Not all cells are recovered after straining, as some remain trapped within the pellet of tissue fragments. An additional washing and straining of the detritus can be conducted to increase cell yield, but this will extend the isolation time.
9. A final volume of 15 cc is arbitrary and can vary based on personal preference. When mixing SVF with fat for Cell-assisted Lipotransfer (CAL), final volume is not as important because there is usually a larger volume of fat, but when directly injecting SVF, a smaller volume (5 cc–10 cc) may be desired.
10. Samples with <80% viability will not be approved for clinical use. Samples which detect any contamination during the initial gram stain will not be approved for clinical use. A patient whose sample has a positive aerobic/anaerobic culture will be closely monitored for infection. A positive culture does not mean that a patient will develop an infection.
11. A dilution of the final SVF product may be needed to achieve a cell concentration in the readable range for the cell counting method used. Typically a 1:10 dilution in LR was sufficient, but in a few rare cases a 1:100 dilution may be required.

Disclosures

None of the authors have a financial interest in any of the products, devices, or drugs mentioned in this chapter.

References

1. Ude CC, Sulaiman SB, Min-Hwei N et al (2014) Cartilage regeneration by chondrogenic induced adult stem cells in osteoarthritic sheep model. PLoS One 9(6):e98770
2. Planat-Bernard V, Silvestre JS, Cousin B et al (2004) Plasticity of human adipose lineage cells towards endothelial cells: physiological and therapeutic perspectives. Circulation 109:656–663

3. Naderi N, Wilde C, Haque T et al (2014) Adipogenic differentiation of adipose-derived stem cells in a 3-dimensional spheroid culture (microtissue): implications for the reconstructive surgeon. J Plast Reconstr Aesthet Surg 67(12):1726–1734
4. Zuk PA, Zhu M, Mizuno H et al (2001) Multilineage cells from human adipose tissue: implications for cell-based therapies. Tissue Eng 7:211–229
5. Yoshimura K, Sato K, Aoi N et al (2008) Cell-assisted lipotransfer for cosmetic breast augmentation: supportive use of adipose-derived stem/stromal cells. Aesthet Plast Surg 32(1):48–55
6. Rehmam J, Traktuev D, Li J et al (2004) Secretion of angiogenic and antiapoptotic factors by human adipose stromal cells. Circulation 109:1292–1298
7. Kapur SK, Katz AJ (2013) Review of the adipose derived stem cell secretome. Biochimie 95:2222–2228
8. Matsumoto D, Sato K, Gonda K et al (2006) Cell-assisted lipotansfer: supportive use of human adipose-derived stem cells for soft tissue augmentation with lipoinjection. Tissue Eng 12(12):3375–3383
9. Lee HC, An SG, Lee HW et al (2012) Safety and effect of adipose tissue-derived stem cell implantation in patients with critical limb ischemia. Circ J 76:1750–1760
10. Salibian AA, Widgerow AD, Abrouk M et al (2013) Stem cells in plastic surgery: a review of current clinical and translational applications. Arc Plast Surg 40:666–675
11. Michalek J, Moster R, Lukac L et al (2015) Autologous adipose tissue-derived stromal vascular fraction cells application in patients with osteoarthritis. Cell Transplant. https://doi.org/10.3727/096368915X686760
12. Gautam M, Fujita D, Kimura K et al (2015) Transplantation of adipose tissue-derived stem cells improves cardiac contractile function and electrical stability in a rat myocardial infarction model. J Mol Cell Cardiol 81:139–149
13. Shah FS, Wu X, Dietrich M et al (2013) A non-enzymatic method for isolating human adipose-derived stromal stem cells. Cytotherapy 15:979–985
14. Markarian FM, Frey GZ, Silveira MD et al (2014) Isolation of adipose-derived stem cells: a comparison among different methods. Biotechnol Lett 36:693–702
15. Baptista LS, do Amaral RJ, Carias RB et al (2009) An alternative method for the isolation of mesenchymal stromal cells derived from lipoaspirate samples. Cytotherapy 11(6):706–715
16. Yoshimura K, Shiguera T, Matsumoto D et al (2006) Characterization of freshly isolated and cultured cells derived from the fatty and fluid portions of liposuction aspirates. J Cell Physiol 208:64–76
17. Wünsch E, Heidrich HG (1963) On the qualitative determination of collagenase. Z Physiol Chem 333:149
18. McCarthy RC, Spurlin B, Wright MJ et al (2008) Development and characterization of a collagen degradation assay to assess purified collagenase used in islet isolation. Transplant Proc 40:339
19. Aronowitz JA, Ellenhorn JD (2013) Adipose stromal vascular fraction isolation: a head-to-head comparison of four commercial cell separation systems. Plast Reconstr Surg 132(6):932e–939e

Chapter 3

Scaffold-Free, Size-Controlled Three-Dimensional Culture of Rabbit Adipose-Derived Stem Cells

Christina L. Rettinger, Kai P. Leung, and Rodney K. Chan

Abstract

Adipose-derived stem cells are capable of self-renewal and differentiation along multiple cell lineages, and have potential applications in a wide range of therapies. ASCs are commonly cultured as monolayers on tissue culture plastic, but there are indications that they may lose their cell-specific properties with time in vitro. There has been a growing interest in culturing adherent cells using three-dimensional techniques based on the understanding that growing cells on plastic surfaces cannot truly recapitulate 3D in vivo conditions. Here we describe a novel method for generating and culturing rabbit ASCs as scaffold-free 3D cell aggregates using micropatterned wells via a forced aggregation technique.

Key words Cell aggregates/spheroids, Rabbit adipose-derived stem cells, 3D culture

1 Introduction

Much of the work conducted on adult stem cells has focused on mesenchymal stem cells (MSCs) found within the bone marrow stroma. Bone marrow-derived mesenchymal stem cells (BM-MSCs) have been shown to differentiate into a variety of cell types, including adipocytes, chondrocytes, myoblasts, and osteoblast [1–5]. However, traditional bone marrow procurement procedures are painful, often require general anesthesia, and result in low cell yield upon processing. Over the past decade, adipose tissue has emerged as an attractive, alternative source of multipotent adult stem cells termed "adipose-derived stem cells (ASCs). ASCs can be easily extracted from discarded liposuction or resected adipose tissue via collagenase digestion and differential centrifugation. They can be maintained in vitro for extended periods of time with stable population doubling and low levels of senescence [6, 7]. These characteristics, along with their easy accessibility, low donor-site morbidity, and their broad differentiation potential, have made ASCs an excellent candidate for a wide range of cell-based therapeutics.

Bruce A. Bunnell and Jeffrey M. Gimble (eds.), *Adipose-Derived Stem Cells: Methods and Protocols*, Methods in Molecular Biology, vol. 1773, https://doi.org/10.1007/978-1-4939-7799-4_3, © Springer Science+Business Media, LLC, part of Springer Nature 2018

ASCs are typically cultured as monolayers using conventional tissue culture techniques. These methods have proven adequate but there are reports demonstrating a loss in cell replication ability, colony-forming efficiency, and differentiation capacity in MSCs over time [8–10]. As a result, there is growing interest in culturing adherent cells using three-dimensional (3D) cell culture techniques, with evidence showing significant differences between the cellular phenotype and biological response of cells cultured in monolayers vs. 3D culture.

Approaches used to provide cells a 3D environment have included a variety of porous scaffolds, hydrogels, and multicellular aggregates, each of which has been successfully employed with a variety of cell types [11–16]. Among these techniques, the use of multicellular aggregates is attractive due to their self-assembly nature and the lack of exogenous substrates. Three-D cell aggregates have been widely used in cancer research to mimic the tumor microenvironment [13, 17]. They have also been applied to stem cell cultures to emulate embryogenesis [11, 18].

Methods used to generate cell aggregates have involved culturing cells on nonadherent surfaces or exposing them to gravity to encourage self-assembly of cell spheroids. However, these methods are labor-intensive and produce cell aggregates that are heterogeneous in size [17, 18]. Recently, Ungrin et al. [19] developed a microfabrication-based nonadhesive surface (AggreWell™, STEMCELL Technologies, Vancouver, BC) for the culture of embryonic stem cells to improve embryoid body homogeneity and differentiation. A number of groups have since extended this technology to mesenchymal stem cell research, with the majority of these in vitro studies focused on the differentiation potential of 3D cell aggregates generated from murine or human bone marrow-derived mesenchymal stem cells (BMSCs) [20–23].

Since the methods used to extract cells from adipose tissue were first described by Rodbell and colleagues in the 1960s [24], ASCs have been isolated not only from human adipose tissue but also from adipose tissue of many other species, ranging from small rodents, rabbits, to large animals such as pigs and Rhesus monkeys. Rabbits are phylogenetically closer to primates than rodents and are invaluable at bridging the gap between small animal models and large animal models required for preclinical, translational research. The New Zealand White (NZW) rabbits are commonly used to model various human injury and disease conditions. Here, we describe methods for the isolation and culture of rabbit adipose-derived stem cells (rbASCs), in addition to techniques used to generate scaffold-free, sized-controlled 3D rbASC aggregates.

2 Materials

2.1 Tissue

1. Inguinal fat pads obtained from young female New Zealand White (NZW) rabbits under general anesthesia or postmortem (3–4 months old, 2.5–3.5 kg, Harlan Laboratories, Inc.).

2.2 General Supplies

1. Personal protective equipment.
2. 0.2-μm filter units.
3. 15-mL conical tubes.
4. 50-mL conical tubes.
5. 75-cm^2 tissue culture (T75) flasks.
6. Cryovials (Nalgene).
7. Sterile, disposable scalpels.
8. Sterile forceps and scissors, autoclaved.
9. 100-μm cell strainers (Falcon) or sheets of sterile gauze.
10. Freezing apparatus (Mr. Frosty™ Freezing Container).

2.3 Equipment

1. Inverted light microscope (Zeiss Axio Vert).
2. Confocal microscope (Zeiss LSM 710).
3. Shaking water bath (Precision™ Reciprocal Shaking Bath).
4. Orbital shaker (Thermolyne Roto Mix Type 50800).
5. Centrifuge (Eppendorf Centrifuge 5810R).
6. Tissue culture incubator set to 37 °C, 5% CO_2 (Thermo).
7. Automated cell counter (Countess™).
8. Tissue processor (Leica ASP 6025).
9. Tissue embedding station (Thermo Histostar).
10. Automatic slide stainer (Leica Autostainer XL).

2.4 Cell Isolation, Culture, and Cryopreservation

1. Phosphate buffered saline (PBS).
2. Hank's buffered salt solution (HBSS).
3. Collagenase solution: 0.075% type II collagenase (Gibco) prepared in HBSS, filtered through a 0.2-μm membrane.
4. Erythrocyte lysis buffer: 10 mM $KHCO_3$, 150 mM NH_4Cl, 0.1 mM EDTA prepared in distilled water, filtered through a 0.2-μm membrane.
5. Cell culture medium: Dulbecco's Modified Eagle Medium: Nutrient Mixture F12 (DMEM/F12) supplemented with 10% fetal bovine serum (FBS) and 1% penicillin/streptomycin.
6. Trypsin–EDTA.
7. Trypan blue.

8. Freezing medium: complete culture medium supplemented with 10% dimethyl sulfoxide (DMSO).
9. Isopropanol.

2.5 Aggregate Formation

1. AggreWell™ 400/400Ex/800 plates (STEMCELL Technologies).
2. AggreWell™ rinsing solution (STEMCELL Technologies).
3. 6-well ultralow adherence plates (Corning).

2.6 Cell Viability and Proliferation

1. LIVE/DEAD Viability/Cytotoxicity Kit: Calcein AM and ethidium homodimer-1 (Invitrogen).
2. Click-iT EdU Imaging Assay Kit with Alexa Fluor 488 (Invitrogen).
3. 4% paraformaldehyde.
4. 10% buffered formalin.
5. Bovine serum albumin (Sigma).
6. Fibrinogen (Sigma).
7. Thrombin (Sigma).

3 Methods

All media, tubes, flasks, and cells should be kept sterile. All work should be performed in a biosafety hood and sterile techniques should be applied.

Place PBS, collagenase solution, erythrocyte lysis buffer, and cell culture media in a 37 °C water bath prior to use. Collagenase solution should be made fresh and used within 1 h.

3.1 Rabbit ASC Isolation and Culture

1. Begin with a central midline incision through the NZW rabbit's dermis in the inguinal region, followed by incisions from flank to flank at the rostral and caudal aspects of the midline incision. Separate the skin from the underlying adipose tissue and carefully remove the inguinal fat pads. Place the dissected fat pads in sterile, prewarmed PBS. Remove any visible blood vessels.
2. Weigh tissue and place 5–10 *g* of tissue in each 50 mL conical tube.
3. Wash tissue extensively in PBS. PBS can be supplemented with antibiotics such as penicillin/streptomycin, gentamicin, and amphotericin b if contamination is a concern. Mince tissue manually to the consistency of applesauce, and digest in 0.75% type II collagenase solution for 1 h at 37 °C in a shaking water bath. Vortex sample (or manually invert the conical tube) from time to time speeds up the digestion process.

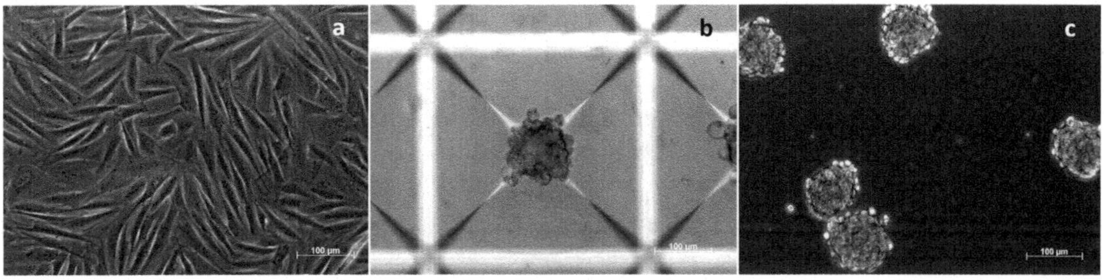

Fig. 1 Phase-contrast images of rbASCs as confluent 2D monolayers on tissue culture plastic (**a**), as 3D cell aggregates in micropatterned wells (**b**), or in suspension culture (**c**). Magnification of images is 100×

4. After 1 h, following inactivation of the collagenase, filter the digested mixture through a sterile nylon mesh filter and centrifuge at 1000 rpm (~164 × *g*) for 10 min.
5. After carefully aspirating mature adipocytes in the supernatant, resuspend the pellet in 10 mL of erythrocyte lysis buffer, and allowed to sit in room temperature for 10 min. Centrifuge for 10 min at 1000 rpm. This pellet consists of the stromal vascular fraction (SVF), containing a heterogeneous population of cells, which includes fibroblasts, pericytes, endothelial cells, preadipocytes, and ASCs.
6. After centrifugation, resuspend the resultant pellet in complete ASC culture media with 10% FBS and plated in a T75 flask.
7. After overnight culture, wash the flask several times with PBS and replenish with fresh complete ASC culture media. Change the medium every 3 days. Primary culture rabbit ASCs become confluent within 7 days (Fig. 1a).
8. Cryopreservation of rbASCs: rbASCs should be harvested at 75–80% confluence for cryopreservation/freezing. Harvest and count the cells. After counting, resuspend the pellet in freezing media at a concentration of 1–2 × 10^6 cells/mL. Place cryovials in a freezing apparatus in a −80 °C freezer. Transfer cells to liquid nitrogen for long term storage the next day.

3.2 ASC Aggregate Formation and Maintenance

1. Pretreat AggreWell™ plates with 1 mL of AggreWell™ rinsing solution, centrifuge at 2000 × *g* for 5 min, and wash repeatedly with PBS.
2. Dissociate ASC monolayers with Trypsin-EDTA, count, add to the pretreated AggreWell™ plates and centrifuge at 300 × *g* for 5 min to sediment cells into the microwells. Overnight incubation at 37 °C and 5% CO_2 is required to promote aggregate formation.
3. ASC aggregates of different sizes can be generated by simply varying the number of cells plated per well (*see* **Note 1**).

4. ASC aggregates can be maintained in AggreWell™ plates in complete growth medium with 10% FBS for extended periods of time (as shown in Fig. 1b) or they can be harvested from the microwells and cultured as free-floating cell aggregates (Fig. 1c) in six-well ultralow adherence plates with agitation via the use of an orbital shaker (*see* **Note 2**).
5. Surface characteristics of rbASC aggregates can be examined via scanning electron microscopy (*see* **Note 3**).

3.3 Viability of ASC Aggregates

Viability of ASC aggregates in culture can be assessed qualitatively via the LIVE/DEAD Viability/Cytotoxicity Assay Kit (Invitrogen, Carlsbad, CA). The LIVE/DEAD assay is a two-color fluorescence cell viability assay that is based on the simultaneous determination of live and dead cells with two probes that measure parameters of cell viability: intracellular esterase activity in living cells as measured by Calcein AM and compromised plasma membrane integrity in dead cells (as measured by ethidium homodimer-1 (EthD-1)).

1. Harvest ASC aggregates from AggreWell™ plates via repetitive pipetting. Wash several times with PBS to remove cell culture media. Wait for ASC aggregates sediment at the bottom of the conical tube before removing the supernatant carefully.
2. Prepare a solution containing 2 μM Calcein AM and 4 μM EthD-1 in PBS and shield it from light.
3. Incubate ASC aggregates with the solution containing Calcein AM and EthD-1 for 30 min in RT. Wash several times with PBS.
4. Pipet stained ASC aggregates onto a glass slide, coverslip, and imaged immediately via confocal microscopy (e.g., Zeiss LSM 710). Viable cells will stain green, whereas non-viable cells will stain red (*see* **Note 4**).

3.4 Proliferation of ASC Aggregates

The proliferative capability of ASC aggregates can be assessed qualitatively on the basis of EdU (5-ethynyl-2′-deoxyuridine) incorporation using Invitrogen's Click-iT EdU Imaging Assay Kit with Alexa Fluor 488 (Invitrogen, Carlsbad, CA). Similar to BrdU (bromo-deoxyuridine), EdU is a nucleoside analog of thymidine, and is incorporated into DNA during active DNA synthesis. Detection is based on a "click reaction," which is a copper-catalyzed covalent reaction between an azide and alkyne. The EdU contains an alkyne and Alexa Fluor dye contains an azide. Standard fixation and detergent permeabilization are sufficient for click-iT detection reagent to gain access to the DNA (no harsh treatments such as DNA denaturation to expose BrdU are required).

1. ASC aggregates in culture are pulsed with 10 μM of EdU for 24 h.
2. Harvest ASC aggregates from microwells via repetitive pipetting. After allowing the aggregates to sediment to the

bottom of the 15 mL conical tube and forming a visible pellet, wash several times with PBS and fix for 30 min in 4% paraformaldehyde.

3. Permeabilize ASC aggregates with 0.5% Triton X-100 and wash several times with 3% BSA solution. Incubate in a click-iT reaction cocktail containing 1× click-iT reaction buffer, CuSO4, Alexa Fluor azide, and reaction buffer additive for 30 min in RT, followed by a 30 min incubation with a 1× Hoechst 33,342 solution for DNA counter-staining.
4. A suspension of stained cell aggregates is added to a glass slide and imaged immediately via fluorescence or confocal microscopy (*see* **Note 5**).
5. Pulsed ASC aggregates can also be encapsulated in fibrin gel and processed for paraffin embedding as "microtissues" (*see* **Note 6**). Histological sections of the aggregates can be stained for EdU following the same protocol outlined above.
6. Quantification of cell proliferation or growth over time is also feasible (*see* **Note** 7).

4 Notes

1. Rabbit ASCs seeded at different densities compact to form cell aggregates after overnight culture in plates containing micropatterned wells. Cell aggregates generated using this method appear to be highly uniform in size. Further compaction of the cell aggregates can be observed when they are maintained in micropatterned wells for extended periods of time. Cross-sectional area or diameter of rbASC aggregates can be determined using the ImageJ image analysis software on phase contrast images taken at different time points.
2. Monolayer ASCs can also be derived from rbASC aggregates by plating intact cell aggregates onto tissue culture treated plates or flasks. Cells will grow out from plated aggregates within 1–2 days.
3. Scanning electron microscopy can reveal the surface characteristics of rbASC aggregates. Each cell aggregate is composed of many tightly compacted single cells, some of which lack clearly defined cell-to-cell boundaries (*see* Fig. 2).
4. Studies conducted in our laboratory indicate that rbASCs were able to form compact cell aggregates and remain viable in 3D static culture for up to 7 days [25]. However, nonviable cells were evident in cell aggregates with diameters greater than 200 μm (generally aggregates composed of more than 4000 cells/aggregate).

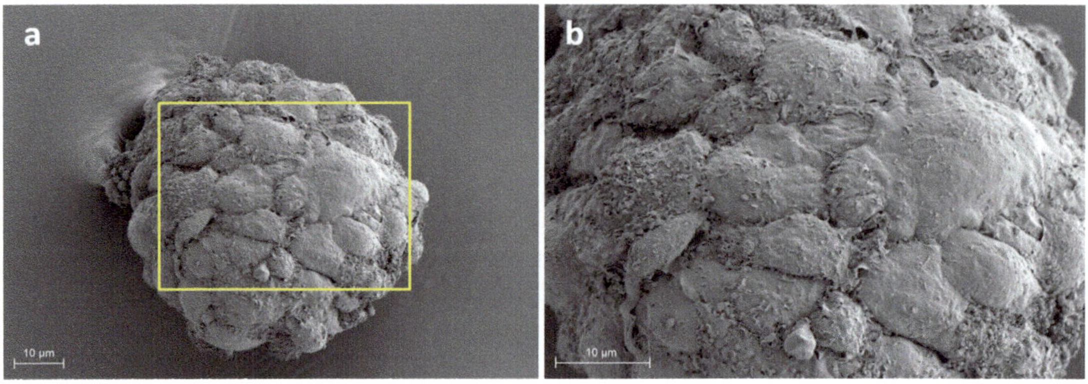

Fig. 2 Scanning electron microscopy images of rbASCs as 3D cell aggregates at 500 cells/aggregate (**a**). (**b**) is a close-up of the ROI highlighted in (**a**)

5. Limited cell proliferation was observed in cell aggregates of all sizes based on experiments performed in our laboratory. Our data suggest that the majority of cells within an aggregate appear to be dormant or quiescent in the G_0 phase of the cell cycle [25].
6. ASC aggregates can be encapsulated in fibrin gel, fixed in 10% buffered formalin and processed for paraffin embedding via the use of a tissue processor. Prepare stock solutions of 17.3 mg/mL fibrinogen and 167 U/mL thrombin. Harvest ASC aggregates from micropatterned wells via repeated pipetting, resuspend aggregates in fibrinogen solution in a metal mold and add an equal volume of thrombin. Fix fibrin gel encapsulated ASC aggregates in 10% buffered formalin for 30 min in RT. Carefully retrieve the gel from the metal mold and place in a tissue cassette lined with biopsy paper. These samples can also be processed manually. The dehydrated and paraffinized samples are set in hot paraffin and embedded for histological sectioning. Sections of ASC aggregates in fibrin gel can be stained with Hematoxylin and Eosin (H&E) or other markers of interest via standard immunohistochemistry techniques.
7. ASC aggregates can be dissociated via incubation in 0.25% trypsin-EDTA at 37 °C for 15–30 min (depending on the size of the aggregate) with mechanical agitation to obtain a single-cell suspension. The resultant cell suspension can be counted with a hemocytometer or an automated cell counter.

Acknowledgments

This work was supported by the US Army Medical Research and Material Command (W81XWH-10-2-0054).

Disclaimers

The opinions or assertions contained herein are the private views of the authors and are not to be construed as official or reflecting the views of the Department of Defense or US Government. One of the authors (KPL) is an employee of the US Government, and this work was prepared as part of his official duties.

References

1. Beresford JN, Bennett JH, Devlin C, Leboy PS, Owen ME (1992) Evidence for an inverse relationship between the differentiation of adipocytic and osteogenic cells in rat marrow stromal cell cultures. J Cell Sci 102(Pt 2):341–351
2. Caplan AI (1991) Mesenchymal stem cells. J Orthop Res 9(5):641–650. https://doi.org/10.1002/jor.1100090504
3. Cheng SL, Yang JW, Rifas L, Zhang SF, Avioli LV (1994) Differentiation of human bone marrow osteogenic stromal cells in vitro: induction of the osteoblast phenotype by dexamethasone. Endocrinology 134(1):277–286. https://doi.org/10.1210/endo.134.1.8275945
4. Johnstone B, Hering TM, Caplan AI, Goldberg VM, Yoo JU (1998) In vitro chondrogenesis of bone marrow-derived mesenchymal progenitor cells. Exp Cell Res 238(1):265–272. https://doi.org/10.1006/excr.1997.3858
5. Pittenger MF, Mackay AM, Beck SC, Jaiswal RK, Douglas R, Mosca JD, Moorman MA, Simonetti DW, Craig S, Marshak DR (1999) Multilineage potential of adult human mesenchymal stem cells. Science 284(5411):143–147
6. Zuk PA, Zhu M, Ashjian P, De Ugarte DA, Huang JI, Mizuno H, Alfonso ZC, Fraser JK, Benhaim P, Hedrick MH (2002) Human adipose tissue is a source of multipotent stem cells. Mol Biol Cell 13(12):4279–4295. https://doi.org/10.1091/mbc.E02-02-0105
7. Zuk PA, Zhu M, Mizuno H, Huang J, Futrell JW, Katz AJ, Benhaim P, Lorenz HP, Hedrick MH (2001) Multilineage cells from human adipose tissue: implications for cell-based therapies. Tissue Eng 7(2):211–228. https://doi.org/10.1089/107632701300062859
8. Banfi A, Muraglia A, Dozin B, Mastrogiacomo M, Cancedda R, Quarto R (2000) Proliferation kinetics and differentiation potential of ex vivo expanded human bone marrow stromal cells: implications for their use in cell therapy. Exp Hematol 28(6):707–715
9. Baxter MA, Wynn RF, Jowitt SN, Wraith JE, Fairbairn LJ, Bellantuono I (2004) Study of telomere length reveals rapid aging of human marrow stromal cells following in vitro expansion. Stem Cells 22(5):675–682. https://doi.org/10.1634/stemcells.22-5-675
10. Reiser J, Zhang XY, Hemenway CS, Mondal D, Pradhan L, La Russa VF (2005) Potential of mesenchymal stem cells in gene therapy approaches for inherited and acquired diseases. Expert Opin Biol Ther 5(12):1571–1584. https://doi.org/10.1517/14712598.5.12.1571
11. Baharvand H, Hashemi SM, Shahsavani M (2008) Differentiation of human embryonic stem cells into functional hepatocyte-like cells in a serum-free adherent culture condition. Differentiation 76(5):465–477. https://doi.org/10.1111/j.1432-0436.2007.00252.x
12. Bittinger F, Brochhausen C, Skarke C, Kohler H, Kirkpatrick CJ (1997) Reconstruction of peritoneal-like structure in three-dimensional collagen gel matrix culture. Exp Cell Res 236(1):155–160. https://doi.org/10.1006/excr.1997.3724
13. Kelm JM, Timmins NE, Brown CJ, Fussenegger M, Nielsen LK (2003) Method for generation of homogeneous multicellular tumor spheroids applicable to a wide variety of cell types. Biotechnol Bioeng 83(2):173–180. https://doi.org/10.1002/bit.10655
14. Takahashi T, Ogasawara T, Asawa Y, Mori Y, Uchinuma E, Takato T, Hoshi K (2007) Three-dimensional microenvironments retain chondrocyte phenotypes during proliferation culture. Tissue Eng 13(7):1583–1592. https://doi.org/10.1089/ten.2006.0322
15. Tong JZ, Sarrazin S, Cassio D, Gauthier F, Alvarez F (1994) Application of spheroid culture to human hepatocytes and maintenance of their differentiation. Biol Cell 81(1):77–81
16. Wang Y, Kim UJ, Blasioli DJ, Kim HJ, Kaplan DL (2005) In vitro cartilage tissue engineering with 3D porous aqueous-derived silk scaffolds and mesenchymal stem cells. Biomaterials 26(34):7082–7094. https://doi.org/10.1016/j.biomaterials.2005.05.022

17. Timmins NE, Dietmair S, Nielsen LK (2004) Hanging-drop multicellular spheroids as a model of tumour angiogenesis. Angiogenesis 7(2):97–103. https://doi.org/10.1007/s10456-004-8911-7
18. Kurosawa H (2007) Methods for inducing embryoid body formation: in vitro differentiation system of embryonic stem cells. J Biosci Bioeng 103(5):389–398. https://doi.org/10.1263/jbb.103.389
19. Ungrin MD, Joshi C, Nica A, Bauwens C, Zandstra PW (2008) Reproducible, ultra high-throughput formation of multicellular organization from single cell suspension-derived human embryonic stem cell aggregates. PLoS One 3(2):e1565. https://doi.org/10.1371/journal.pone.0001565
20. Baraniak PR, McDevitt TC (2012) Scaffold-free culture of mesenchymal stem cell spheroids in suspension preserves multilineage potential. Cell Tissue Res 347(3):701–711. https://doi.org/10.1007/s00441-011-1215-5
21. Cook MM, Futrega K, Osiecki M, Kabiri M, Kul B, Rice A, Atkinson K, Brooke G, Doran M (2012) Micromarrows--three-dimensional coculture of hematopoietic stem cells and mesenchymal stromal cells. Tissue Eng Part C Methods 18(5):319–328. https://doi.org/10.1089/ten.TEC.2011.0159
22. Kabiri M, Kul B, Lott WB, Futrega K, Ghanavi P, Upton Z, Doran MR (2012) 3D mesenchymal stem/stromal cell osteogenesis and autocrine signalling. Biochem Biophys Res Commun 419(2):142–147. https://doi.org/10.1016/j.bbrc.2012.01.017
23. Markway BD, Tan GK, Brooke G, Hudson JE, Cooper-White JJ, Doran MR (2010) Enhanced chondrogenic differentiation of human bone marrow-derived mesenchymal stem cells in low oxygen environment micropellet cultures. Cell Transplant 19(1): 29–42. https://doi.org/10.3727/096368909X478560
24. Rodbell M (1966) Metabolism of isolated fat cells. II. The similar effects of phospholipase C (Clostridium Perfringens Alpha toxin) and of insulin on glucose and amino acid metabolism. J Biol Chem 241(1):130–139
25. Rettinger CL, Fourcaudot AB, Hong SJ, Mustoe TA, Hale RG, Leung KP (2014) In vitro characterization of scaffold-free three-dimensional mesenchymal stem cell aggregates. Cell Tissue Res 358(2):395–405. https://doi.org/10.1007/s00441-014-1939-0

Chapter 4

Differentiation of Brown Adipocyte Progenitors Derived from Human Induced Pluripotent Stem Cells

Anne-Laure Hafner, Tala Mohsen-Kanson, and Christian Dani

Abstract

In this chapter, we describe a protocol to induce at a high rate the differentiation of brown/brown like adipocyte progenitors (BAPs) derived from human induced pluripotent stem cells (hiPSCs). We also describe culture conditions to maintain hiPSCs and to derive hiPSC-BAPs.

This novel culture system provides an unlimited source of human brown adipocytes and a unique means for studying events regulating the generation and recruitment of human BAPs.

Key words Human induced pluripotent stem cells, Brown adipocytes, Adipocyte, Differentiation, Adipocyte progenitors

1 Introduction

In mammals, three types of adipocytes exist, i.e., brown, beige, and white adipocytes, which are all involved in regulation of the energy balance while having opposite functions. White adipose tissue (WAT) is dispersed throughout the body and is mainly involved in energy storage. In contrast to WAT, brown adipose tissue (BAT) is specialized in energy expenditure. Activated BAT consumes metabolic substrate and burns fat to produce heat thanks to the Uncoupling Protein (UCP)-1. Beige adipocytes have been recently described as brown-like adipocytes and represent a third type of adipocytes that are recruited in WAT [1]. Activation of BAT or recruitment of beige adipocytes leads to powerful antiobesity and antidiabetic effects in mice. It was previously reported that BAT implants were able to restore normoglycemia in diabetic mice [2, 3]. Therefore, the notion of transplanting brown or beige adipocyte progenitors (APs) in obese patients as a therapeutic approach to counteract obesity and its associated metabolic complications recently emerged. However, BAT represents a minor fraction of adipose tissue in humans and disappears from most areas with age, persisting only around deeper organs [4].

Bruce A. Bunnell and Jeffrey M. Gimble (eds.), *Adipose-Derived Stem Cells: Methods and Protocols*, Methods in Molecular Biology, vol. 1773, https://doi.org/10.1007/978-1-4939-7799-4_4, © Springer Science+Business Media, LLC, part of Springer Nature 2018

Human brown APs are hard to isolate in this regard. Therefore, a cellular source is urgently needed to amplify and characterize human BAPs and to identify factors involved in their recruitment.

Following the pioneer work of Yamanaka's group on the generation of patient-specific induced pluripotent stem cells (hiPSCs) by reprogramming fibroblasts [5], hiPSCs emerged as an unlimited source of APs for autologous cell-based therapy to treat obesity. The capacity of hiPSCs to generate functional adipocytes was first reported by Nakao's group. Interestingly, adipocytes generated from hiPSCs can maintain their functional properties for several weeks after transplantation into nude mice [6, 7]. These data revealed that hiPSC-adipocytes could potentially be used to correct metabolic parameters in lipodystrophic patients. In these experiments, differentiated hiPSCs, but not purified APs, were transplanted into mice. Indeed, differentiated hiPSC cultures are enriched with adipocytes after adipogenic induction, but contain also several other cell types that are undesirable for transplantation, including undifferentiated iPSCs that can form teratomas. This indicates that purification of hiPSC-APs with a high adipogenic capacity is required prior proposing hiPSC-based therapeutic approaches. Nishio et al. [8] developed a procedure to generate functional brown adipocytes at a high frequency from hiPSCs using a hematopoietic cocktail to induce hiPSC differentiation. Remarkably, hiPSC-brown adipocytes were able to improve glucose tolerance after transplantation in mice. This report indicated that hiPSCs could potentially be used to generate brown adipocytes with therapeutic properties. Finally, Ahfeldt et al. [9] were able to generate pure brown and white adipocytes from hiPSCs, but only following transduction with master genes of adipogenesis. The need to genetically modify hiPSC-APs to generate adipocytes clearly illustrates the low adipogenic potential of hiPSC-derived APs and hampers their clinical use.

We recently reported the potential of hiPSCs to generate both brown and white APs [10]. However, as we recently discussed [11], a bottleneck is the weak adipogenic potential of hiPSC-derived APs compared to adult adipose tissue-derived APs. This feature has been observed by us and others using different approaches to derive mesenchymal cells from hESCs and hiPSCs. Therefore, the differentiation of hiPSC-BAPs requires a protocol different than those used to differentiate APs derived from adult tissues. We describe in this chapter a protocol to differentiate hiPSC-BAPs at a high rate.

2 Materials

2.1 Preparation of STO Feeder Cells

1. STO feeder cells, which are mouse embryonic fibroblasts, are available from European Collection of Cell Cultures; ECACC (https://www.phe-culturecollections.org.uk/collections/ecacc.aspx).

2. Feeder cell growth medium. DMEM High glucose, supplemented with ((final concentrations) FCS (10%), L-glutamine (2 mM), sodium pyruvate (1 mM), HEPES buffer (10 mM), penicillin (50 units), and streptomycin (50 mg/mL).
3. Solution of 0.1% gelatin.
4. Feeder cell dissociation buffer. Solution of trypsin (0.25%) and ethylenediaminetetraacetic acid (EDTA) (1 mM).
5. Feeder cell mitomycin C-inactivation buffer (*see* **Note 1**): Mitomycin C (10 μg/mL).

2.2 Culture/ Maintenance of hiPSCs

1. hiPSC Growth medium (*see* **Note 2**): DMEM/F12 supplemented with (final concentrations) KSR (20%), FGF2 (10 ng/mL), L-glutamine (2 mM), nonessential amino acids (0.1 mM), 2-mercaptoethanol (0.1 mM), penicillin (50 units), and streptomycin (50 mg/mL).
2. hiPSC dissociation buffer (*see* **Note 3**): Trypsin (0.25%), collagenase IV (1 mg/mL), KSR (20%), and Cacl2 (1 mM).
3. Six-well plates tissue culture grade.

2.3 hiPSCs Freezing Buffer

1. Solution of Y-27632 (10 mM).
2. Freezing medium (*see* **Note 4**): DMSO (12%), acetamine (0.6 M), and 1.2 propanediol (0.7 M), hiPSC growth medium (5%).

2.4 hiPSC Differentiation

1. Differentiation medium: DMEM/F12 medium supplemented with 20% KSR.
2. Six-well plates, ultralow attachment.
3. Solution of 0.5% gelatin.

2.5 BAPs Derivation and Expansion

1. Mesenchymal cell derivation and expansion medium: DMEM Low glucose supplemented with ((final concentrations) FCS (10%), FGF2 (5 ng/mL), L-glutamine (2 mM), sodium pyruvate (1 mM), HEPES buffer (10 mM), penicillin (50 units), and streptomycin (50 mg/mL).
2. Freezing medium: solution of DMSO/FCS (10/10%) in growth medium.
3. Antibodies to identify BAPs by FACS: CD105, CD90, CD146, CD73, CD31, and CD45.
4. Sequences of primers to mark BAPs by real-time PCR:

   hDIO2: Fw GTCACTGGTCAGCGTGGTTTT; Rev. TTCTTCACATCCCCCAATCCT

   hPAX3: Fw ACACCGTGCCGTCAGTGAGT; Rev. TCGCTTTCCTCTGCCTCCTT

hHOXC8: Fw GTCTCCCAGCCTCATGTTTC; Rev. TCTGATACCGGCTGTAAGTTTGC

hHOXC9: Fw GCAGCAAGCACAAAGAGGA; Rev. CGTCTGGTACTTGGTGTAGGG

hHOXA5: Fw CCCAGATCTACCCCTGGATG; Rev. CAGGGTCTGGTAGCGCGTGT

5. Sequences of primers to mark BAP-differentiated progenies:

hUCP1: Fw GTGTGCCCAACTGTGCAATG, Rev. CCAGGATCCAAGTCGCAAGA

hPLN1: Fw ACCATCTCCACCCGCCTC; Rev. GATGGGAACGCTGATGCTGT

2.6 BAP Differentiation

1. Differentiation medium I: EBM-2 medium supplemented with (final concentrations) FCS (0.1%), SB431542 (5 μM), ascorbic acid (25.5 μg/mL), hydrocortisone (4 μg/mL), EGF (10 ng/mL), IBMX (0.5 mM), dexamethasone (0.25 μM), T3 (0.2 nM), insulin (1 μg/mL), and rosiglitazone (1 μM).
2. Differentiation medium II (*see* **Note 5**): EBM-2 medium supplemented with (final concentrations) FCS (0.1%), SB431542 (5 μM), ascorbic acid (25.5 μg/mL), hydrocortisone (4 μg/mL), EGF (10 ng/mL), T3 (0.2 nM), insulin (1 μg/mL), and rosiglitazone (1 μM).

3 Methods

We routinely maintain undifferentiated hiPSC on mitomycin C-inactivated STO feeder cell layers. BAPs are derived from human iPS cells, named hiPS NOK6, -NOK6-tat, and -O27-tat, that were generated by reprogramming neural stem cells [12] with PiggyBac technology [13] and from hiPS cells, named hiPS-10, that were generated by reprogramming adipose-derived stem cells [14]. These hiPSCs were previously described in [10].

3.1 Preparation of Mitomycin C-Inactivated STO Feeder Cells

1. Place 10^6 STO cells per gelatin-coated 100 mm dish in feeder cell growth medium. Incubate overnight at 37 °C, 5% CO_2.
2. The day after, change the feeder cell growth medium with feeder cell growth medium supplemented with mitomycin C. Incubate for 2 h at 37 °C, 5% CO_2.
3. After incubation, aspirate the mitomycin-C medium and wash the cells three times with 10 mL of PBS.
4. Aspirate off PBS and add 1 mL of feeder cell dissociation buffer. Incubate at 37 °C, 5% CO_2 for 2–3 min. Check under an inverted microscope that cells are detached from the bottom of the dish.

5. Add 10 mL of growth medium to stop trypsinization and resuspend the cells by pipetting. Transfer the cells to a sterile tube and centrifuge at 250 × *g* for 5 min at room temperature.
6. Aspirate the medium off and resuspend the cell pellet with 5 mL of growth medium by pipetting up and down two or three times. Count the cells (*see* **Note 6**).
7. Seed the cells at 0.3 × 10^6 cells per gelatin-coated well of six-well plate.
8. Before seeding hiPSCs, aspirate the feeder growth medium and wash twice with PBS.
9. Seed hiPSCs and maintain in hiPSC growth medium.

3.2 Expansion of hiPSCs

1. One hour before passaging, add 10 mM of Y-27632 in each dish containing hiPSCs. This step dramatically increases hiPSC viability.
2. Aspirate the medium and wash twice with PBS.
3. Add 0.3 mL of hiPSC dissociation buffer per well and incubate at 37 °C for about 2 min. Check under an inverted microscope that feeder cells are detached from the well while iPS colonies remain attached.
4. Aspirate off the dissociation buffer. Add 0.5 mL of PBS per well.
5. Aspirate off the PBS. Add 2 mL of hiPSC growth medium and resuspend the cells to small clumps by pipetting up and down with a 5 mL sterile pipette. Collect in a 15 mL tube.
6. Add 3 mL of hiPSC growth medium to harvest the rest of cell aggregates. Pool with the 2 mL.
7. Centrifuge at 250 × *g* for 1 min at room temperature.
8. Wash the plate containing mitomycin C treat-STO, thawed the day before, twice with PBS. Aspirate the PBS and add 1 mL of hiPSC growth medium.
9. Suspend the hiPSC pellet with 6 mL of hiPSC growth medium. Pipet up and down with a blue tip.
10. Transfer 1 mL of hiPSC suspension to a well of 6-well plates containing mitomycin C-treated STO feeder cells prefilled with 1 mL of hiPSC growth medium
11. Incubate at 37 °C, 5% CO2 until cells become 80% confluent in six-well plates.
12. Change the hiPSC growth medium every day.

3.3 hiPSC Freeze Stock

1. Before freezing, add 10 μM of Y-27632 in each dish containing hiPSCs and incubate for 1 h.
2. Aspirate the medium and wash with 2 mL of PBS.

3. Aspirate the PBS and add 0.3 mL of hiPSC dissociation buffer per well. Proceed as describe Subheading 3.2, **steps 3–8**.
4. Suspend the hiPSC pellet with 0.2 mL of hiPSC freezing buffer.
5. Transfer the cell suspension to freeze vials.
6. Put the vials quickly into liquid nitrogen.

3.4 Thawing hiPSCs from Freezing Tubes in Liquid Nitrogen

1. Prepafre 10 mL of prewarmed hiPSC growth medium supplemented with 10 μM of Y-27632.
2. Take frozen stocks of hiPSCs from liquid nitrogen.
3. Add 0.8 mL of hiPSC growth medium prepared at the **step 1** and thaw cells quickly by shaking the vials gently.
4. Transfer the cell suspension to the tube prepared at the **step 1** and centrifuge at 250 × *g* for 2 min.
5. Aspirate the supernatant and add 2 mL of hiPSC growth medium.
6. Transfer cell suspension to a well of 6-well plate seeded with mitomycin C-treated STO cells. Do not move the plate for 48 h. Change the medium every day beginning at day 2 (*see* **Note 7**).

3.5 hiPSC Differentiation

Differentiation of hiPSC is initiated by floating cultivation to form embryoid bodies (EBs).

The experimental scheme with timeline for hiPSCs and BAPs differentiation is given in Fig. 1.

1. Seed hiPSCs from a well of 6-well plate onto a well of ultralow attachment 6-well plate in hiPSC growth medium without FGF2. This is considered as the day 0 of EB formation.
2. Do not move the plate for 3 days. Change the medium every day from day 3.
3. Ten days after EB formation, EBs are plated onto one gelatin-coated well of 6-well plate tissue culture grade and maintained in hiPSC growth medium without FGF2.
4. Do not move the plate for 3 days. Change the medium every other day from day 3 to day 10 after EB plating.

3.6 BAP Derivation

Derivation of BAPS is performed 10 after EB plating, i.e., 20 days after EB formation. The derivation is mainly carried out as previously described by Brown and colleagues for mesenchymal stem cell derivation from human embryonic stem cells [15].

1. Remove the hiPSC growth medium from the wells and wash once with PBS.
2. Remove PBS and add 2 mL of mesenchymal cell derivation and expansion medium. Change the medium every other day.
3. A week after, trypsinize cells as described Subheading 3.1, **steps 4–5**.

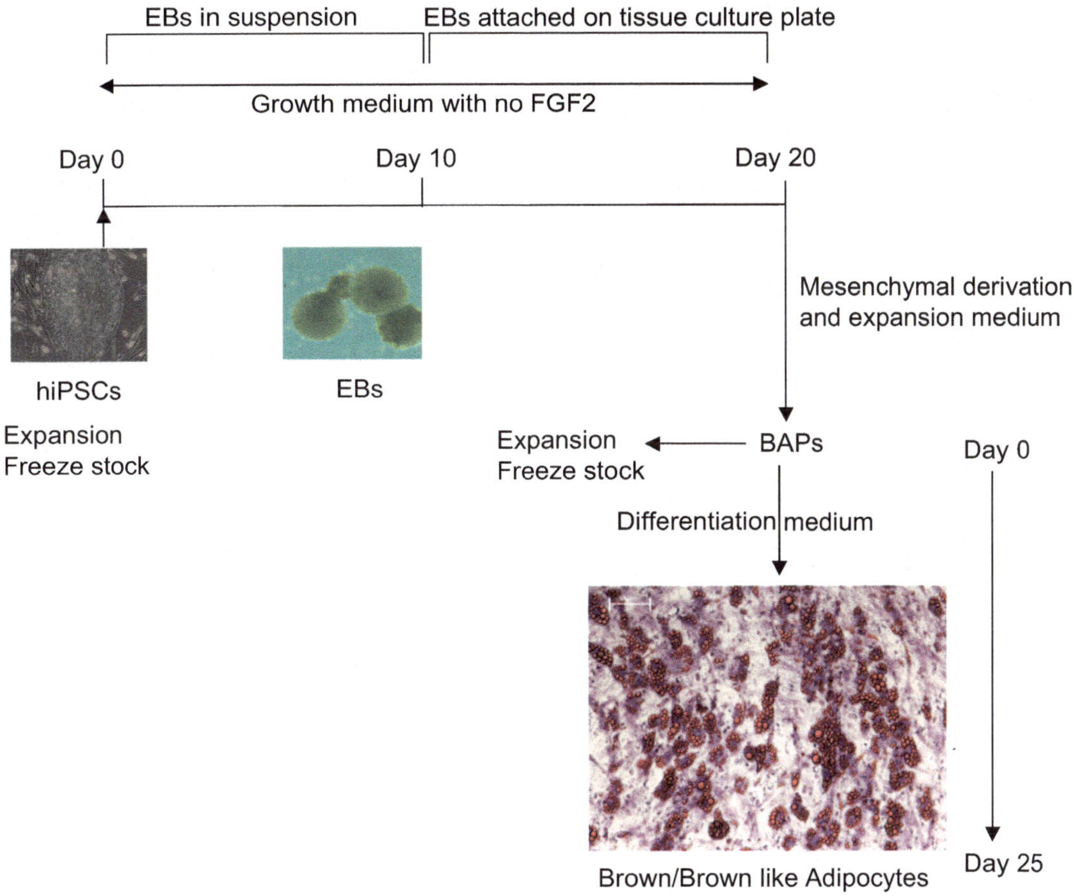

Fig. 1 Schematic diagram for hiPSC-BAP derivation and differentiation. BAPs are derived after 20 days of hiPSC differentiation. Adipocytes, stained with Oil Red O for lipid droplets, are detectable after the maintenance of BAPs in the adipogenic cocktail

4. Seed the cells with a 1:3 split ratio and maintain cells in mesenchymal cell derivation and expansion medium.
5. Repeat **step 4** when cells reach 80% confluence.
6. Repeat **step 5** until achieving homogenous CD73 labeling determined by classical FACS analysis. Isolated BAPs can be characterized in more details as following: Expression of CD105, CD90, CD146 and CD73 at the cell surface and the absence of CD31 and CD45 determined by FACS marks the mesenchymal origin of isolated cells. Expression of Pax3 and Dio2 and no expression of HOXC8, HOXC9, and HOXA5 determined by real-time PCR mark BAPs as we previously described [10].
7. Expand and freeze BAPs in freezing medium.

3.7 BAP Differentiation

1. Plate BAPs at a high density, i.e., 5×10^4 cells/cm^2 in mesenchymal cell derivation and expansion medium.
2. One or 2 days after, when cells reach confluence, wash cells once with PBS.
3. Remove PBS and add BAP differentiation medium I.
4. Three days later, switch cells into BAP differentiation medium II. Change medium once a week.
5. Adipocytes should appear 10 days after and can be maintained up to 35 days. Figure 1 shows BAPs visualized after Oil Red O staining 25 days after induction of the differentiation.
6. Adipocytes differentiated from isolated BAPs can be further characterized by monitoring expression of perilipin 1 (PLN1) gene (as an adipocyte marker) and of UCP1 gene (as a marker of brown/brown-like adipocytes) (*see* **Note 8**).

4 Notes

1. To prepare 200 mL of mitomycin C at 10 μg/mL: Dissolve 2 mg of mitomycin C into 5 mL of PBS, then into 195 mL of feeder cell growth medium (*see* Subheading 2, **item 1**). Aliquot into 10 mL and store at −20 °C. Mitomycin C is light sensitive.
2. To prepare 2-mercaptoethanol 1000×: add 100 μL of 2-mercaptoethanol 14 M into 14 mL H_2O and filter sterilized. Store at 4 °C up to 2 weeks. FGF2 is added into the growth medium before use. Store the growth medium at 4 °C up to a week.
3. To prepare 100 mL of dissociation buffer, mix 10 mL of 2.5% trypsin, 10 mL of 10 mg/mL collagenase IV, 20 mL of KSR, 1 mL of 100 mM $CaCl_2$, and 59 mL of PBS. Aliquot and store at −20 °C.
4. To prepare 10 mL of hiPSC freezing buffer, mix 1.42 mL of DMSO, 6 mL of acetamide 1 M, 2.2 mL of 1.2 propanediol 3 M, and 0.5 mL of hiPSC growth medium. Aliquot into 0.4 mL and store at −80 °C.
5. Differentiation medium II is the differentiation medium I without dexamethasone and IBMX.
6. Mitomycin-inactivated STO can be used the day after or can be frozen at −80 °C or in liquid nitrogen to make a stock. This stock should be thawed in a gelatin-coated plate with the 3 days before use.
7. The viability of hiPSCs after freezing is a critical point. It is important to perform steps of Subheading 3.3, **steps 3–6** and of Subheading 3.4, **steps 2–4** as quick as possible. The use of Y-27632 dramatically improves hiPSC viability.

8. At this stage, the level of UCP1 gene expression is similar to that observed upon differentiation of BAPs isolated from human chin fat depot, as described in [16].

References

1. Wu J, Bostrom P, Sparks LM, Ye L, Choi JH, Giang AH, Khandekar M, Virtanen KA, Nuutila P, Schaart G, Huang K, Tu H, van Marken Lichtenbelt WD, Hoeks J, Enerback S, Schrauwen P, Spiegelman BM (2012) Beige adipocytes are a distinct type of thermogenic fat cell in mouse and human. Cell 150(2):366–376
2. Stanford KI, Middelbeek RJ, Townsend KL, An D, Nygaard EB, Hitchcox KM, Markan KR, Nakano K, Hirshman MF, Tseng YH, Goodyear LJ (2013) Brown adipose tissue regulates glucose homeostasis and insulin sensitivity. J Clin Invest 123(1):215–223
3. Gunawardana SC, Piston DW (2010) Reversal of type 1 diabetes in mice by brown adipose tissue transplant. Diabetes 61(3):674–682
4. Frontini A, Cinti S (2010) Distribution and development of brown adipocytes in the murine and human adipose organ. Cell Metab 11(4):253–256
5. Takahashi K, Tanabe K, Ohnuki M, Narita M, Ichisaka T, Tomoda K, Yamanaka S (2007) Induction of pluripotent stem cells from adult human fibroblasts by defined factors. Cell 131(5):861–872
6. Taura D, Noguchi M, Sone M, Hosoda K, Mori E, Okada Y, Takahashi K, Homma K, Oyamada N, Inuzuka M, Sonoyama T, Ebihara K, Tamura N, Itoh H, Suemori H, Nakatsuji N, Okano H, Yamanaka S, Nakao K (2009) Adipogenic differentiation of human induced pluripotent stem cells: comparison with that of human embryonic stem cells. FEBS Lett 583(6):1029–1033
7. Noguchi M, Hosoda K, Nakane M, Mori E, Nakao K, Taura D, Yamamoto Y, Kusakabe T, Sone M, Sakurai H, Fujikura J, Ebihara K, Nakao K (2013) In vitro characterization and engraftment of adipocytes derived from human induced pluripotent stem cells and embryonic stem cells. Stem Cells Dev 22(21):2895–2905
8. Nishio M, Yoneshiro T, Nakahara M, Suzuki S, Saeki K, Hasegawa M, Kawai Y, Akutsu H, Umezawa A, Yasuda K, Tobe K, Yuo A, Kubota K, Saito M, Saeki K (2012) Production of functional classical brown adipocytes from human pluripotent stem cells using specific hemopoietin cocktail without gene transfer. Cell Metab 16(3):394–406
9. Ahfeldt T, Schinzel RT, Lee YK, Hendrickson D, Kaplan A, Lum DH, Camahort R, Xia F, Shay J, Rhee EP, Clish CB, Deo RC, Shen T, Lau FH, Cowley A, Mowrer G, Al-Siddiqi H, Nahrendorf M, Musunuru K, Gerszten RE, Rinn JL, Cowan CA (2012) Programming human pluripotent stem cells into white and brown adipocytes. Nat Cell Biol 14(2):209–219
10. Mohsen-Kanson T, Hafner AL, Wdziekonski B, Takashima Y, Villageois P, Carriere A, Svensson M, Bagnis C, Chignon-Sicard B, Svensson PA, Casteilla L, Smith A, Dani C (2014) Differentiation of human induced pluripotent stem cells into brown and white adipocytes: role of pax3. Stem Cells 32(6):1459–1467
11. Hafner AL, Dani C (2014) Human induced pluripotent stem cells: a new source for brown and white adipocytes. World J Stem Cells 6(4):467–472
12. Sun Y, Pollard S, Conti L, Toselli M, Biella G, Parkin G, Willatt L, Falk A, Cattaneo E, Smith A (2008) Long-term tripotent differentiation capacity of human neural stem (NS) cells in adherent culture. Mol Cell Neurosci 38(2):245–258
13. Guo G, Yang J, Nichols J, Hall JS, Eyres I, Mansfield W, Smith A (2009) Klf4 reverts developmentally programmed restriction of ground state pluripotency. Development 136(7):1063–1069
14. Rodriguez A-M, Pisani D, Dechesne CA, Turc-Carel C, Kurzenne J-Y, Wdziekonski B, Villageois A, Bagnis C, Breittmayer J-P, Groux H, Ailhaud G, Dani C (2005) Transplantation of a multipotent cell population from human adipose tissue induces dystrophin expression in the immunocompetent mdx mouse. J Exp Med 201(9):1397–1405
15. Brown SE, Tong W, Krebsbach PH (2009) The derivation of mesenchymal stem cells from human embryonic stem cells. Cells Tissues Organs 189(1–4):256–260
16. Kouidhi M, Villageois P, Mounier CM, Ménigot C, Rival Y, Piwnica D, Aubert J, Chignon-Sicard B, Dani C (2015) Characterization of human knee and chin adipose-derived stromal cells. Stem Cells Int 2015:592090

Chapter 5

Methylcellulose Based Thermally Reversible Hydrogels

Anoosha Forghani and Ram Devireddy

Abstract

The creation of single and multilayered adult stem cells (ASCs) sheets is presented. The stem cell sheets preserve the cell–cell and cell–extracellular matrices and are developed by utilizing a thermally reversible methylcellulose (MC) coated tissue culture polystyrene (TCPS) dish. This technique is an improvement and a simplification of earlier noninvasive cell retrieval methods based on the use of a temperature-responsive poly(N-isopropylacrylamide) (PIPAAm) coated TCPS dishes. The optimal combination of MC–water–salt was determined to be 12–14% of MC (mol. wt. of 15,000) in water with 0.5× PBS (~150 mOsm). This solution exhibited a gel formation temperature of ~32 °C. The addition (evenly spread) of 1 ml of 3 mg/ml rat tail type-I (pH adjusted to 7.5) over the MC coated surface at 37 °C improves ASC adhesion and proliferation on the methylcellulose system. Upon confluence, a continuous monolayer ASC sheet was formed on the surface of the MC hydrogel system. When the grown cell sheet was removed from the incubator and exposed to room temperature (~30 °C), it spontaneously and gradually detached from the surface of the thermoresponsive hydrogel, creating an ASC sheet.

Key words Cell sheet engineering, Temperature-responsive polymers, Adult stem cells, Tissue engineering

1 Introduction

Traditional tissue engineering methods have generally focused on one of two strategies: either the injection of isolated cell suspensions or the use of biodegradable scaffolds to support tissue formation. There has been considerable progress in developing cellularized scaffolds to regenerate various tissues including bone, cartilage, blood vessel, nerve, liver, myocardium, and many other tissues [1–10]. However, most of the conventional scaffolds do not allow sufficient cell migration to establish adequate cell–extracellular matrix (ECM) and cell–cell interaction. Therefore, high density cell seeding is commonly utilized in static and dynamic cultures inside bioreactors to establish adequate cell–extracellular matrix (ECM) and cell–cell interactions. The process of cell seeding plays a crucial role in determining the progression of tissue formation to establish 3D cell culture [11–14]. Unfortunately, the

Bruce A. Bunnell and Jeffrey M. Gimble (eds.), *Adipose-Derived Stem Cells: Methods and Protocols*, Methods in Molecular Biology, vol. 1773, https://doi.org/10.1007/978-1-4939-7799-4_5, © Springer Science+Business Media, LLC, part of Springer Nature 2018

cells located in the interior of the scaffold rely on diffusion for solute transport, and thus, the current methods of seeding are compromised by the depletion of nutrients [12, 15–17].

To overcome some of the drawbacks associated with traditional scaffold based tissue engineering, cell sheet engineering has been developed as an alternative approach for tissue engineering [18–25]. Cell sheet engineering has the advantage of eliminating the use of biodegradable scaffolds. Using temperature-responsive poly(N-isopropylacrylamide) (PIPAAm) coated tissue culture polystyrene (TCPS) dishes, cultured cells were successfully harvested as intact sheets by simple temperature changes, thereby avoiding the use of proteolytic enzymes [26]. In a more recent study, a novel method, using a thermoresponsive hydrogel coated on TCPS dishes, was developed for harvesting living cell sheets [27]. The hydrogel was prepared by pouring aqueous methylcellulose blended with distinct salts on TCPS dishes at room temperature and subsequently gelled at 37 °C for cell culture. Methylcellulose (MC) with a viscosity of 4000 cP at 2% solution was used to form gels at physiologically relevant temperatures [27]. This chapter presents a modified MC–collagen hydrogel method to create cell sheets, specifically tailored to work with adipose tissue derived adult stem cells (ASCs) and for the creation of multidimensional cell sheets [28].

2 Materials and Major Equipment

Where appropriate, materials should be sterile and all manipulations performed using a laminar flow bench (fume hood). Adequate safety equipment (gloves, face shield) will be used during all liquid nitrogen handling procedures.

1. Sterile plastic dishes/pipettes/micropipettes.
2. Cell digestion/isolation, culture, and differentiation media (*see* text below for further details).
3. 37 °C CO_2 incubator.
4. 4 °C refrigerator.
5. Heater.
6. Uniform temperature control stirrer.
7. pH meter.
8. Osmometer.
9. Differential scanning calorimeter (DSC).
10. Light microscope.
11. PBS, iso-osmotic or 0.3 M.

12. Pure distilled water.
13. Tissue culture polystyrene (TCPS) dish.
14. Diluent: DMEM, DMEM-F12, fetal calf serum, and human serum.
15. Methylcellulose (M7140; Mol. wt of 15,000) obtained from Sigma Chemicals (Sigma Chemical Company, USA).
16. Type I collagen solution (3 mg/ml rat tail).

3 Methods

3.1 Collection and Isolation of ASCs

The process of adult stem cell isolation (and culture/expansion/differentiation) from human liposuction aspirate (adipose tissue) is described elsewhere [28–34] and is only briefly described below:

1. Obtain liposuction aspirates from the plastic surgeon.
2. Wash tissue and collagenase digest at 37 °C with continuous shaking for 1 h.
3. Centrifuge digest to obtain "stromal vascular fraction (SVF)" Pellet; *see* **Note 1**.
4. Suspend SVF in culture medium containing 10% FBS at equivalent of 1 ml of tissue per ml of media—*see* **Note 2**.
5. Seed flasks at a density of 0.156 ml tissue digest/sq. cm surface area as "Passage 0 (P0)." Incubate at 37 °C, 5% CO_2.
6. Wash cultures to remove nonadherent cells after 24 h of incubation.
7. Feed cultures every 2–3 days.
8. Harvest P0 by trypsin–EDTA digestion when it reaches 80–90% confluency (mean, 5.5 days).
9. Seed "Passage 1 (P1)" and subsequent passages of undifferentiated ASCs at a density of 5×10^3 cells per sq. cm and maintain according to **steps 6** and **7**.
10. Harvest P1 by trypsin/EDTA digestion when it reaches 80–90% confluency (mean, 5.5 days)—*see* **Note 3**.

3.2 Preparation of MC Solutions

MC solutions were formed by a dispersion technique, as described in the product manual. Table 1 lists the composition of the various MC–water mixtures investigated. Note that as the molecular weight of the MC increases the % of MC that can be dissolved in water decreases to a maximum of 6% for MC M0512 (Mol. wt of 88,000) to a maximum of 10% for MC M0262 (Mol. wt of 41,000) to a maximum of 16% for MC M7140 (Mol. wt of 15,000). Additional experiments were also conducted by adding salts (increasing the osmolality) to the MC–water system—Fig. 1.

Table 1
Gelation temperature of the methylcellulose (MC)–water solutions as a function of molecular weight and concentrations

MC concentration (%)	Gel point from calorimetry experiments (°C) M0512 (mol. wt. = 88,000)	M0262 (mol. wt. = 41,000)	M7140 (mol. wt. = 15,000)
2	53.1 (±0.8)	52.5 (±0.3)	50.8 (±0.7)
4	52.2 (±1.3)	48.2 (±1.1)	46.4 (±1.9)
6	48.2 (±1.0)	45.8 (±1.4)	44.9 (±1.5)
8	–	43.2 (±1.4)	44.1 (±0.9)
10	–	41.7 (±2.0)	40.1 (±1.2)
12	–	–	37.6 (±0.8)
14	–	–	34.6 (±1.9)
16	–	–	31.1 (±1.5)

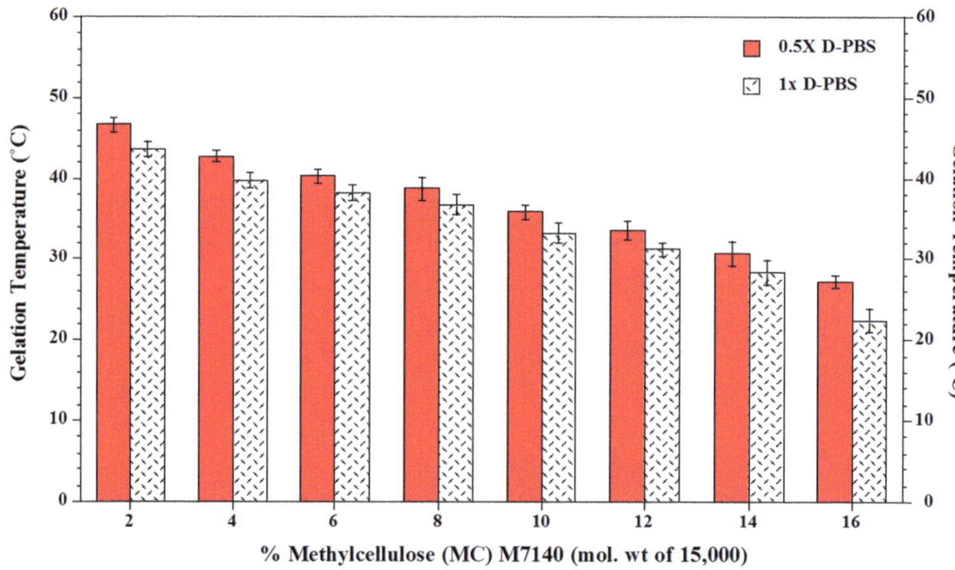

Fig. 1 Thermal gelation temperatures of MC M7140 (mol. wt. of 15,000) in 0.5× and 1× D-PBS solutions obtained using a calorimeter. MC–PBS solutions were prepared in different concentrations by dispersing the weighed MC powders in heated D-PBS solution at 90 °C. As shown in the figure increasing the concentration of MC in solution and increasing the osmolality of the D-PBS solutions from 0.5× to 1× leads to a lower gelation temperature. For example, a gelation temperature of ~25 °C can be achieved by using 16% MC in 1× D-PBS as opposed to a value of ~40 °C with 8% MC in 0.5× D-PBS. Since, lowering the concentration of MC leads to solutions with lower viscosity (and hence, increased ease of use), the optimal MC solution was chosen to be 14% MC with 0.5× D-PBS. This optimal solution of MC had a gelation temperature of ~32 °C. The gelation temperature is shown on the y-axis while the concentration of MC in solution (either 0.5× D-PBS or 1× D-PBS) is shown on the x-axis

1. Mix equal volume of 1× PBS with pure distilled water. This will create the solvent, 0.5× PBS solution.
2. Heat half of the required volume of the solvent (0.5× PBS) to 80 °C.
3. Stir in the MC powder and wet thoroughly.
4. Add in the remaining volume of the solvent (0.5× PBS) at ~30 °C to the heated MC–solvent mixture and gently stir.
5. Equilibrate the MC–solvent mixture to ~30 °C and stir/agitate for a further 30 min.
6. Equilibrate for 12–18 h in a 4 °C refrigerator, to obtain a clear viscous solution without any bubbles or inclusions.
7. Prior to use, slowly equilibrate (~couple of hours) the solutions with the room temperature by placing them in a sterile environment.

The optimal combination of MC–water–salt was found to be 12–14% of MC (mol. wt. of 15,000) in water with 0.5× PBS (~150 mOsm). This solution exhibited a gel formation temperature of ~32 °C.

3.3 Calorimetric Characterization of MC Solutions

The gelation temperature of the various MC–aqueous solutions was investigated using a differential scanning calorimeter (DSC), as described earlier [28, 35–42]. DSC has been routinely used to detect endothermic heat of dehydration of polymeric materials.

1. Place 15–20 μl of the MC–water solution in the DSC sample pan.
2. Place 15–20 μl of deionized or distilled water in the DSC reference pan; this will increase the sensitivity of the measurements. Alternatively use an empty DSC reference pan.
3. Heat the solution in the sample pan using the DSC from 10 to 90 °C at a heating rate of 10 °C/min.
4. The transition temperature in the DSC thermograms can be obtained using the DSC software—Fig. 2.

3.4 MC Coated Tissue Culture Polystyrene (TCPS) Dish

1. Place approximately 1 ml of the chosen MC–aqueous solution into the center of each TCPS dish (35 mm by 10 mm) at ~30 °C.
2. Distribute the aqueous solution onto the TCPS dish to evenly spread a thin transparent layer.
3. Incubate the MC coated TCPS dish at 37 °C for 1 h; at which time an opaque and gelled MC hydrogel layer should be formed on the dish—Fig. 3.

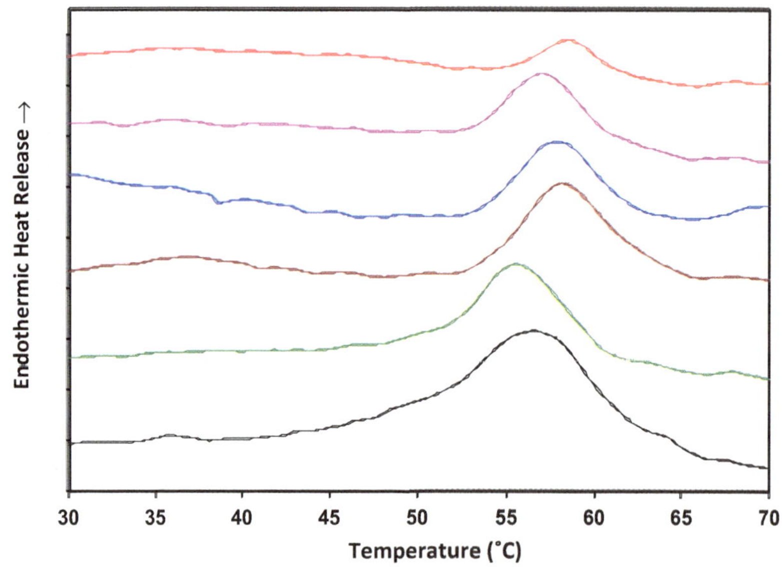

Fig. 2 A series of differential scanning calorimeter (DSC) thermograms showing the incipient gelation temperature of MC M0512 (mol. wt. of 88,000) at various concentrations (1, 2, 3, 4, 5, and 6%) in 1× PBS solution. The thermograms from top to bottom represent increasing concentrations of MC in solution. For the ease of visualization, the thermograms for all the concentrations were displaced by a stepwise amount to prevent overlapping of the curves. In this figure, increasing the amount of MC in solution not only increases the endothermic peak height but it also makes it broader. This suggests that the total energy absorbed during heating of the MC solutions is dependent upon the concentration of MC. The temperature is shown on the x-axis, while the endothermic heat release is shown on the y-axis

Fig. 3 Photomicrograph of a detached intact MC cell sheet. Approximately 1 ml of the aqueous MC solution was poured into the center of each TCPS dish at room temperature (30 °C) and evenly spread into a thin transparent layer. The TCPS dish was placed in a convective drying at 37 °C for 1 h and subsequently, coated with 1 ml aqueous type I collagen (3 mg/ml, Rat tail, Gibco by Life Technologies) at 37 °C followed by incubation at 37 °C in humidified incubator for a further 30 min. The scale bar in the figure represents 1 cm

3.5 Assessment of Degradation, Swelling and Osmotic Stability of MC Gels

1. Measure the initial weight of the MC hydrogel—"MC weight."
2. Add 2 ml of cell culture media, (DMEM-F12 and 10% fetal bovine serum) to the TCPS dishes and measure the weight—"Wet MC weight @ $t = 0$".
3. Assess the weight of the MC hydrogel at various time points (1, 4, 7 and 10 days)—"Wet MC weight @ t".
4. Dry the MC hydrogels overnight in a convective oven at 70 °C. Record the dried weight as—"Dried MC weight @ t".
5. Quantity Swelling and Degradation, as a function of time (t), using the following formulae:

$$\%\text{Swelling}(t) = \frac{\text{Wet MC weight} @ t = 0}{\text{Wet MC weight} @ t}$$

$$\%\text{Degradation}(t) = \frac{(\text{Dried weight} @ t = 0) - (\text{Dried weight} @ t)}{(\text{Dried weight} @ t = 0)}$$

6. Monitor the osmolality of the loaded media was monitored with time using an osmometer to evaluate whether the salts blended in the MC hydrogel would leach out with time during in vitro culture; utilize an uncoated TCPS dish loaded with the same media as control.

3.6 Culture of ASCs and NIH/3T3 Cells on MC Coated TCPS Dishes

1. Adjust the pH of aqueous type I collagen to 7.5.
2. Coat approximately 1 ml aqueous type I collagen (3 mg/ml rat tail collagen type-I; pH adjusted to 7.5) over the MC coated TCPS dish surface at 37 °C. *See* **Note 4**.
3. Add ASCs (~200,000 cells/ml in the cell culture media) to the MC coated TCPS dish or 3T3 cells (1×10^6 cells/dish). Make certain that the temperature of the media is 37 °C (and not below the gel transition temperature of 32 °C). *See* **Note 5**.
4. Observe cell attachment and growth using a light microscope; utilize an uncoated TCPS dish loaded with the same media as control—Figs. 4 and 5.

3.7 Thermal Reversibility of Attachment

1. Upon confluence, a continuous monolayer ASC sheet will be formed on the surface of the MC hydrogel system—Fig. 3.
2. Remove the TCPS dish from the incubator and equilibrate with an environment at 30 °C (or at a temperature below the transition temperature obtained earlier using the DSC).
3. The ASC sheet will spontaneously and gradually detach from the TCPS dish within 1 min. Detachment should be complete within 2–3 min. *See* **Note 6**.

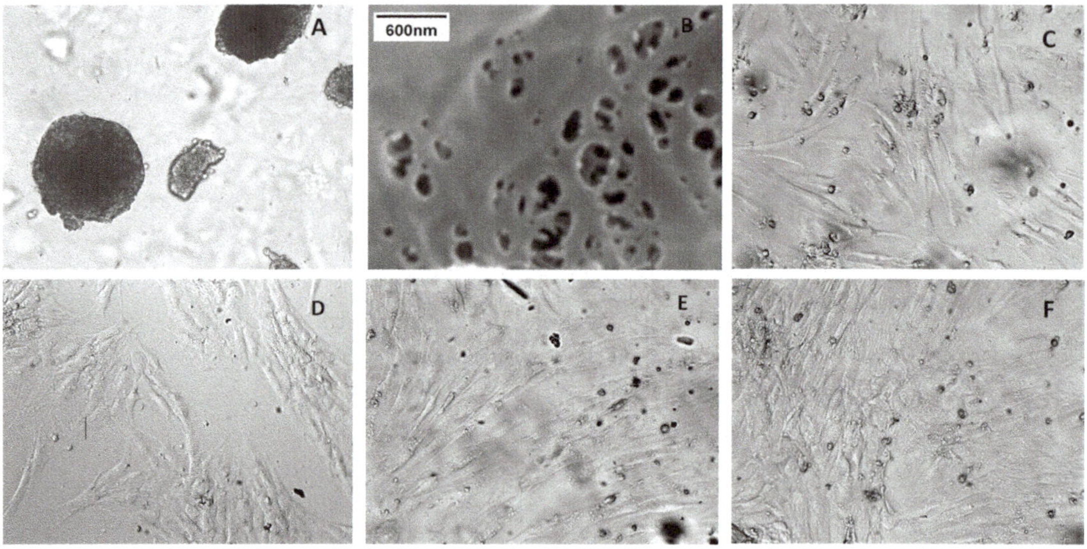

Fig. 4 (**a**) ASCs grown on MC hydrogel system for 1 day (**b**) SEM image of the collagen network formed onto the MC coated TCPS dish (**c**) ASCs grown on MC hydrogel system coated with collagen for 1 day (**d**) Cell attached and grown on a noncoated TCPS dish (control) for 1 day. The ASC were completely unreceptive to the hydrogel surface and generated embryoid-like structures (**a**). The addition (evenly spread) of 200 μl of 2 mg/ml bovine collagen type-I (pH adjusted to 7.5) over the MC coated surface at 37 °C (**b**), significantly improved the ASC cell adhesion on the hydrogel system (**c**). The cells grown on a control uncoated TCPS dish are also shown for reference (**d**). Also shown are ASCs cultured and grown for 3 days (**e**) and 5 days (**f**) days on the MC hydrogel system

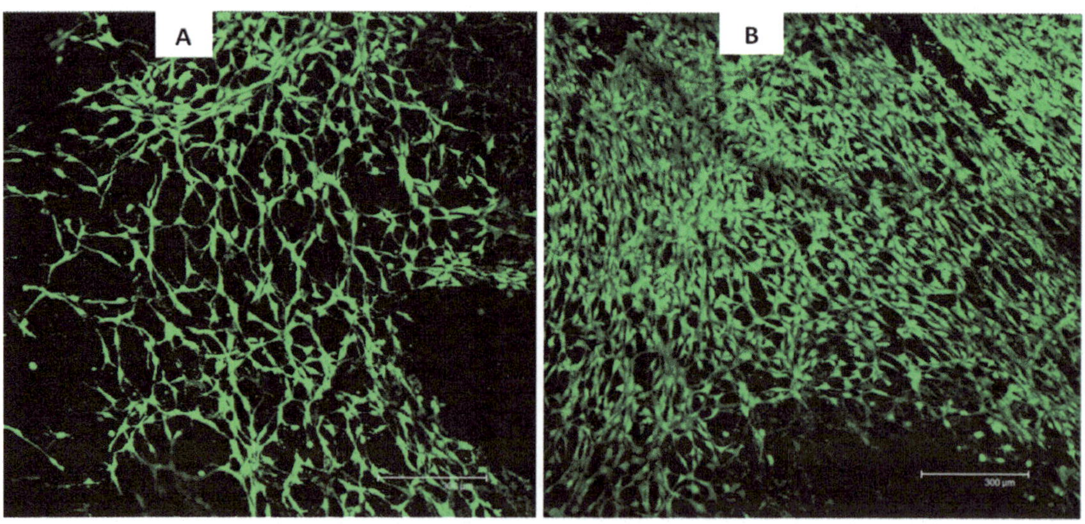

Fig. 5 Live 3T3 cells stained with Calcein-AM grown on MC hydrogel system for 1 day at 20× (**a**) and 10× (**b**). The scale bar represents 300 μm in both the figures

4 Notes

1. Centrifuge the SVF at room temp for 5 min with the speed of 1500 rpm.
2. Culture media is DMEM F-12 containing 1% triple antibiotics and10% FBS.
3. Detach the cells by adding trypsin-EDTA and incubation at 37 °C CO_2 incubator for 7–10 min.
4. The addition (evenly spread) of 200 μl of 2 mg/ml bovine collagen type-I (pH adjusted to 7.5) over the MC coated surface at 37 °C, significantly improves the ASC cell adhesion and proliferation on the hydrogel system.
5. Care should be taken to add the media at temperature above the gel-transition temperature (>32 °C). Otherwise, the MC gel will disintegrate and the integrity of the cell sheet will be lost.
6. Upon confluence, a continuous monolayer ASC sheet should be formed on the surface of the MC hydrogel system. When the grown cell sheet is removed from the incubator and exposed to room temperature (~30 °C), it spontaneously and gradually detaches from the surface of the thermoresponsive hydrogel.

Acknowledgments

The authors thank Dr. Elizabeth Clubb and Dr. James Wade at the Pennington Biomedical Research Center (PBRC) for supplying the liposuction aspirates and their many patients for consenting to participate in this protocol. Acknowledgements are also due to Dr. Gimble, a longtime colleague and friend without whose help this work would never have been initiated. And finally, acknowledgements are also due to Dr. S. Thirumala for performing the seminal experiments reported in this chapter. This work was supported in part by funding from the Louisiana Board of Regents and the Department of Mechanical Engineering at the Louisiana State University (LSU). Anoosha Forghani was supported the NSF EPSCoR LA-SiGMA project under award #EPS-1003897.

References

1. Hubbell JA, Massia SP, Desai NP, Drumheller PD (1991) Endothelial cell-selective materials for tissue engineering in the vascular graft via a new receptor. Biotechnology (N Y) 9:568–572
2. Hutmacher DW (2000) Scaffolds in tissue engineering bone and cartilage. Biomaterials 21:2529–2543
3. Griffith LG (2002) Emerging design principles in biomaterials and scaffolds for tissue engineering. Ann N Y Acad Sci 961:83–95
4. Griffith LG, Naughton G (2002) Tissue engineering–current challenges and expanding opportunities. Science 295:1009–1014
5. Grignon J, Scallan A (1980) Effect of pH and neutral salts upon the swelling of cellulose gels. J Appl Polym Sci 25:2829–2843
6. Bauer TW, Muschler GF (2000) Bone graft materials. An overview of the basic science. Clin Orthop Relat Res:10–27

7. Muschler GF, Nakamoto C, Griffith LG (2004) Engineering principles of clinical cell-based tissue engineering. J Bone Joint Surg Am 86-A:1541–1558
8. Sikavitsas VI, Bancroft GN, Mikos AG (2002) Formation of three-dimensional cell/polymer constructs for bone tissue engineering in a spinner flask and a rotating wall vessel bioreactor. J Biomed Mater Res 62:136–148
9. Akizuki T, Oda S, Komaki M, Tsuchioka H, Kawakatsu N, Kikuchi A, Yamato M, Okano T, Ishikawa I (2005) Application of periodontal ligament cell sheet for periodontal regeneration: a pilot study in beagle dogs. J Periodontal Res 40:245–251
10. Galaev IY, Mattiasson B (1999) Smart polymers and what they could do in biotechnology and medicine. Trends Biotechnol 17:335–340
11. Sistino JJ (2003) Bioreactors for tissue engineering–a new role for perfusionists? J Extra Corpor Technol 35:200–202
12. Martin I, Wendt D, Heberer M (2004) The role of bioreactors in tissue engineering. Trends Biotechnol 22:80–86
13. Janssen FW, Oostra J, Oorschot A, van Blitterswijk CA (2006) A perfusion bioreactor system capable of producing clinically relevant volumes of tissue-engineered bone: in vivo bone formation showing proof of concept. Biomaterials 27:315–323
14. Dvir T, Benishti N, Shachar M, Cohen S (2006) A novel perfusion bioreactor providing a homogenous milieu for tissue regeneration. Tissue Eng 12:2843–2852
15. Khong YM, Zhang J, Zhou S, Cheung C, Doberstein K, Samper V, Yu H (2007) Novel intra-tissue perfusion system for culturing thick liver tissue. Tissue Eng 13:2345–2356
16. Avgoustiniatos ES, Colton CK (1997) Effect of external oxygen mass transfer resistances on viability of immunoisolated tissue. Ann N Y Acad Sci 831:145–167
17. Mikos AG, McIntire LV, Anderson JM, Babensee JE (1998) Host response to tissue engineered devices. Adv Drug Deliv Rev 33:111–139
18. Yamato M, Utsumi M, Kushida A, Konno C, Kikuchi A, Okano T (2001) Thermo-responsive culture dishes allow the intact harvest of multilayered keratinocyte sheets without dispase by reducing temperature. Tissue Eng 7:473–480
19. Matsuda N, Shimizu T, Yamato M, Okano T (2007) Tissue engineering based on cell sheet technology. Adv Mater 19:11
20. Akiyama Y, Kushida A, Yamato M, Kikuchi A, Okano T (2007) Surface characterization of poly(N-isopropylacrylamide) grafted tissue culture polystyrene by electron beam irradiation, using atomic force microscopy, and X-ray photoelectron spectroscopy. J Nanosci Nanotechnol 7:796–802
21. Shimizu T, Sekine H, Yang J, Isoi Y, Yamato M, Kikuchi A, Kobayashi E, Okano T (2006) Polysurgery of cell sheet grafts overcomes diffusion limits to produce thick, vascularized myocardial tissues. FASEB J 20:708–710
22. Yang J, Yamato M, Kohno C, Nishimoto A, Sekine H, Fukai F, Okano T (2005) Cell sheet engineering: recreating tissues without biodegradable scaffolds. Biomaterials 26:6415–6422
23. Hayashida Y, Nishida K, Yamato M, Yang J, Sugiyama H, Watanabe K, Hori Y, Maeda N, Kikuchi A, Okano T, Tano Y (2006) Transplantation of tissue-engineered epithelial cell sheets after excimer laser photoablation reduces postoperative corneal haze. Invest Ophthalmol Vis Sci 47:552–557
24. Harimoto M, Yamato M, Hirose M, Takahashi C, Isoi Y, Kikuchi A, Okano T (2002) Novel approach for achieving double-layered cell sheets co-culture: overlaying endothelial cell sheets onto monolayer hepatocytes utilizing temperature-responsive culture dishes. J Biomed Mater Res 62:464–470
25. Hirose M, Kwon OH, Yamato M, Kikuchi A, Okano T (2000) Creation of designed shape cell sheets that are noninvasively harvested and moved onto another surface. Biomacromolecules 1:377–381
26. Yamao M (2003) Cell sheet engineering: from temperature-responsive culture surfaces to clinics. In: European cells and materials: BioSurf V–functional polymeric surfaces in biotechnology. ETH Zürich, Switzerland
27. Chen CH, Tsai CC, Chen W, Mi FL, Liang HF, Chen SC, Sung HW (2006) Novel living cell sheet harvest system composed of thermoreversible methylcellulose hydrogels. Biomacromolecules 7:736–743
28. Thirumala S, Gimble JM, Devireddy RV (2013) Methylcellulose based thermally reversible hydrogel system for tissue engineering applications. Cell 2:460–475
29. Zuk PA, Zhu M, Mizuno H, Huang J, Futrell JW, Katz AJ, Benhaim P, Lorenz HP, Hedrick MH (2001) Multilineage cells from human adipose tissue: implications for cell-based therapies. Tissue Eng 7:211–228
30. Gimble J, Guilak F (2003) Adipose-derived adult stem cells: isolation, characterization, and differentiation potential. Cytotherapy 5:362–369

31. Gimble JM, Katz AJ, Bunnell BA (2007) Adipose-derived stem cells for regenerative medicine. Circ Res 100:1249–1260
32. Thirumala S, Gimble JM, Devireddy RV (2005) Transport phenomena during freezing of adipose tissue derived adult stem cells. Biotechnol Bioeng 92:372–383
33. Thirumala S, Zvonic S, Floyd E, Gimble JM, Devireddy RV (2005) Effect of various freezing parameters on the immediate post-thaw membrane integrity of adipose tissue derived adult stem cells. Biotechnol Prog 21:1511–1524
34. Devireddy RV, Thirumala S, Gimble JM (2005) Cellular response of adipose derived passage-4 adult stem cells to freezing stress. J Biomech Eng 127:1081–1086
35. Li L, Shan H, Yue C, Lam Y, Tam K, Hu X (2002) Thermally induced association and dissociation of methylcellulose in aqueous solutions. Langmuir 18: 7291–7298
36. Zheng P, Li L, Hu X, Zhao X (2004) Sol-gel transition of methylcellulose in phosphate buffer saline solutions. J Polym Sci B 42:1849–1860
37. Sarkar N, Walker L (1995) Hydration-dehydration properties of methylcellulose and hydroxypropylmethylcellulose. Carbohydr Polym 27:177–185
38. Sarkar N (1979) Thermal gelation properties of methyl and hydroxypropyl methylcellulose. J Appl Polym Sci 27:177–185
39. Wang Q, Li L (2005) Effects of molecular weight on thermoreversible gelation and gel elasticity of methylcellulose in aqueous solution. Carbohydr Polym 62:232–238
40. Xu Y, Wang C, Tam KC, Li L (2004) Salt-assisted and salt-suppressed sol-gel transitions of methylcellulose in water. Langmuir 20:646–652
41. Li L (2002) Thermal gelation of methylcellulose in water: scaling and thermoreversibility. Macromolecules 35:5990–5998
42. Liua R, Cellesic F, Tirellic N, Saundersa B (2009) A study of thermoassociative gelation of aqueous cationic poly(N-isopropyl acrylamide) graft copolymer solutions. Polymer 50:1456–1462

Chapter 6

Decellularized Adipose Tissue Scaffolds for Soft Tissue Regeneration and Adipose-Derived Stem/Stromal Cell Delivery

Pascal Morissette Martin, Arthi Shridhar, Claire Yu, Cody Brown, and Lauren E. Flynn

Abstract

Surgically discarded adipose tissue is not only an abundant source of multipotent adipose-derived stem/stromal cells (ASCs) but can also be decellularized to obtain a biomimetic microenvironment for tissue engineering applications. The decellularization methods involve processing excised fat through a series of chemical, mechanical, and enzymatic treatment stages designed to extract cells, cellular components, and lipid from the tissues. This process yields a complex 3D bioscaffold enriched in collagens that mimics the biochemical and biomechanical properties of the native extracellular matrix (ECM). For ASC culture and delivery, decellularized adipose tissue (DAT) provides a cell-supportive platform that is conducive to adipogenesis. While DAT can be applied in its intact form as an off-the-shelf adipogenic matrix, it can also be used as an ECM source for the fabrication of an array of other scaffold formats including adipose ECM-derived microcarriers and porous foams. In this chapter, we describe the methods developed in our lab to decellularize human adipose tissue and to further process it into a variety of scaffolding materials for a range of applications in soft tissue regeneration, wound healing, and cell culture.

Key words Adipose tissue, Adipose-derived stem/stromal cells, Cell delivery, Collagen, Decellularization, Extracellular matrix (ECM), Scaffold, Soft tissue regeneration, Tissue engineering

1 Introduction

The extracellular matrix (ECM) plays a pivotal role in mediating cellular functions such as adhesion, survival, proliferation, and differentiation. Matrix interactions can modulate cell behavior via integrin binding and through the effects of small bioactive components including matricellular proteins and matrikines [1–4]. Further, the ECM can also serve as a natural reservoir for cell-secreted growth factors, binding and regulating their activity [5]. With growing evidence of the critical importance of the ECM microenvironment in directing cellular responses, ECM-derived bioscaffolds are emerging as a highly promising platform for tissue

Bruce A. Bunnell and Jeffrey M. Gimble (eds.), *Adipose-Derived Stem Cells: Methods and Protocols*, Methods in Molecular Biology, vol. 1773, https://doi.org/10.1007/978-1-4939-7799-4_6,

regeneration and cell delivery, as well as for the development of complex 3D cell culture models.

Recent advances in decellularization technologies have provided a practical means of obtaining tissue-specific ECM sourced from a diverse range of mammalian organs and tissues [6–10]. Ideally, these strategies should be designed to extract immunogenic cellular components using chemical, physical, and/or enzymatic means, while retaining the structural and bioactive components of the native ECM as much as possible [7]. To ensure effective cell extraction, decellularization protocols should be designed based on the specific physical and biochemical properties of the target tissue source.

Recognizing the importance of the complex ECM in the context of tissue regeneration, our group is interested in harnessing the potential of adipose-derived ECM as a cell instructive matrix for volume augmentation applications in reconstructive and cosmetic plastic surgery. Moving toward this goal, we pioneered a detergent-free decellularization protocol for generating scaffolds from excised human fat discarded as surgical waste [7, 11]. In our experience, avoiding the use of detergents helps to preserve the complex structure and composition of the adipose ECM, and also circumvents cytotoxicity concerns associated with residual detergent that can be difficult to extract at the end of processing [7]. Historically regarded as largely cellular, adipose tissue is now known to contain a complex ECM network including collagen types I, III, IV, V, and VI [12, 13]; laminin, fibronectin, and proteoglycans [14]; matricellular proteins [15–18]; and bioactive soluble factors [19]. The prevalence of an intact basement membrane in DAT is also postulated to be advantageous for tissue regeneration based on its roles in organogenesis and wound healing [7]. DAT processed using our methods has similar mechanical properties to native human adipose tissue [20], is easy to handle, and can be stored for extended periods of time in a lyophilized format.

Supporting a tissue-specific approach to adipose regeneration, cell culture studies have demonstrated that bioscaffolds derived from adipose ECM have (a) "adipo-conductive" and (b) "adipo-inductive" properties [7, 21–24]. More specifically, human ASCs seeded on DAT scaffolds (a) readily differentiate into mature adipocytes when cultured in adipogenic differentiation medium and (b) express adipogenic markers, including the master regulatory genes PPARγ and C/EBPα, as well as low levels of adipogenic enzyme activity and intracellular lipid accumulation, in the absence of exogenous differentiation factors [7, 21–23]. In vivo studies have demonstrated that adipose ECM-derived bioscaffolds provide a biocompatible microenvironment that stimulates soft tissue regeneration, including new adipocyte development [22, 24, 25].

In addition to using DAT directly as an implantable scaffolding material, the extracted ECM can also be further processed into a variety of forms including particulates, microcarriers, and porous foams to produce scaffolds with properties suitable for other applications in soft tissue regeneration and cell delivery (Fig. 1a) [7, 21–23]. DAT particles can be injected as a filler via a fine gauge needle, or combined with polymer carriers to form injectable or implantable composite gels [23]. For cell culture applications, DAT-based microcarriers have demonstrated promise as an ASC expansion platform, and can also be applied as a biocompatible and biodegradable injectable cell delivery vehicle [21, 26]. Porous foams and bead foams synthesized from enzymatically digested DAT are stable without the need for synthetic additives or chemical cross-linking. DAT foams have been shown to strongly promote angiogenesis in vivo, supporting rapid host cell infiltration, followed by scaffold remodeling and resorption [22]. Interestingly, the inductive effects of the DAT on ASC adipogenesis have been observed in all of the scaffold formats.

In this chapter, we describe in detail our detergent-free decellularization protocol for human adipose tissue, as well as the further processing of the DAT through cryo-milling to generate fine particulates or through enzymatic digestion with α-amylase to fabricate DAT foams, microcarriers, and bead foams.

2 Materials

2.1 Decellularization of Adipose Tissue

1. Human adipose tissue (*see* **Note 1**).
2. Solution A: Prepare 10× stock solution with 10 mM tris(hydroxymethyl)aminomethane (Tris), 5 mM ethylenediaminetetraacetic acid (EDTA). Add 900 mL of deionized water to a graduated cylinder. Weigh 12.1 g Tris and 14.6 g EDTA and transfer to the cylinder. Add deionized water to a final volume of 1 L and stir until dissolved. Store at room temperature. To prepare a 1× working solution, add 100 mL of 10× stock solution to a graduated cylinder and dilute with deionized water to a final volume of 1 L. Adjust pH to 8 (*see* **Note 2**).
3. Sorensen's phosphate buffer rinsing solution (SPB rinse): Prepare 10× stock SPB rinse with 0.55 M sodium phosphate dibasic heptahydrate ($Na_2HPO_4 \cdot 7H_2O$), 0.17 M potassium phosphate (KH_2PO_4). Add 900 mL of deionized water to a graduated cylinder. Weigh 147 g $Na_2HPO_4 \cdot 7H_2O$ and 23 g KH_2PO_4 and transfer to the cylinder. Add deionized water to a final volume of 1 L and stir until dissolved. Store at room temperature. To prepare a 1× working solution, add 100 mL of 10× stock solution to a graduated cylinder and dilute with deionized water to a final volume of 1 L. Adjust pH to 7.3.
4. Sorensen's phosphate buffer digest solution (SPB digest): Prepare 10× stock SPB digest with 0.55 M sodium phosphate

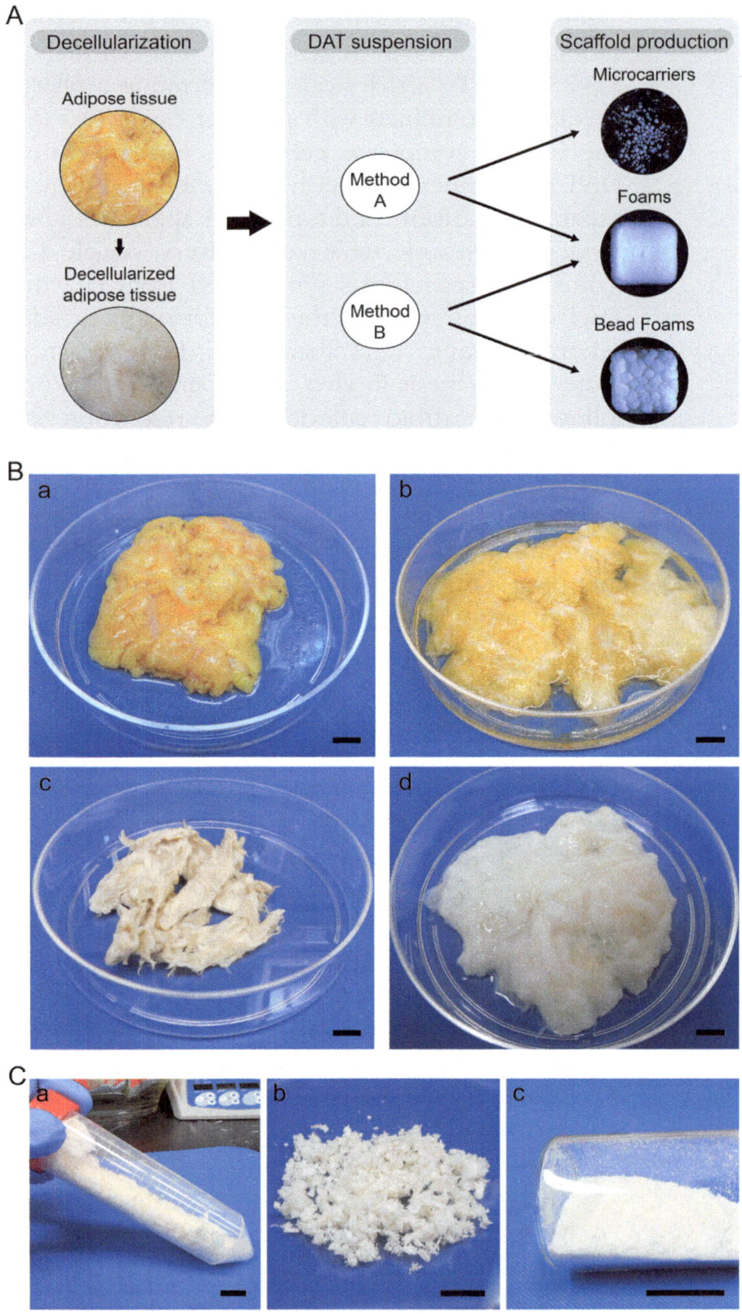

Fig. 1 Overview of adipose decellularization and further processing steps for DAT suspension preparation. (**A**) Flow chart highlighting the key stages for producing the various DAT scaffold formats. (**B**) Macroscopic tissue appearance during decellularization. (**a**) Excised adipose tissue prior to decellularization. (**b**) Tissue following the first polar solvent extraction. (**c**) Tissue after lipid extraction by mechanical pressing. (**d**) DAT hydrated in PBS. (**C**) Representative images of the DAT during further processing steps. (**a**) Lyophilized DAT. (**b**) Minced lyophilized DAT. (**c**) Cryo-milled DAT. Scale bars represent 1 cm

dibasic heptahydrate ($Na_2HPO_4 \cdot 7H_2O$), 0.17 M potassium phosphate (KH_2PO_4), 0.049 M magnesium sulphate heptahydrate ($MgSO_4 \cdot 7H_2O$). Add 900 mL of deionized water to a graduated cylinder. Weigh 147 g $Na_2HPO_4 \cdot 7H_2O$, 23 g KH_2PO_4 and 12 g $MgSO_4 \cdot 7H_2O$ and transfer to the cylinder. Add deionized water to a final volume of 1 L and stir until dissolved. Store at room temperature. To prepare a 1× working solution, add 100 mL of 10× stock solution to a graduated cylinder and dilute with deionized water to a final volume of 1 L. Adjust pH to 7.3.

5. Deoxyribonuclease (DNase) stock solution: Weigh out 0.175 g NaCl and dissolve in 10 mL deionized water. Carefully adjust pH to 7.3. Add 10 mL of glycerol and mix well. Add 15 mL of the NaCl–glycerol solution to 200,000 Units of DNase (Sigma-Aldrich CAT# D4527) and mix thoroughly. Aliquot 1 mL in cryovials and store at −20 °C. Thaw at room temperature immediately before use and do not refreeze.
6. Ribonuclease (RNase) stock solution: Weigh out 0.06 g Tris, 0.44 g NaCl and dissolve in 50 mL deionized water. Adjust pH to 7.5. Add 40 mL of the Tris–NaCl solution to 500 mg of RNase (Sigma-Aldrich CAT# R5125). Aliquot 1 mL in cryovials. Store at −20 °C. Thaw at room temperature immediately before use and do not refreeze.
7. Lipase stock solution: Prepare 5 mM calcium chloride ($CaCl_2$) solution. Weigh out 36.75 mg of calcium chloride dihydrate and transfer into a 50 mL volumetric flask. Add deionized water to a final volume of 50 mL and mix well to dissolve. Chill $CaCl_2$ solution to 4 °C. Add 50 mL of chilled $CaCl_2$ solution to 100,000 Units lipase (Sigma-Aldrich CAT# L0382). Aliquot 2 mL in cryovials and store at −20 °C. Thaw at room temperature immediately before use and do not refreeze.
8. Phenylmethylsulfonyl fluoride (PMSF) solution: Add 28.7 mL of absolute isopropanol to 1 g of PMSF (Sigma-Aldrich CAT# P7626) in a foil-wrapped glass bottle with a tightly sealing lid to protect from exposure to moisture and light (*see* **Note 3**). Stir until clear and store at room temperature.
9. Trypsin-EDTA: 0.25% (1×) with EDTA 4Na (Life Technologies CAT# 25200–072).
10. 100× Antibiotic/Antimycotic (ABAM) (Life Technologies CAT# 15240–062).
11. Phosphate buffered saline (PBS), pH 7.4.
12. Bovine serum albumin (BSA).
13. Ethanol: 70%.
14. Isopropanol, absolute (*see* **Note 4**).
15. Sterile deionized water.

16. Hydrochloric acid (HCl): 6 M (For pH adjustment).
17. Sodium hydroxide (NaOH): 6 M (For pH adjustment).
18. 500 mL polypropylene specimen jars (autoclavable).
19. Stainless steel mesh filters (1 mm pore size).
20. Forceps.
21. Surgical scissors.
22. Pipette aid and serological pipettes.
23. Micropipettors (P20 and P1000).
24. Agitating incubator (capable of 100 rpm and maintaining 37 °C).
25. Freezer (−80 °C).

2.2 Lyophilization and Cryo-Milling of DAT

1. Liquid nitrogen.
2. Scissors.
3. Glass vials.
4. Stainless steel test sieves with desired mesh size for particle size separation (e.g., 200 μm, 100 μm, 50 μm).
5. Ball milling system [e.g., Retsch mixer mill MM 400 ball mill grinder with 25 mL stainless steel milling chamber and stainless steel milling balls (2 × 10 mm)].
6. Lyophilization system.

2.3 Preparation of DAT Suspension

1. Minced or cryo-milled DAT.
2. NaH_2PO_4 solution: 0.22 M, pH 5.4.
3. Acetic acid solution: 0.2 M in deionized water.
4. NaCl solution: 5% (w/v) in deionized water.
5. α-amylase from *Aspergillus oryzae* (Sigma-Aldrich CAT# 10065).
6. Distilled water.
7. 15 mL centrifuge tubes.
8. 50 mL centrifuge tubes.
9. Electronic tissue homogenizer (10 mm saw-tooth probe recommended).
10. Centrifuge.

2.4 DAT Foam Preparation

1. DAT suspension.
2. Molds for desired scaffold size and shape.
3. Freezer.
4. Lyophilization system.
5. Desiccator.

2.5 Electrospraying DAT Suspension/Microcarrier Preparation

1. DAT suspension.
2. Distilled water.
3. Liquid nitrogen.
4. Aluminum foil.
5. 3 mL syringe.
6. 25G winged needle (beveled).
7. 15 mL centrifuge tubes.
8. Stainless steel scoopula.
9. 1 mm pore stainless steel strainer.
10. 500 mL Dewar flask.
11. Electrospinning apparatus (power supply, syringe pump, etc.).
12. Freezer (−80 °C).
13. Lyophilization system.
14. Desiccator.

2.6 DAT Bead Foam Preparation

1. DAT microcarriers.
2. Petri dishes.
3. Stainless steel scoopula.
4. Molds for desired scaffold size and shape.
5. Freezer (−20 or −80 °C).
6. Lyophilization system.
7. Desiccator.

2.7 DAT Foam/Microcarrier/Bead Foam Rehydration

1. Ethanol, absolute.
2. Ethanol solution: 75% in deionized water.
3. Ethanol solution: 50% in deionized water.
4. Ethanol solution: 25% in deionized water.
5. Sterile PBS.

3 Methods

3.1 Decellularization of Adipose Tissue

The decellularization protocol established by our lab involves freeze–thaw cycles, agitation, gentle mechanical manipulation, and chemical/biochemical treatments to extract cells and lipid, while striving to preserve the bioactivity and mechanical integrity of the ECM. Figure 1b outlines the appearance of the tissue at key stages in the decellularization process.

Day 1:

1. With forceps and surgical scissors, remove cauterized tissue and large blood vessels from the freshly isolated adipose tissue sample. Cut the tissue into ~2 cm × 2 cm × 2 cm (~25 g) pieces to increase the surface area in contact with the decellularization reagents (*see* **Note 1**).
2. Transfer approximately 100 g (4 x ~25 g pieces) of adipose tissue to a new sterile specimen jar. Add 99 mL of 1× Solution A, 7 μL of PMSF solution, and 1 mL of 100× ABAM (*see* **Note 5**).
3. Store sample at −80 °C until frozen.
4. Remove the sample from the freezer and thaw in an agitating incubator at 37 °C and 100 rpm.
5. Once thawed, replace liquid contents with 99 mL of fresh 1× Solution A, 7 μL of PMSF solution, and 1 mL of 100× ABAM.
6. Repeat **steps 3–5** an additional two times for a total of three freeze–thaw cycles.
7. At the end of the third freeze–thaw cycle, transfer the tissue to a new sterile specimen jar and add 99 mL of 0.25% trypsin-EDTA and 1 mL 100× ABAM. Incubate for 16 h at 37 °C under agitation at 100 rpm.

Day 2:

8. Transfer the tissue to a new sterile specimen jar and add 99 mL of absolute isopropanol supplemented with 7 μL of PMSF solution and 1 mL of 100× ABAM. Incubate at 37 °C under agitation at 100 rpm for 8 h.
9. Replace the solution with 99 mL of fresh absolute isopropanol, 7 μL of PMSF solution, and 1 mL of 100× ABAM and incubate at 37 °C under agitation at 100 rpm for 16 h (*see* **Note 6**).

Day 3:

10. Transfer the tissue to a new sterile specimen jar and add 99 mL of fresh absolute isopropanol, 7 μL of PMSF solution and 1 mL of 100× ABAM and incubate at 37 °C under agitation at 100 rpm for 8 h (*see* **Note 6**).
11. Replace solution with 99 mL of fresh absolute isopropanol, 7 μL of PMSF solution, and 1 mL of 100× ABAM and incubate at 37 °C under agitation at 100 rpm for 16 h (*see* **Note 7**).

Day 4:

12. Transfer the tissue to a new sterile specimen jar and rinse in 99 mL of SPB rinse solution for 30 min under agitation at 37 °C and 100 rpm. Supplement SPB rinse solution with 7 μL of PMSF solution and 1 mL of 100× ABAM (*see* **Note 8**).

13. Repeat **step 12** twice more for a total of three rinses (*see* **Note 9**).
14. Following rinses, incubate the sample in 99 mL of 0.25% trypsin–EDTA supplemented with 1 mL of 100× ABAM for a total of 6 h under agitation at 37 °C and 100 rpm.
15. Transfer the tissue to a new sterile specimen jar and rinse in 99 mL of SPB rinse solution supplemented with 7 μL of PMSF and 1 mL of 100× ABAM for 30 min under agitation at 37 °C and 100 rpm.
16. Repeat **step 15** twice more for a total of three rinses.
17. Following rinsing, incubate the tissue in 194 mL of SPB digest solution with 1 mL of DNase stock solution, 1 mL of RNase stock solution, and 2 mL of lipase stock solution. Supplement with 14 μL of PMSF solution and 2 mL of 100× ABAM. Incubate at 37 °C under agitation at 100 rpm for 16 h (*see* **Note 10**).

Day 5:

18. Transfer the tissue to a new sterile specimen jar and add 99 mL of absolute isopropanol supplemented with 7 μL of PMSF solution and 1 mL of 100× ABAM. Incubate at 37 °C under agitation at 100 rpm for 8 h.
19. Rinse the tissue sample in 99 mL of SPB rinse solution supplemented with 7 μL of PMSF solution and 1 mL of 100× ABAM for 30 min under agitation at 37 °C and 100 rpm.
20. Repeat **step 19** twice more for a total of three rinses.
21. Rinse tissue in 70% ethanol for 30 min.
22. Repeat **step 21** twice more for a total of three rinses.
23. Transfer the tissue to a new sterile specimen jar and rinse the DAT in sterile deionized water three times for 30 min under agitation at room temperature to remove residual ethanol (*see* **Notes 11** and **12**).

3.2 Cryo-Milling DAT

Cryo-milling is used to mechanically process the DAT into small particles under conditions that can help to preserve bioactivity. The cryo-milled DAT can be used as an injectable scaffold, incorporated into a composite scaffold, or further processed into a homogenous DAT suspension as a first step in the fabrication of DAT foams or microcarriers (Fig. 1c).

1. Place the rinsed DAT in a 50 mL centrifuge tube. Do not exceed 30 mL per centrifuge tube (*see* **Note 13**). Freeze overnight at −20 °C.
2. Keep samples frozen while transporting to the lyophilizer. Lyophilize samples for 48 h or until completely dry (sample size dependent).

3. Store lyophilized DAT in a desiccator until further use (*see* **Note 14**).
4. Finely mince the DAT with sharp scissors (1–2 mm^3 pieces).
5. Transfer the minced DAT into the milling chamber.
6. Place two stainless steel milling balls (10 mm diameter) into the chamber.
7. Loosely seal the chamber and submerge the milling chamber in liquid nitrogen for 3 min.
8. Secure the frozen chamber to the milling system and mill for 3 min continuously at 30 Hz or 1800 rpm (*see* **Note 15**).
9. Store the cryo-milled DAT in a glass container.
10. Stainless steel test sieves of different pore sizes may be used to separate the cryo-milled DAT based on particle size.

3.3 Preparation of DAT Suspension

Two methods that can be used to produce the DAT suspension are described in detail in Table 1 (*see* **Notes 16–18**). In Method A, the DAT is first processed into a dried particulate form through mechanical mincing or cryo-milling (*see* Subheading 3.2 for detailed procedures), while Method B utilizes larger fragments of hydrated DAT as the starting material. These processing methods can influence the ECM structure and affect the final characteristics of the DAT suspension. Importantly, the methods presented allow for flexibility in creating a range of porous scaffolds with different geometries and physical properties. For instance, the DAT suspension generated with Method A with cryo- milling yields a more homogeneous porous foam as compared to the less processed DAT suspension prepared using either Method A with mechanical mincing or Method B. Notably, for the fabrication of small diameter (300–500 μm) microcarriers, a less viscous DAT suspension prepared from cryo-milled DAT following Method A should be used for electrospraying (*see* Subheading 3.5). In contrast, Method B results in a more intact collagen fibril network that can be used to create DAT “bead foams” comprised of fused networks of porous DAT microcarriers that are stabilized through fibril entanglement between neighboring beads (*see* Subheading 3.7) [27]. A flow chart outlining the key steps in scaffold fabrication is shown in Fig. 1a, while representative images of the DAT foams, microcarriers, and bead foams are shown in Fig. 2.

3.4 Fabrication of DAT Foams

Porous DAT foams can be made using DAT suspensions prepared using either Method A or B (Table 1). Altering the preprocessing conditions through mincing versus cryo-milling will affect the physical and mechanical properties of the resultant foams. More specifically, foams prepared from minced DAT suspensions generally have greater heterogeneity and increased stiffness as compared to those fabricated from cryo-milled DAT suspensions.

Table 1
Methods of DAT suspension preparation

Method A	Method B
1. Weigh 250 mg of lyophilized DAT that has been cryo-milled or minced finely with scissors (1 mm^3) and transfer into a 15 mL centrifuge tube. *See* Subheading 3.2 for further details on cryo-milling methods.	1. Weigh 2 g wet weight of PBS hydrated DAT (0.5–1 cm^3 pieces) that has been manually pressed to remove excess liquid and then transfer into a 15 mL centrifuge tube.
2. Add 5 mL of 0.22 M NaH_2PO_4 solution and carefully mark the liquid volume on the outside of the centrifuge tube (*see* **Note 16**).	2. Add 10 mL of 0.22 M NaH_2PO_4 solution to the centrifuge tube.
3. Add 0.75 mg of α-amylase (equivalent to 0.3% (w/w) of dry tissue).	3. Add 6 mg of α-amylase (equivalent to 0.3% (w/w) of wet tissue).
4. Add 0.22 M NaH_2PO_4 solution to a final volume of approximately 10 mL.	4. Allow the mixture to digest at room temperature for 72 h under continuous agitation (300 rpm).
5. Allow the mixture to digest at room temperature for 72 h under continuous agitation (300 rpm).	5. After digestion, carefully remove the digested tissue mass and manually press to extract excess liquid.
6. Centrifuge at 1500 × *g* for 10 min.	6. Place the pressed DAT back into the tube, add 10 mL of 5% NaCl solution, and agitate at room temperature for 10 min.
7. Remove the supernatant without disturbing the insoluble DAT pellet.	7. Carefully remove the digested tissue mass and manually press to extract excess liquid.
8. Add 10 mL of 5% NaCl solution to the DAT pellet. Disrupt the pellet by vortexing and agitate at room temperature for 10 min.	8. Repeat **steps 6–7** for a total of 2 rinses with 5% NaCl solution.
9. Repeat **steps 6–8** once more with 5% NaCl solution (for a total of 2 rinses with 5% NaCl solution).	9. Repeat **steps 6–7** with distilled water (for a total of 1 rinse with distilled water).
10. Repeat **steps 6–8** with distilled water (for a total of 1 rinse with distilled water).	10. Add 10 mL of 0.2 M acetic acid solution and mince the tissue finely with scissors (2–5 mm^3 pieces).
11. Centrifuge at 1500 × *g* for 10 min and remove the supernatant.	11. Place the sample in an agitating incubator overnight at 120 rpm and 37 °C.
12. Add 0.2 M acetic acid solution to the 5 mL mark on the tube made on day 1. Vortex tube until the DAT pellet is suspended in the acetic acid solution.	12. Homogenize the DAT suspension with a homogenizer. Ensure that the suspension does not overheat during homogenization (*See* **Note 17**).
13. Place the sample in an agitating incubator overnight at 120 rpm and 37 °C.	13. Centrifuge at 1500 × *g* for 10 min to promote settling of remaining DAT fragments. Transfer the liquefied fraction (supernatant) into a 50 mL centrifuge tube.
14. Homogenize the DAT suspension with a homogenizer until no large pieces are visible. Ensure that the suspension does not overheat during homogenization (*see* **Note 17**).	14. Add an additional 10 mL of 0.2 M acetic acid solution to the fragments and homogenize until no large pieces remain.
15. Store the final DAT suspension at 4 °C until further use (stock suspension is 50 mg of DAT/mL by starting dry weight) (*see* **Note 18**).	15. Pool the homogenized DAT suspensions and store at 4 °C until further use (stock suspension is 100 mg of DAT/mL by starting wet weight) (*see* **Note 18**).

1. Warm the DAT suspension, produced using either Method A or Method B (*see* Subheading 3.3), to 37 °C and dilute the stock solution with distilled water as required (recommended range: 10 mg/mL to 50 mg/mL) (*see* **Note 19**).

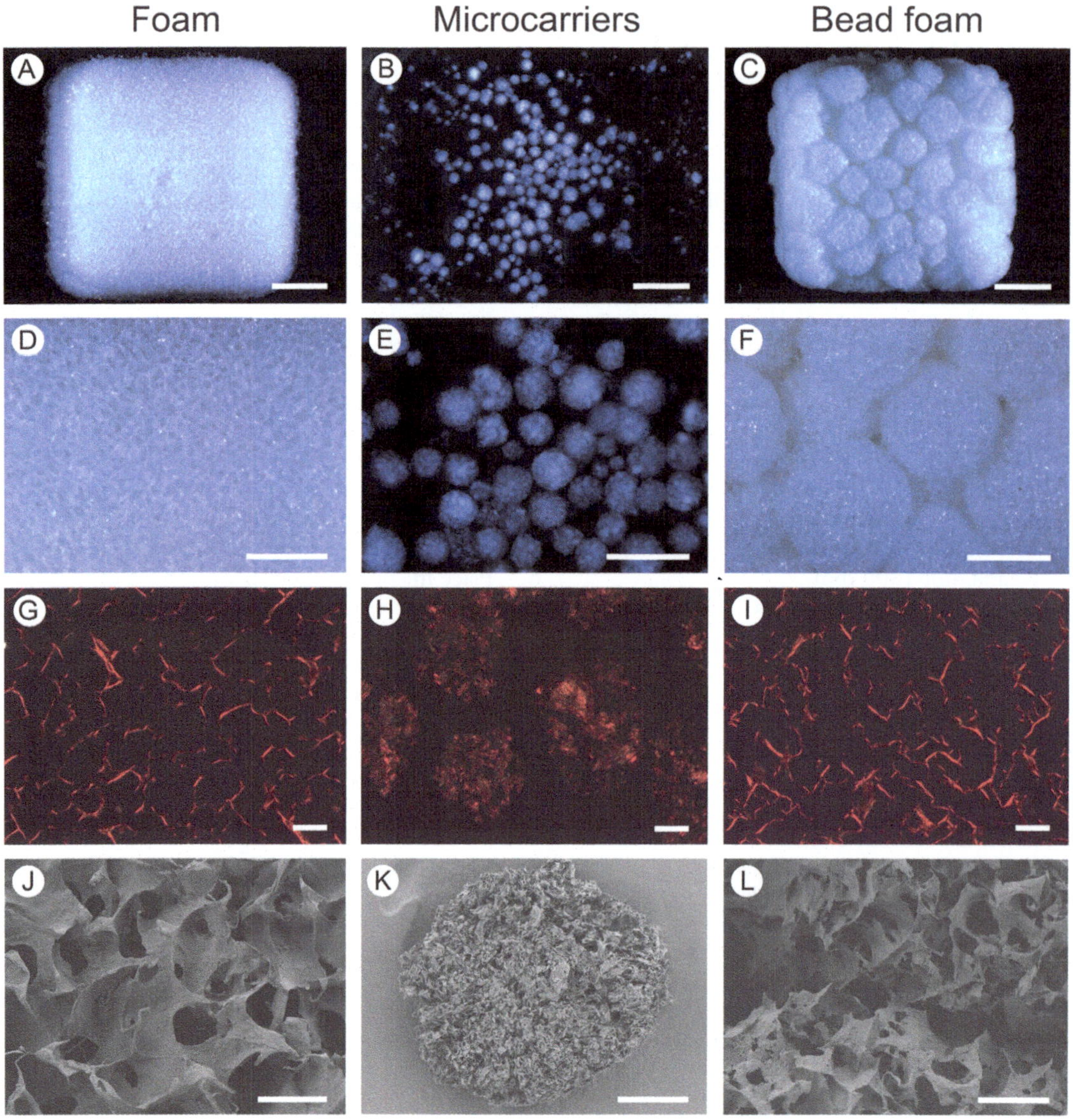

Fig. 2 Representative images of the DAT foams, microcarriers, and bead foams. (**a**–**f**) Macroscopic structure; Scale bars: (**a**–**c**) = 2 mm, (**d**–**f**) = 1 mm. (**g**–**i**) Picrosirius red staining of collagen visualized by polarized light microscopy. (**j**–**l**) Scanning electron microscopy images showing the porous scaffold architecture in the various DAT-based scaffold formats. Scale bars: (**g**–**l**) = 100 μm

2. Transfer the diluted DAT suspension into the desired mold and freeze overnight (*see* **Note 20**).
3. Place the mold inside the lyophilizer for 24 h (size dependant).
4. Following freeze-drying, remove the scaffolds from the lyophilizer and store in a desiccator until further use (*see* **Note 21**).

3.5 Electrospraying DAT Suspension

DAT suspension produced using either Method A or Method B (Table 1) can be electrosprayed as the first step in the production of both DAT microcarriers and bead foams. Subsequent processing stages in the fabrication of the microcarriers and bead foams are discussed in detail in Subheadings 3.6 and 3.7 respectively.

1. Warm the DAT suspension to 37 °C and dilute the stock solution as needed with distilled water to achieve the desired DAT concentration (recommended range: 10–50 mg/mL).
2. Set up the electrospraying apparatus as illustrated in Fig. 3.
3. Check that the distance between the tip of the needle and the liquid nitrogen inside the Dewar flask is between 4 and 6 cm (*see* **Note 22**).
4. Ensure that the aluminum foil is submerged in the liquid nitrogen and connected to the ground.
5. Connect the needle tip to the cathode and carefully position the opening of the bevelled tip relative to the foil as shown in detail in Fig. 3.
6. Varying the voltage and the viscosity of the DAT suspension controls the size of the electrosprayed beads (suggested voltage range: 15–25 kV). For example, it is recommended to set the voltage to 17 kV for a 10 mg/mL cryo-milled DAT suspension and 20 kV for a 50 mg/mL cryo-milled DAT suspension to generate beads in the size range of 300–500 μm.
7. The extrusion rate of the DAT suspension should be controlled using a syringe pump (suggested range: 25–35 mL/h).
8. Electrospray the DAT suspension directly into the liquid nitrogen (*see* **Note 23**).

3.6 DAT Microcarrier Fabrication

For the microcarriers, cryo-milled DAT suspensions prepared using Method A (Table 1) are recommended.

1. Immediately following electrospraying (*see* Subheading 3.5), remove the frozen microcarriers from the Dewar flask with a stainless steel scoopula.
2. Place the microcarriers into a precooled 15 mL centrifuge tube (*see* **Note 24**). Do not overfill the tubes beyond the 3 mL mark to ensure proper lyophilization (*see* **Note 25**).
3. Immediately lyophilize the microcarriers overnight (*see* **Note 21**).
4. Store lyophilized microcarriers in a desiccator.
5. When ready to use, resuspend the dried microcarriers in absolute ethanol and sieve through a 1 mm stainless steel strainer to remove any aggregates. Continue with scaffold rehydration (*see* Subheading 3.8) (*see* **Note 26**).

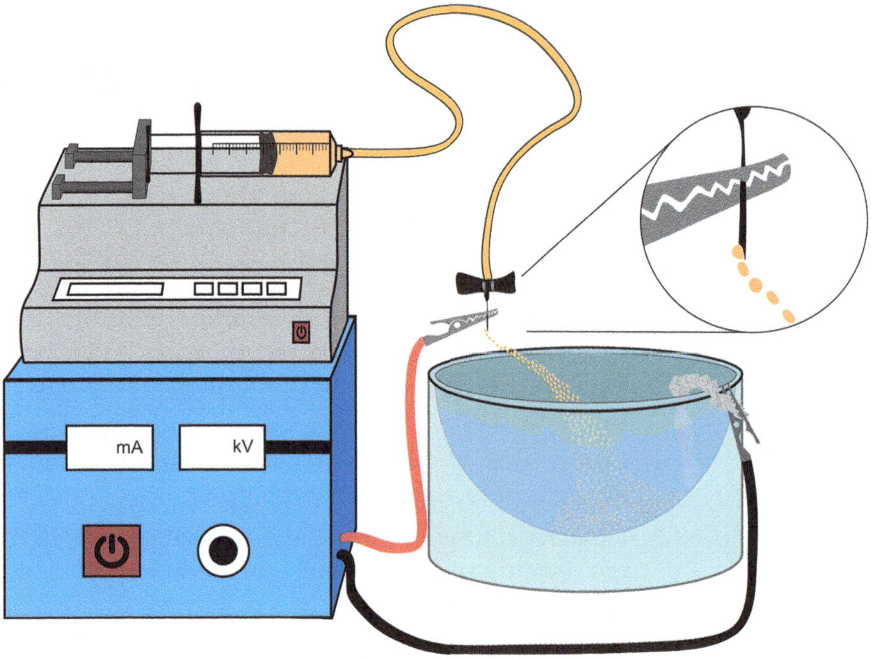

Fig. 3 Diagram of the electrospraying apparatus used for the production of DAT microcarriers and bead foams. In this setup, the positive pole of the power supply is connected to the 25G beveled needle with an alligator clip, and the aluminum foil strip is connected to the ground. The other end of the foil is submerged into liquid nitrogen in the Dewar flask. A syringe pump is used to control the flow rate of the DAT suspension. Variables that can influence the size and geometry of the beads include the viscosity of the DAT suspension, the extrusion rate, the voltage applied, and the needle height. A magnification of the needle position is shown in the inset. The needle should be oriented parallel to the foil strip, on the opposite side of the Dewar, with the opening of the bevelled edge positioned toward the flask to aid in droplet generation

3.7 DAT Bead Foam Fabrication

For the bead foams, it is recommended to use DAT suspensions prepared using Method B (Table 1) to achieve a robust bead network.

1. Immediately following electrospraying (*see* Subheading 3.5), remove the frozen microcarriers (beads) from the Dewar flask with a stainless steel scoopula.
2. Allow the beads to thaw completely at room temperature in a petri dish and carefully remove any large or irregular beads.
3. Carefully transfer the thawed beads into a mold and freeze overnight at −20 or −80 °C.
4. Lyophilize the samples within the molds overnight (size dependant).
5. Once freeze-dried, remove the scaffolds from the lyophilizer and store them in a desiccator until further use (*see* **Note 21**).

3.8 DAT Foam/Bead Foam/Microcarrier Rehydration

1. Immerse the lyophilized scaffolds in absolute ethanol at room temperature overnight (*see* **Note 27**).
2. Replace the solution with 75% ethanol and incubate overnight at room temperature.
3. Replace the solution with 50% ethanol and incubate overnight at room temperature.
4. Replace the solution with 25% ethanol and incubate for 3 h at room temperature.
5. Replace the solution with PBS and incubate for 3 h at room temperature.
6. Rinse with sterile PBS for a total of 3 rinses for 30 min each at room temperature.
7. Use the scaffolds immediately in subsequent in vitro or in vivo experiments.

4 Notes

1. While our protocol was designed for excised tissue blocks, lipoaspirates can also be processed. A recommended modification of the approach for lipoaspirates is to use stainless steel strainers with every solution change to minimize sample loss. Based on our experience, processing adipose tissue *en bloc* results in a higher yield of ECM. The adipose tissue sample should be transported to the lab on ice in sterile PBS supplemented with 2% BSA immediately following extraction.
2. All pH adjustments should be done using HCl or NaOH. The pH of solution A is recommended to be pH 8 to minimize protease activity [28].
3. PMSF is a serine protease inhibitor that deactivates proteases released during cell lysis. PMSF is highly unstable if exposed to water and the stock solution must be prepared with dry solvents and protected from both water and light during storage.
4. Isopropanol is miscible in both water and oil and is used to extract lipids from the tissue.
5. For all processes, scale the reagents as required based on the original amount of tissue. When processing larger tissue volumes, it is critical to ensure that the samples are well mixed during mechanical agitation stages and that the tissue fragments do not become aggregated over time.
6. Before placing tissue in fresh isopropanol, gently press the tissue between two stainless steel mesh filters to aid in the removal of lipids; mechanical extraction is crucial for lipid separation.

If stainless steel mesh filters are unavailable, manual massaging and manipulation can be used as an alternative.

7. The tissue will shrink after the initial isopropanol washes (Fig. 1b).
8. Since day 4 involves a relatively long procedure of consecutive rinses, it is recommended to start the protocol as early in the day as possible.
9. Each rinse will require the use of fresh SPB rinse solution, supplemented with ABAM and PMSF solution.
10. SPB digest solution contains a high salt concentration, which may precipitate during storage. Ensure that the solution is well mixed prior to use.
11. For short term storage (up to 1 month), DAT should be rinsed in sterile PBS three times for 30 min each at room temperature to remove residual ethanol and then kept in sterile PBS supplemented with 1% ABAM at 4 °C until further use. For longer-term storage, it is recommended that the DAT be stored at room temperature in a lyophilized state, following the lyophilization methods described in Subheading 3.2.
12. The DAT samples should have a light yellowish-white appearance at the end of processing. A darker yellow or brown appearance is indicative of incomplete lipid and/or cell extraction.
13. Place the 50 mL centrifuge tube on its side while freezing to allow for greater surface area during lyophilization in order to reduce drying times.
14. Lyophilized DAT is highly hygroscopic and moisture from the environment will reduce milling efficacy, producing flakes instead of fine powder.
15. The DAT particle size can be reduced by increasing the milling time and/or frequency.
16. Since the DAT swells when hydrated, the final volume of the solution will be more than 5 mL after the addition of the NaH_2PO_4 solution.
17. To prevent overheating of the DAT suspension during homogenization, the tube can be cooled through incubation in cold water in between short (4–5 s) homogenizing cycles. Since the solution may be highly viscous after homogenizing, examine it carefully to ensure that there are no large pieces of tissue remaining. Stirring the solution gently with a stirring rod can help to detect any remaining unhomogenized pieces.
18. The viscosity of the DAT stock solution may vary to some extent due to tissue donor variability. To ensure consistency between the batches, it is recommended to pool DAT from at

least five donors and use the same DAT suspension stock to prepare all scaffolds for each experiment.

19. Warming the solution reduces the viscosity of the DAT suspension. Due to the inherent viscosity, a syringe should be used to measure the required volumes when diluting. The DAT suspension concentration will have an impact on the properties of the scaffold. A lower concentration will result in a less viscous DAT suspension and a softer foam scaffold with a higher porosity.
20. For cell culture studies, cryomolds or the wells of multiwell plates can be used as molds. The geometry of the resulting foam will be defined by the shape of the mold used for freezing the DAT suspension. The freezing temperatures prior to lyophilization can impact the porosity and pore sizes of the resultant foams. Lower freezing temperatures will typically yield scaffolds with smaller pore sizes.
21. It is recommended that the scaffolds be stored in their lyophilized state in a desiccator at room temperature until further use.
22. The low temperatures associated with the liquid nitrogen below the needle may cause clogging of the DAT suspension within the needle. To avoid this, it is recommended to ensure that the DAT suspension is warm and electrospray only small volumes at a time, pooling multiple batches to obtain a larger number of microcarriers.
23. A fine uniform stream of DAT suspension should be seen spraying from the needle tip with little to no disruption. If the syringe becomes clogged, the DAT suspension may be too cool or may not be properly homogenized. If large DAT pieces are visible, homogenize the DAT suspension again prior to electrospraying.
24. Precool the tubes using liquid nitrogen or by placing them in the −80 °C freezer.
25. To prevent adhesion during lyophilization, do not allow the microcarriers to thaw at any point during the drying process. Remove the microcarriers from the Dewar in small batches and store the filled tubes in the −80 °C freezer until ready to lyophilize all samples.
26. Microcarriers are best handled when immersed in absolute ethanol since they readily adhere to plastic surfaces in the hydrated state.
27. It is critical to rehydrate the scaffolds slowly to prevent structural collapse and the formation of bubbles. Ensure that the scaffolds completely sink to the bottom of the container before proceeding to the next rehydration step. The scaffolds may be placed inside an evacuated chamber to speed the process.

References

1. Kyriakides TR, Bornstein P (2003) Matricellular proteins as modulators of wound healing and the foreign body response. Thromb Haemost 90(6):986–992. https://doi.org/10.1267/THRO03060986
2. Bellon G, Martiny L, Robinet A (2004) Matrix metalloproteinases and matrikines in angiogenesis. Crit Rev Oncol Hematol 49(3):203–220. https://doi.org/10.1016/j.critrevonc.2003.10.004
3. Maquart FX, Bellon G, Pasco S, Monboisse JC (2005) Matrikines in the regulation of extracellular matrix degradation. Biochimie 87(3–4):353–360. https://doi.org/10.1016/j.biochi.2004.10.006
4. Walker JT, Kim SS, Michelsons S, Creber K, Elliott CG, Leask A, Hamilton DW (2015) Cell–matrix interactions governing skin repair: matricellular proteins as diverse modulators of cell function. Res Rep Biochem 2015(5):73–88
5. Vlodavsky I, Miao HQ, Medalion B, Danagher P, Ron D (1996) Involvement of heparan sulfate and related molecules in sequestration and growth promoting activity of fibroblast growth factor. Cancer Metastasis Rev 15(2):177–186
6. Flynn LE, Prestwich GD, Semple JL, Woodhouse KA (2008) Proliferation and differentiation of adipose-derived stem cells on naturally derived scaffolds. Biomaterials 29(12):1862–1871. https://doi.org/10.1016/j.biomaterials.2007.12.028
7. Flynn LE (2010) The use of decellularized adipose tissue to provide an inductive microenvironment for the adipogenic differentiation of human adipose-derived stem cells. Biomaterials 31(17):4715–4724. https://doi.org/10.1016/j.biomaterials.2010.02.046
8. Moutos FT, Estes BT, Guilak F (2010) Multifunctional hybrid three-dimensionally woven scaffolds for cartilage tissue engineering. Macromol Biosci 10(11):1355–1364. https://doi.org/10.1002/mabi.201000124
9. Frohlich M, Grayson WL, Marolt D, Gimble JM, Kregar-Velikonja N, Vunjak-Novakovic G (2010) Bone grafts engineered from human adipose-derived stem cells in perfusion bioreactor culture. Tissue Eng A 16(1):179–189. https://doi.org/10.1089/ten.TEA.2009.0164
10. Lee HY, Yang HJ, Rhie JW, Han KT (2014) Adipose tissue regeneration in vivo using micronized acellular allogenic dermis as an injectable scaffold. Aesthetic Plast Surg 38(5):1001–1010. https://doi.org/10.1007/s00266-014-0379-2
11. Flynn LE (2015) Decellularized adipose tissue. U.S. Patent No. 9,034,386 (B2); Issue date May 19, 2015
12. Nakajima I, Yamaguchi T, Ozutsumi K, Aso H (1998) Adipose tissue extracellular matrix: newly organized by adipocytes during differentiation. Diff Res Biol Div 63(4):193–200. https://doi.org/10.1111/j.1432-0436.1998.00193.x
13. Kubo Y, Kaidzu S, Nakajima I, Takenouchi K, Nakamura F (2000) Organization of extracellular matrix components during differentiation of adipocytes in long-term culture. In Vitro Cell Dev Biol Anim 36(1):38–44. https://doi.org/10.1290/1071-2690(2000)036<0038:OOEMCD>2.0.CO;2
14. Pierleoni C, Verdenelli F, Castellucci M, Cinti S (1998) Fibronectins and basal lamina molecules expression in human subcutaneous white adipose tissue. Eur J Histochem 42(3):183–188
15. Przybyt E, van Luyn MJ, Harmsen MC (2015) Extracellular matrix components of adipose derived stromal cells promote alignment, organization, and maturation of cardiomyocytes in vitro. J Biomed Mater Res A 103(5):1840–1848. https://doi.org/10.1002/jbm.a.35311
16. Kong P, Cavalera M, Frangogiannis NG (2014) The role of thrombospondin (TSP)-1 in obesity and diabetes. Adipocytes 3(1):81–84. https://doi.org/10.4161/adip.26990
17. Tartare-Deckert S, Chavey C, Monthouel MN, Gautier N, Van Obberghen E (2001) The matricellular protein SPARC/osteonectin as a newly identified factor up-regulated in obesity. J Biol Chem 276(25):22231–22237. https://doi.org/10.1074/jbc.M010634200
18. Van Hul M, Frederix L, Lijnen HR (2012) Role of thrombospondin-2 in murine adipose tissue angiogenesis and development. Obesity 20(9):1757–1762. https://doi.org/10.1038/oby.2011.260
19. Fain JN (2006) Release of interleukins and other inflammatory cytokines by human adipose tissue is enhanced in obesity and primarily due to the nonfat cells. Vitam Horm 74:443–477. https://doi.org/10.1016/S0083-6729(06)74018-3
20. Omidi E, Fuetterer L, Reza Mousavi S, Armstrong RC, Flynn LE, Samani A (2014) Characterization and assessment of hyperelastic and elastic properties of decellularized human adipose tissues. J Biomech 47(15):3657–3663.

https://doi.org/10.1016/j.jbiomech.2014.09.035

21. Turner AE, Yu C, Bianco J, Watkins JF, Flynn LE (2012) The performance of decellularized adipose tissue microcarriers as an inductive substrate for human adipose-derived stem cells. Biomaterials 33(18):4490–4499. https://doi.org/10.1016/j.biomaterials.2012.03.026
22. Yu C, Bianco J, Brown C, Fuetterer L, Watkins JF, Samani A, Flynn LE (2013) Porous decellularized adipose tissue foams for soft tissue regeneration. Biomaterials 34(13):3290–3302. https://doi.org/10.1016/j.biomaterials.2013.01.056
23. Cheung HK, Han TT, Marecak DM, Watkins JF, Amsden BG, Flynn LE (2014) Composite hydrogel scaffolds incorporating decellularized adipose tissue for soft tissue engineering with adipose-derived stem cells. Biomaterials 35(6):1914–1923. https://doi.org/10.1016/j.biomaterials.2013.11.067
24. Wu I, Nahas Z, Kimmerling KA, Rosson GD, Elisseeff JH (2012) An injectable adipose matrix for soft-tissue reconstruction. Plast Reconstr Surg 129(6):1247–1257. https://doi.org/10.1097/PRS.0b013e31824ec3dc
25. Adam Young D, Bajaj V, Christman KL (2014) Award winner for outstanding research in the PhD category, 2014 Society for Biomaterials annual meeting and exposition, Denver, Colorado, April 16-19, 2014: Decellularized adipose matrix hydrogels stimulate in vivo neovascularization and adipose formation. J Biomed Mater Res A 102(6):1641–1651. https://doi.org/10.1002/jbm.a.35109
26. Yu C, Kornmuller A, Brown CFC, Hoare T, and Flynn LE (2017) Decellularized adipose tissue microcarriers as a dynamic culture platform for human adipose-derived stem/stromal cell expansion. Biomaterials 120:66–80. https://doi.org/10.1016/j.biomaterials.2016.12.017
27. Flynn LE, Yu C (2015) Collagenous foam materials. U.S. Patent No. 9,629,939; Issue date April 25, 2017
28. Gilbert TW, Sellaro TL, Badylak SF (2006) Decellularization of tissues and organs. Biomaterials 27(19):3675–3683. https://doi.org/10.1016/j.biomaterials.2006.02.014

Chapter 7

Induction of Skin Allograft Transplantation Tolerance in Mice Using Human Adipose Derived Stromal Cells

Anthony D. Foster, Nicholas Clark, and Thomas A. Davis

Abstract

Murine models of allograft transplantation are valuable for understanding the immunological mechanisms of allograft acceptance and rejection, the evaluation of immunosuppressive drugs and strategies, and the restoration of functional defects. Herein, we describe methods to create a skin murine allograft surgical model and how to administer adipose-derived stromal cells (ASC) with limited numbers of donor bone marrow to create stable multilineage donor cell chimerism and indefinite immunological tolerance.

Key words Skin allograft transplantation, Tolerance induction, Donor cell chimerism, In vivo immune regulation, Adipose stem cells

1 Introduction

Adipose-derived stromal cells (ASC) exhibit a wide variety of applications in vivo, including wound repair, angiogenesis, modulation of inflammatory responses, and regulation of autoimmune diseases [1–4]. Importantly, ASCs are immunologically tolerated cells, and lack expression of MHC class II antigens and costimulatory molecules [5, 6]. The intrinsic trophic and paracrine actions, anti-inflammatory and immunomodulatory properties of these cells have been shown to be key in stimulating the recruitment and mobilization of endogenous progenitor cells and tissue regeneration, dampening immune-mediated responses, inhibiting activated T cell proliferation, and suppression of dendritic cell activation and differentiation within various inflammatory niches [3, 5, 7]. Recently, we demonstrated that the coinfusion of ASCs in combination with limited donor bone marrow cells in immunologically conditioned recipient mice support stable multilineage donor cell chimerism and indefinite allogeneic graft survival [8]. ASCs are an ideal source of allogeneic donor cells as they are well characterized phenotypically and functionally and can be easily obtained using relatively minimal invasive methods from normal individuals in high numbers, reducing the need for repeated

Bruce A. Bunnell and Jeffrey M. Gimble (eds.), *Adipose-Derived Stem Cells: Methods and Protocols*, Methods in Molecular Biology, vol. 1773, https://doi.org/10.1007/978-1-4939-7799-4_7, © Springer Science+Business Media, LLC, part of Springer Nature 2018

ex vivo expansion in culture [9, 10]. Collectively, these features make ASCs attractive cells for therapeutic tolerance induction applications. This chapter highlights a generalized protocol to establish long-term multilineage donor cell chimerism and indefinite allograft graft survival using ASC as immune regulatory cells in a murine skin allograft transplantation model.

2 Materials

1. Mouse strains.
   (a) BALB/c donor mice (The Jackson Laboratory, Bar Harbor, ME).
   (b) C57BL/6 recipient mice (The Jackson Laboratory, Bar Harbor, ME).
   (c) C3H/HeJ mice (The Jackson Laboratory, Bar Harbor, ME).
2. Human adipose-derived stromal cells (ASC; LaCell LLC, New Orleans, LA).
3. Equipment:
   (a) FACSAria II (BD Biosciences, San Jose, CA).
   (b) Refrigerated rate controlled benchtop centrifuge (Thermo Scientific, Sorvall Legend XTR).
   (c) Inverted phase contrast microscope.
   (d) Electric clippers (Wahl, Model: 9757 with size 50 clippers (0.4 mm) Sterling, IL).
   (e) Mark I Cesium-137 irradiator.
   (f) 250 watt heat lamp.
   (g) Digital scale (Mettler-Toledo, Model: MLI5026).
   (h) Automated cell counter: (Nexcelom Cellometer Auto 2000, Lawrence, MA).
   (i) Analog vortex mixer (VWR, Radnor PA).
   (j) VWR Analog rocking platform shaker (VWR, Radnor PA).
   (k) Mouse restrainer (Mouse Tail Illuminator Restrainer, Braintree Scientific, Braintree, MA).
   (l) Ice bucket/tray.
   (m) 4–6″ forceps.
   (n) 4–6″ straight iris scissors.
   (o) Scalpel handle and #10 blades.

4. Medias/Buffers.
   (a) Dulbecco's phosphate-buffered saline (D-PBS) without calcium chloride and magnesium chloride.
   (b) Heat-inactivated fetal bovine serum (FBS; HyClone, Logan, UT).
   (c) FACS staining buffer (2.5% FBS, 0.1% sodium azide in D-PBS).
      - Sodium azide (Sigma-Aldrich, St. Louis, MO Cat# 1101497037).
   (d) ASC growth media (10% heat-inactivated FBS and 1% penicillin/streptomycin in DMEM/F-12).
      - Penicillin (10,000 IU) and streptomycin (10,000 μg/mL) (100×) (Cat#15140–22; Invitrogen, Frederick, MD).
      - Dulbecco's Modified Eagle Medium: Nutrient Mixture F-12 (DMEM/F-12).
   (e) Mixed lymphocyte reaction (MLR) culture media:
      - Roswell Park Memorial Institute 1640 (RPMI-1640) supplemented with 10% heat-inactivated FBS, 10 mM HEPES, 1% nonessential amino acids, 2 mM L-glutamine, 100 U/mL penicillin, 100 mg/mL streptomycin (all from Invitrogen, Carlsbad, CA, http://www.invitrogen.com), and 5 mg/mL 2-mercaptoethanol (Sigma-Aldrich, St. Louis, MO).
   (f) Injection Vehicle: D-PBS with 100 U/mL preservative-free heparin and 100 U/mL DNase containing 5 mM Mg^{2+}.
      - DNase containing 5 mM Mg^{2+} (Worthington Biochemical Corp., Lakewood Township, NJ, http://www.worthington-biochem.com).
      - Preservative-free heparin (Sigma-Aldrich, St. Louis, MO, http://www.sigmaaldrich.com).
   (g) ACK (ammonium-chloride-potassium) lysing buffer (Lonza, Allendale, NJ, Cat # 10-548E).
5. Treatment/medications.
   (a) Ketamine HCL 100 mg/mL (Pfizer, New York, NY NDC Code 0856–2013-01).
   (b) Xylazine HCL 100 mg/mL (Lloyd Inc., Shenandoah, IA NDC Code 61311–482-10).
   (c) Busulfan (Sigma-Aldrich, St. Louis, MO, Cat# B2635-10G).
   (d) Buprenex (Reckitt Benckiser Pharmaceuticals Inc., Richmond, VA NDC Code 12496–0757-1.
   (e) Buprenex SR (ZooPharm, Windsor, CO Cat# BZ8069317).

(f) T cell depletion antibodies.
 - Rat anti-mouse CD4 antibody (clone YTS 191.1, 10 mg/kg; BioXCell, West Lebanon NH, Cat# BE0119).
 - Rat anti-mouse CD8 antibody (clone YTS 169.4; BioXCell, West Lebanon NH, Cat# BE0117).

6. Reagents/consumables:
   (a) EDTA-treated blood collection tubes (BD Microtainer Cat# 365973).
   (b) 100 mm sterile petri dish.
   (c) 10 cc syringe.
   (d) 1 cc syringes.
   (e) 27.5-gauge needles.
   (f) Transfer pipette.
   (g) 50 cc polystyrene conical centrifuge tubes.
   (h) 70 micron cell strainer (Falcon Corning).
   (i) Cellometer slides (Cat# CHT4-SD100–014; Nexcelom).
   (j) AOPI Staining Solution (Cat# CS2–0106-25 mL; Nexcelom).
   (k) Providone–iodine swabstick (Medline, Mundelein, IL).
   (l) 70% isopropyl alcohol.
   (m) Ice.
   (n) Dry ice.
   (o) Ink pad.
   (p) 4 × 4 gauze.
   (q) One inch wide sheer adhesive bandage (2.5 cm × 7.6 cm).
7. FACS Staining Reagents.
   (a) 12 × 75 mm 5 mL round bottom polystyrene FACS tubes (Falcon Corning).
   (b) Purified rat anti-mouse CD16/CD32 (Clone 2.4G2) (BD Pharmigen, San Jose, CA Cat# 553142).
   (c) Flurochrome labeled antibodies: FITC H2-D^d BD Bioscience Cat# 553579, APC Foxp3 eBioscience Cat# 17–5773-82, APC Cy7 CD8 BD Biosciences Cat# 557654, PerCP EFluor 710 CD4 eBioscience Cat# 46–0042-82, PETr CD45, Invitrogen, Frederick, MD Cat# MCD4517, BV510 CD3 Biolegend, San Diego, CA Cat# 100234.
   (d) BioLegend Fluorofix (BioLegend, San Diego, CA Cat# 422101).
8. Cell proliferation BrdU ELISA kit (chemiluminescent; Roche).

3 Methods (See Note 1)

3.1 Harvest and Preparation of Skin Grafts

1. Purchase pathogen-free (10–12 weeks of age) BALB/c (H-2^d) mice (white hair) from a certified vendor (The Jackson Laboratory; Bar Harbor, ME, USA). We use female BALB/c mice as tissue donors for trunk skin allografts (*see* **Note 2**).
2. Place the donor mice (BALB/c) or the home cage in an approved transparent CO_2 euthanasia chamber (*see* **Note 3**). Animals are placed in 100% air atmosphere and gradually transitioned to 100% CO_2 atmosphere over 2–3 min (2–3 L of CO_2/min) for a standard shoe-box mouse cage. Mice should be left in the chamber for 2–3 min after they become unconscious (~30 s).
3. Using electric clippers (blade #50) carefully shave all the hair on the dorsum and abdomen areas between the fore and hind limbs of the mice.
4. Generously wet the clipped hair area of the entire mouse with 70% isopropyl alcohol.
5. Lay the mouse on its dorsum (back). Using sterile scissors and forceps, carefully excise the trunk skin circumferentially as a single piece between the fore and hind limbs. Loosen and gently peel the skin from the underlying connective tissue using the blunt end of a forceps or hemostat (*see* **Note 4**).
6. Place the removed trunk skin in a sterile petri dish containing enough cold (4 °C) D-PBS and float the excised skin internal side down. Place the dish on ice. Proceed to remove the skin from all the mice. After all tissues have been collected, properly dispose of the animal carcasses in accordance with your institute's animal disposal policy or SOP.
7. Using a scalpel and blade, carefully scrape away all the adipose (yellow fat) and subcutaneous tissue from the excised trunk skin pieces. Following the removal of the adipose tissue the dermal side of the skin graft should appear shinny and somewhat translucent.
8. Using sterile forceps and scalpel, cut the trunk skin into $3 \times 3\ cm^2$ pieces, gently submerge trunk grafts in sterile cold D-PBS and then float trunk grafts internal side down in a sterile petri dish containing fresh cold D-PBS.

3.2 Skin Grafting

1. We use pathogen-free female (10–12 weeks of age) BALB/c (H-2^d) and C57BL/6 (H-2^b), mice purchased from The Jackson Laboratory (Bar Harbor, ME, USA) as skin graft donors (white hair) and recipients (black hair), respectively (*see* **Note 5**).
2. Weigh C57BL/6 recipient mice using a digital scale and record individual weights.

3. After the induction of deep anesthesia in C57BL/6 recipients with an intraperitoneal injection (*see* **Note 6**) of ketamine (100 mg/kg; Pfizer, New York, NY) and xylazine (10 mg/kg; Lloyd Inc., Shenandoah), use an electric shaver (blade #50) to remove hair from midline dorsum area below the shoulder blades to the base of the tail, including the lateral chest area (*see* **Note 7**).

   Disinfect the shaved skin area (dorsal midline) with providone–iodine solution followed by three gentle skin scrubs using sterile gauze soaked with 70% isopropyl alcohol.

   (a) A graft bed (2 cm in diameter on the mid-dorsum) is prepared using sterile forceps and fine curved scissors by removing the area of epidermis and dermis down to the level of the intrinsic muscle (panniculus carnosus) (*see* **Note 8**).

   (b) A graft of donor skin, slightly larger than the recipient's graft bed, is fitted to the prepared bed without suturing (internal surface of donor skin placed on graft bed). Carefully, trim away excess donor skin so the graft lies flat over the recipient graft bed. Graft is secured with an adhesive sheer plastic bandage (2.5 cm × 7.6 cm). Make sure that the bandage is secure around the abdomen but does not interfere with breathing and mobility (*see* **Note 9**).

   (c) Transfer each mouse to a separate clean cage. For the first 2–3 weeks post skin grafting mice should be individually housed in microisolator cages.

   (d) Place cage under a heat lamp (85–95 °F) until the mouse has completely recovered from the anesthesia and is mobile.

   (e) For pain management, mice are treated with a single intraperitoneal dose of Buprenex SR (3.25 mg/kg) or buprenorphine (Reckitt Benckiser Pharmaceuticals, Richmond, VA). Buprenex SR is the preferred method of pain treatment as it provides a sustained release over 3 days with a single injection of 3.25 mg/kg. Alternatively, multiple injections of buprenorphine at 0.25–1.0 mg/kg BW q/12 h. (bid) may be given for 3 days post skin transplantation for pain management.

   (f) At day 7 post skin grafting, carefully manually scruff the mouse and cut through the adhesive bandage with blunt scissors (*see* **Note 10**).

3.3 Conditioning Protocol

1. Following skin grafting, mice need to undergo a short-term immunosuppression conditioning regimen consisting of: intraperitoneal (i.p using a 1 mL syringe fitted with a 25 gauge needle; total injection volume 0.5 mL) injection consisting of a cocktail of a rat anti-mouse CD4 depletion antibody and a rat anti-mouse CD8 depletion antibody (anti-CD4 clone YTS 191.1, 10 mg/kg; anti-CD8 clone YTS 169.4, 10 mg/kg;

BioXCell, West Lebanon, NH) at days 0, 2, 5, 7, and 14 post transplantation and i.p. injection of busulfan (5 mg/kg, Sigma-Aldrich, St. Louis, MO) at day 5 post grafting. The busulfan treatment acts as a mild form of myelo-ablation to make space within the bone marrow compartment for donor marrow cells to seed. Donor chimerism is a critical component of long-term graft tolerance.

3.4 Preparation of Bone Marrow Cells

1. Prelabel a 50 cc conical centrifuge tube with approximately 10–15 mL of sterile D-PBS (depending on the number of donor animals to be used) and place on ice.
2. Using BALB/c donor animals euthanized as described in Subheading 3.2 (Skin Grafting), collect both hind legs as follows. Place the animal on a hard surface with the dorsal side down. Using surgical scissors and forceps hold one of the legs and cut through the muscle inward toward the hip socket. Cut and separate the leg from the body at the hip socket carefully so as to keep the femur intact. Repeat for the remaining leg. Remove the foot and skin using scissors and forceps and transfer to 50 cc conical centrifuge tube with D-PBS. To aseptically process bone marrow into a single cell suspension the remaining work should be performed inside of a biological safety cabinet. Place the residual leg on a sterile 100 mm petri dish or sterile surgical cutting board. Using a scalpel (#10 blade) and forceps remove the remaining muscle tissue to expose the intact femur and tibia. Cleanly separate the two bones at the knee using the scalpel. Take each bone separately and hold tightly with forceps. Using the scalpel cut approximately 2 mm of bone off of each side. This should expose the medullary cavity on both ends. Prepare a 10 cc syringe with a 25 gauge needle. Fill the syringe with a 1% solution of FBS in PBS. Label a new 50 cc conical tube. Holding the bone over the tube, insert the needle from the syringe into one end, and flush the marrow out of the bone into the tube. Repeat this process for each bone that is remaining. When complete, close the tube and invert several times gently. Centrifuge the tube at 400 × *g* for 5 min at 4 °C. Aseptically remove the supernatant and resuspend cells in 5–10 mL of sterile D-PBS (*see* **Note 11**). Filter the cell suspension through a 70 μM pore size nylon mesh cell strainer (Falcon, BD Biosciences) into a new 50 cc conical tube. The cells are now ready for injection. Determine the number of viable cells using an automated Nexcelom cell counter or a similar device. Alternatively, manual cell counts can be performed using the hemocytometer/Trypan Blue method (*see* **Note 12**). Note that the cell counts should include nucleated cells only, not red blood cells (RBCs). If the counting method used does not exclude red blood cells automatically then an additional step involving the lysis of RBCs using ACK

(hypotonic ammonium-chloride-potassium) lysing buffer (Lonza, Allendale, NJ) will be necessary.

3.5 Adipose Tissue Derived Stromal/Stem Cells

1. Purified human ASCs (passage P-0) can be commercially purchased from LaCell, LLC (New Orleans, LA, http://www.lacell.wpengine.com) or isolated according to a previously described protocol [11]. Cells need to be stored in either the liquid or vapor phase of a liquid nitrogen freezer at < −130 °C until used. P-0 cells can be propagated in culture as described below.
   (a) Initiation of ASC Cultures: Remove cryovials from liquid nitrogen storage and transport to the laboratory on dry ice. Wearing freezer gloves, quickly thaw out each vial by gently swirling the bottom portion of each vial in a clean 37 °C water bath (to prevent contamination issue, do not submerge above the cap threads). Spray the vial with 70% isopropyl alcohol. After drying, open the vial and carefully pipet the cell suspension into a 15 cc conical centrifuge tube. Rinse the empty cryovial once with sterile ASC growth media and add the rinse to the 15 cc conical centrifuge tube. Next, add 2 mL of room temperature ASC growth media for every 1 mL of cell suspension and centrifuge for 5 min at 1000 rpm (218 × *g*) at 20 °C. After decanting the supernatant, gently resuspend the cell pellet in ~ 1–2 mL of fresh ASC culture media by finger flicking the side of the centrifuge tube and/or gentle trituration using a sterile 1 mL pipet. Quantify cell viability and concentration as previously described. Plate/seed ASC into sterile 225 cm^2 uncoated tissue culture flasks at 0.5–1 × 10^6 cells in a total culture volume of 40 mL per flask. Incubate cultures in a fully humidified atmosphere at 37 °C, 5% CO_2 and 95% air.
2. ASC Cell Harvest: After 5–7 days of culture, cells will develop 70–90% confluent monolayers and will be ready for in vivo administration or subculture (P1-passage). Before starting, prewarm sterile trypsin–EDTA solution in a 37 °C in a water bath. Remove flasks from the incubator, stand the flask upright, remove the cap, and remove the spent medium using a 25 cc sterile pipet. Carefully, wash the adherent monolayer of cells with 25 mL of room temperature D-PBS (1×), then remove and discard D-PBS using a 25 cc sterile pipet. Add 15 mL of warm trypsin–EDTA solution to each flask in order to copiously coat the cell monolayers. Return flasks to a 37 °C incubator for ~7–9 min, and when cells start appearing rounded gently tap the side of each flask with the palm of your hand to dislodge any remaining adherent cells. Repeat tapping or incubate further as needed but do not over expose cells to the trypsin–EDTA solution as this will result in reduced viability. Confirm, via light

microscopy using an inverted phase contrast microscope, that >95% of the cells have detached (individual cells should be floating freely). Add fresh ASC culture medium (37 °C) with FBS to inactivate the trypsin. Remove the cells to a 50 cc conical centrifuge tube, centrifuge (5 min at 218 × *g*, 20 °C). Resuspend in 10 mL of sterile D-PBS (1×), and count cells to determine the cell concentration and viability as previously described. Once counted the cells are ready to combine with BMCs for injection using the injection vehicle described below. ASCs should be centrifuged at 1000 rpm (218 × *g*) at room temperature for 5 min and resuspended in 8–10 mL of injection vehicle at room temperature. Cells should be stored at room temperature and injected within 60 min after final preparation (*see* **Note 13**).

3. Injection Vehicle: To minimize cellular aggregation and prevent potential pulmonary embolism, all injections of cells should be performed with the following vehicle: D-PBS with 100 U/mL preservative-free heparin (Sigma-Aldrich, St. Louis, MO, http://www.sigmaaldrich.com) and 100 U/mL DNase containing 5 mM Mg2+ (Worthington Biochemical Corp., Lakewood Township, NJ, http://www.worthington-biochem.com).
4. Preparation of Injection Inoculum: Combine an appropriate number of BMCs and ASCs, depending on the total number of mice to be injected and the goals of the study, (*see* **Note 14**) in a sterile 50 cc conical centrifuge tube. We have found a dose of 5×10^5 donor BMCs and 3×10^6 human ASCs per mouse to be effective. Centrifuge for 5 min at 1000 rpm (218 × *g*) at 20 °C, aseptically remove the supernatant and resuspend in an appropriate volume of vehicle for injection. The recommendation here is a final volume of 0.2 mL per dose.

3.6 Coinfusion of Bone Marrow Cells and Adipose-Derived Stromal Cells

1. Mice need to be appropriately restrained in a Plexiglas mouse restrainer.
2. Place mouse cage under a heat lamp (85–95 °F) for about 10 min or apply warm compresses (< 120 °F) to the tail to increase blood flow.
3. Coinfuse cells intravenously using a 1 cc syringe fitted with a 27 1/2-gauge needle. Injection is accomplished by compression of the lateral tail vein by pressure at the base of the tail (*see* **Note 14**).

3.7 Flow Cytometry for Chimerism and T_{reg} Analysis

At the completion of the experiment (for all tissues) or at an intermediate time point (for peripheral blood only), donor cell chimerism may be determined by flow cytometry analysis according to the following protocol.

1. Tissue Isolation:
   (a) Peripheral Blood: For periodic analysis, small volumes of whole blood (0.2 mL) can be collected under temporary

anesthesia in EDTA collection tubes using tail nick or retro-orbital sinus plexus bleed (*see* **Note 15**). Terminal blood collection by cardiac stick should be performed on an appropriately deep anesthetized mouse (*see* **Note 16**). Following anesthesia, the xyphoid process should be palpated at the caudal aspect of the animal's sternum. Using a 1 inch 25 gauge needle attached to a 1 cc syringe, insert the needle into the animals left xyphoid process notch and direct it toward the heart. Once the needle has been inserted beneath the skin and the heart pierced, apply gentle negative pressure by pulling backward on the plunger. Blood should collect in the syringe. Transfer to a labeled EDTA blood microcollection tube and place on rocker to mix.

(b) Spleen: The entire spleen should be carefully removed using surgical scissors and forceps, so as not to tear the tissue, and placed in D-PBS on ice. To process the spleen into a single cell suspension, place whole spleen in a 100 mm petri dish with approximately 5–8 mL of cold FACS staining buffer as processing media. Take a 10 cc syringe and remove the plunger. Using the flat end of the plunger crush the tissue gently, inside of the petri dish, being careful to process the entire tissue. Rinse the end of the syringe plunger with FACS staining buffer (2.5% FBS, 0.1% sodium azide in D-PBS) into the petri dish. Using a transfer pipette, triturate the suspension several times to break up the cell suspension further. Transfer suspension into a labeled polystyrene 50 cc conical centrifuge tube, using a nylon mesh 70 μm pore cell strainer (Falcon) to remove clumps of cells and yield a single cell suspension. Centrifuge cell suspension for 5 min at 400 × *g* at 4 °C. Discard the supernatant and resuspend cells in an appropriate volume (5–20 mL) of cold FACS staining buffer and store on ice. Calculate the percentage cell viability and viable cell concentration as previously described. Store cells on ice until ready for staining.

(c) Bone Marrow: Collect bone marrow from a single femur (*see* **Note 17**) as described in section 3.4.2. Transfer the cell suspension through a nylon mesh 70 μm pore cell strainer (Falcon) into a labeled polystyrene 50 cc conical centrifuge tube. Centrifuge the cell suspension for 5 min at approximately 400 × *g* at 4 °C, discard the supernatant, and resuspend cells in 2–3 mL of FACS staining buffer. Calculate the percentage cell viability and viable cell concentration as previously described. Store cells on ice until ready for staining.

2. Extracellular Staining for Donor Cell Surface Markers:
   (a) Wash cells: Add 10^6 cells (for BMCs and splenocytes) to an appropriately labeled FACS tube and add 1 mL FACS buffer to each tube to wash. Centrifuge for 5 min at 400 × *g* and 4 °C. Discard the supernatant and resuspend the pellet by vortex.
      - For whole blood place 100 μL of EDTA-treated whole blood into a labeled FACS tube and proceed to the next step.
   (b) Blocking step: add 0.25 μg of FC-block (anti-CD16/anti-CD32 clone 2.4G2, BD Biosciences) in a total volume of 20 μL of FACS buffer to each FACS tube to be stained. Incubate the cells for 10 min at 4 °C.
   (c) Staining Step: While the blocking step is occurring (or beforehand) prepare stock staining solutions containing a cocktail of antibodies for mouse CD3, CD4, CD8, H2-D^d (donor Class I), and CD45 and another stock solution containing a cocktail of irrelevant fluorescently labeled isotype control antibodies. The cocktail should be prepared as follows (*see* **Note 18**): Per sample stained add 1 μL FITC-HDd(Clone 34-2-12), 1 μL PerCP-eFluor 710-CD4 (Clone RM4–5), 1 μL BV510-CD3 (Clone 17A2), 1 μL APC-Cy7-CD8 (Clone 53–6.7), 0.5 μL PE-Texas Red-CD45 (Clone 30-F11), and 95.5 μL FACS buffer for a total volume of 100 μL per sample. When block is complete add 100 μL of staining cocktail to each sample. Incubate covered for 30 min at 4 °C.
   (d) Washing Step: add 1 mL of FACS staining buffer per sample and centrifuge for 5 min at 400 × *g* at 4 °C. Discard the supernatant and resuspend pellet by vortex. Repeat wash step.
   (e) Fixation/Permeabilization step: Using the eBioscience FoxP3 intracellular staining kit (Cat#: 00–5523-00) prepare a solution with 1 part Fixation/Permeabilization Concentrate combined with 3 parts Fixation/Permeabilization Diluent, with a total volume of 1 mL per sample. When the second wash step is complete be certain to vortex the cells after discarding supernatant as adding fixative to pelleted cells will result in a clump of cross-linked cells. Add 1 mL of Fixation/Permeabilization working solution to the resuspended cells and vortex immediately to mix thoroughly. Incubate covered at 4 °C for 45 min to 18 h.
3. Intracellular Staining for FoxP3 Expression:
   (a) Wash Step: prepare a fresh solution of wash buffer from eBioscience Foxp3 intracellular staining kit by diluting 10× Intracellular Wash Buffer concentration with 1 part concentrate plus 9 parts distilled/deionized water (approximately

8 mL of wash buffer will be requires per sample). When fixation/permeabilization is complete add 2 mL of the cold wash buffer to each sample tube and then centrifuge for 5 min at 4 °C and 600 × *g* (*see* **Note 19**). Carefully decant off the supernatant and then vortex to resuspend the pellet. Repeat the wash step.

(b) Intracellular Staining Step: Prior to or during the wash step a working stock dilution of the anti-mouse FoxP3 antibody in the intracellular wash buffer should be prepared combining 0.5 μg of anti-FoxP3 antibody (2.5 μL/sample) with intracellular wash buffer (97.5 μL/sample) for a total volume of 100 μL per sample. Add 100 μL of the working stock solution to each appropriate tube, vortex, and incubate covered for 30 min at 4 °C. It is recommended that you simultaneously prepare an isotype control sample using 0.5 μg of a nonspecific APC labeled isotype control antibody on a cell sample labeled with the same extracellular makers but without APC-FoxP3. This will confirm the staining specificity of the APC-Foxp3 signal in the experimental samples. Further, as with extracellular staining it is highly recommended that the APC-FoxP3 antibody be titrated for optimal staining.

(c) Wash: After incubation, wash cells twice as done previously with 2 mL of intracellular wash buffer followed by centrifugation at 600 × *g* and carefully decant off wash medium and resuspend the cells in the pellet by vortexing.

(d) Fix: Add 0.5 mL of Fluorofix (Biolegend cat# 422101), vortex thoroughly, and incubate covered for 30 min at 4 °C. When complete, add 2 mL of cold FACS buffer to wash, and centrifuge for 5 min at 600 × *g* at 4 °C. Discard supernatant, resuspend cells by vortexing, and add 0.5 mL FACS buffer. Keep samples covered at 4 °C until ready to run on a flow cytometer (*see* **Note 20**).

4. Multiparameter/Multifluorochrome Flow Cytometry (FACS) Analysis of Donor Cell Populations

(a) Chimerism: Using flow cytometry analysis software set up gating for doublet exclusion if desired (*see* **Note 21**). Next, place a gate on the entire $CD45^+$ population to include all hematopoietic derived cells. Then subgate on $CD3^+$, which will include all T cells. Create a two-parameter histogram of CD4 vs. CD8 to allow analysis of these T cell subsets. Separately create two single-parameter histograms gated on (1) $CD45^+CD3^+CD4^+CD8^-$ and (2) $CD45^+CD3^+CD4^-CD8^+$ subsets. Finally, set those histograms to show the expression of H2-D^d, which will be positive on donor cells and negative on host T cells. As such, gate on the H2-D^{d+} cells in each histogram and express the result as a percentage of the total CD4 or CD8 T cell populations, respectively. The resulting percentage will be the rate of donor chimerism for each T cell subset (*see* **Note 22**).

(b) T_{reg} Analysis: Beginning with the CD4 T cell subset, create a 2-parameter histogram showing the expression of CD4 and that of FoxP3. Gate the $CD4^{+}FoxP3^{+}$ events and express as a percentage of the total $CD4^{+}$ T cell population. This will include both host and donor T_{reg} populations. $H2\text{-}D^{d}$ expression may be used to separate host from donor T_{reg} cells.

3.8 Mixed Lymphocyte Reaction

For a direct in vitro analysis of allo-specific T cell responses, a mixed lymphocyte reaction (MLR) may be performed as described below. This assay determines the activity of allo-specific responder T cells in a quantitative manner by measuring proliferation following exposure to allo-mismatched, inactivated target cells in the presence or absence of potentially immunosuppressive agents, such as ASCs. Additionally, this assay may be used to measure the responsiveness of an in vivo treated T cell population using ex vivo stimulation. In this assay, splenocytes from tolerized C57BL/6 recipients are tested for their responsiveness to BALB/c or a third party stimulator.

1. Preparation of splenocytes: To measure the allo-responsiveness of splenocytes from ASC treated recipients perform the following assay. Beginning with skin graft transplanted mice at a time point to be determined by the user, euthanize a C57BL/6 recipient (tolerized) mouse, a naive wild type C57BL/6 mouse, a C3H/HeJ mouse, and a BALB/c mouse using a CO_2 induction chamber as described in section 3.1.2. Aseptically remove the spleen from each animal and place separately into prelabeled collection tubes containing sterile MLR culture medium. Next, working inside a biological safety cabinet, prepare a single cell suspension according to the process described in section 3.7.b, being careful to maintain sterile technique so as to prevent contamination of cells that will be kept in culture for several days. Note that each C57BL/6 spleen tested, as well as the C3H/HeJ and BALB/c spleen tissues should be collected and processed separately. Once complete, centrifuge the labeled 50cc conical tubes containing each preparation of cells for 5 min at 400 × *g* and 20 °C. Aseptically aspirate the supernatant and resuspend the cell pellets in 3–5 mL ACK lysing buffer for 2 min at room temperature to lyse RBCs. Add 40 mL of MLR culture media and centrifuge for 7 min at 400 × *g* and 20 °C. Remove the supernatant and resuspend cells in 5–10 mL of complete MLR culture medium. Calculate the percentage cell viability and viable cell concentration as described above.

2. Radiation of BALB/c and C3H/HeJ target splenocytes: For each reaction, 2×10^7 BALB/c and C3H/HeJ target-stimulatory cells should be irradiated to prevent cellular proliferation with a dose of 30 Gy using a ^{137}Cs or similar irradiator according to manufacturer's guidelines.
3. Allogeneic One-way MLR Culture: In the MLR assay described here the stimulators cells are spleen cells from BALB/c or C3H/HeJ mice and the responder cells are spleen cells from either wild type C57BL/6 mice or allograft tolerant C57BL/6 mice. Culture wild type or tolerized recipient C57BL/6 (responder) splenocytes in triplicate wells at 5×10^6/cells/mL (1×10^5 cells/per well in flat-bottom 96-well plates). To the triplicate cultures add an equal number of irradiated BALB/c target cells, an equal number of C3H/HeJ target cells, or MLR media alone, respectively. Culture in complete MLR media in a humidified atmosphere of 5% CO_2/95% air for 5 days. BrdU Incorporation: After 5 days of culture, the rate of proliferation by BrdU incorporation can be measured using an enzyme linked immunosorbent assay (ELISA) kit (Roche Diagnostics), according to the manufacturer's protocol. The results are then expressed as the net absorbance of the stimulated responder cells minus that of the unstimulated responder cells.
4. Analysis: If ASC treatment was effective in inducing tolerance to BALB/c alloantigen it should be reflected in a decrease in proliferation (measured by reduced BrdU incorporation) in recipient splenocytes cultured with BALB/c target, as compared with naïve C57BL/6 splenocytes cultured with BALB/c target. In contrast, proliferation (measure by increased BrdU incorporation) should be observed in both responder populations when exposed to C3H/HeJ target cells. This result would be indicative not only of successful suppression of the anti-BALB/c allo-response, but also of antigen specific suppression such that suppression is not conferred to a third party target (C3H/HeJ).

4 Notes

1. All procedures involving animals should be performed in accordance with the investigator's Institutional Animal Care and Use Committee guidelines, the Animal Welfare Act and other federal statutes and regulations related to experiments involving animals and adhere to principles stated in the *Guide for the Care and Use of Laboratory Animals.*

2. Female mice tend to be less aggressive than male mice when group housed, and we have noted that trunk grafts from female mice typically have less skin lesions.
3. The chamber must be directly connected to CO_2 gas source (in-house or cylinder) and allow for unobstructed viewing of the mice during the euthanasia procedure.
4. Removal of trunk skin tissue should be conducted under aseptic conditions using sterilized instruments and supplies.
5. BALB/c (H-2^d) and C57BL/6 (H-2^b) mouse strains are both MHC class I and class II mismatched, in addition to minor mismatches, and represent a well-established model of major histocompatibility disparity.
6. For an intraperitoneal injection, the needle should be inserted into the abdominal cavity in the lower left quadrant to avoid the cecum and urinary bladder. Light aspiration should be performed prior to injection, to assure that the tip has not entered the bowel or any vessel
7. It is critical to anesthetize mice individually prior to surgery. Isoflurane anesthesia, which results in rapid induction and recovery can be employed with appropriate continuous monitoring equipment, however we have found it difficult to maintain sedation during application of the skin graft and the dressing.
8. We cut in half the barrel shaft of a 10 cc syringe and use its ink stamped end as a template for cutting uniform circular excisional wounds.
9. It is not uncommon after 4–7 days that the mice will remove the bandage, without affecting the graft placement which typically has fully vascularized by that time.
10. Failure of graft to take or rejection by the host will be evident by tissue contraction, scabbing, necrosis, complete loss of viable tissue and/or lack of new hair growth within 10–14 days post transplantation if not sooner. Evidence of these attributes within a day of removing the bandage suggests the graft failed to vascularize and thus should be considered a technical failure.
11. After the final wash step, BMCs are resuspended in D-PBS alone prior to injection so as to wash out FBS. However, cells should not be left in D-PBS alone without FBS for a long period of time as this will damage the cells. As such, if the injection will not be performed within an hour, BMCs should be kept on ice in D-PBS + 1% FBS and washed with PBS alone prior to injection.
12. To use the Necelom Cellometer Auto 2000:
    (a) Mix 20 μL of sample with 20 μL of AOPI staining solution in a clean centrifuge tube and mix thoroughly.
    (b) Add 20 μL of the mixture to one side of a Cellometer Slide and insert the slide into the front receptor of the machine.

(c) Select the correct assay for the sample (Stem Cell for ASCs, "High RBC Content" for BMCs).

(d) Bring the cells into focus with the arrows on the left of the screen.

(e) Click the count button and results will appear on screen. Use the "live" cell count number for calculations.

(f) *Alternative Cell Count Method:* If BMCs are manually counted using a hemocytometer and the Trypan Blue dye exclusion assay, red blood cells should first be lysed with ACK for 2 min at room temperature, washed with cold PBS + 1% FBS, centrifuged at 400 × *g* for 5 min at 4 °C, resuspended in an appropriate volume of D-PBS, and placed on ice. Next combine 10 μL of cell suspension with 90 mL of a Trypan Blue solution (the cell to stain ratio as well as percentage of Trypan Blue dilution may need to be adjusted depending on the density of cells). Place 10 μL of the mixture (or appropriate volume) on a hemocytometer slide and count using light microscopy. Calculate the concentration of live (Trypan Blue excluding) cells based on the parameters of the hemocytometer chamber in use.

13. ASCs should not be counted when resuspended in the injection vehicle on the Nexelom counter as it interferes with the staining dye.

14. We recommend a total injection volume of 0.2 mL of injection vehicle containing 0.5×10^6 donor bone marrow cells and $1–5 \times 10^6$ ASCs per animal. Control animals should receive tail vein injections of D-PBS + DNAse/heparin as the vehicle control.

15. Periodic blood collection (nonterminal): For periodic assessment of peripheral blood lymphocytes a tail bleed may be performed. Place mouse cage under a heat lamp (85–95 °F) for about 10 min to increase circulation. Transfer mouse to a Plexiglas restrainer as used in the tail vein injection procedure. Have an open EDTA-treated blood collection tube immediately available. Using a #10 scalpel blade make a small shallow cut directly over the lateral tail vein (either side) approximately 1/3rd of the way down from the base of the tail. Immediately place the EDTA-treated blood collection tube in position to receive blood sample. Collected approximately 0.2 mL of blood. Place a sterile 4 × 4 piece of gauze over the cut and hold to stop bleeding and assist the clotting process. After approximately 30 s transfer the animal back to its cage. Care should be taken not to cut too deeply, as well as to avoid collecting an excessive amount of blood as this may be harmful to the animal.

16. Terminal Blood collection and Anesthesia: It is recommended that whole blood be collected either by a tail bleed, or by ter-

minal cardiac stick. In the latter case, mice should be anesthetized with a sodium pentobarbital solution (50 mg/kg) or with a cocktail of ketamine (100 mg/kg; Pfizer, New York, NY) and xylazine (10 mg/kg, Lloyd Inc., Shenandoah), by intraperitoneal injection. Use a 1 cc syringe fitted with a 25-gauge needle. Dilute the anesthetic to a total volume of 200 μL per dose using D-PBS as a vehicle. The use of CO_2 asphyxiation is not recommended as it increases the likelihood of coagulation. Likewise, once the blood is collected into EDTA containing tubes they should be placed on a rocker at room temperature for 30–60 min (as placing samples on ice also increases the changes of coagulation) prior to FACS staining.

17. Depending on the efficiency of bone marrow collection it may be necessary to take both femurs to ensure an adequate number of cells for staining.
18. For FACS staining, the antibody cocktail described here is an example of an effective design with validated antibodies. However it may be necessary to use different fluorochromes or antibody clones depending on the needs of a given study or the flow cytometry equipment available. Regardless of the staining design used (whether this design or another) it is highly recommended that FACS antibodies be titrated to achieve an optimal signal.
19. The process of permeabilization and fixation makes cells more buoyant. To avoid excessive cell loss a higher speed of centrifugation is necessary. A range of 500–700 × *g* should be tested for efficacy. The goal is to limit sample loss while avoiding damage to the cells by excessive centrifugal force.
20. Tandem dyes, such as APC-Cy7, are sensitive to prolonged exposure to paraformaldehyde based fixatives (including Fluorofix). If allowed to remain in Fluorofix or a similar product for an extended period it will decay resulting in an erroneous APC signal. For that reason a final wash and resuspension in FACS buffer following fixation is recommended here. However, per manufacturer's recommendation, if the samples can be run within 4 h of adding the fixative, no final wash step is required.
21. Doublet exclusion is a common practice in FACS analysis. Briefly, in the preparation of a single cell suspension, some cells will remain bound together, creating the possibility of false positive signals. The simplest method of doublet removal during analysis is to first gate all cells as Forward Scatter height (FSC-h) vs. Forward Scatter area (FSC-a). The result is a diagonal population beginning near the origin. Create a narrow gate as shown in Fig. 1. Those cells inside the gate should be considered single cell events. Subgate from this population for all further analysis.
22. Chimerism: Depending on the goals of the study, it may be preferable to analyze the chimerism of the total $CD3^+$ T cell

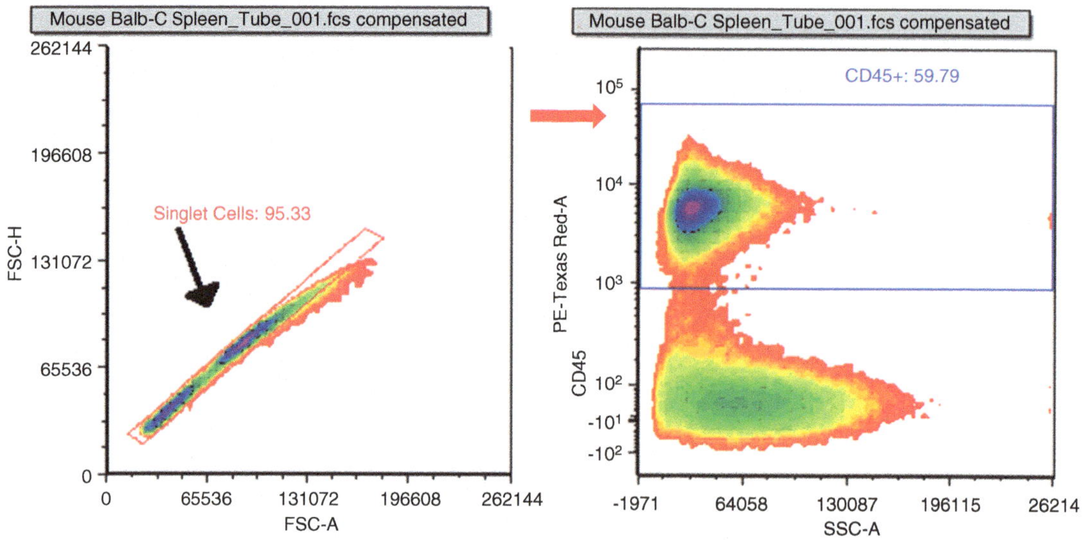

Fig. 1 Doublet exclusion. To remove potential doublet events, first set a two parameter histogram with FSC-height (FSC-h) vs. FSC-area (FSC-a). Set a narrow gate over the population beginning near the origin and going diagonally up and across the histogram as shown here. Close the gate, and use the cells inside as singlet events for subsequent gating

population, rather than the single positive CD4 or CD8 T cell populations. To do so, subgate off of $CD3^+$ events and create a single parameter histogram for the expression of $H2\text{-}D^d$ and analyze the $H2\text{-}D^{d+}$ events as previously indicated.

Disclosures and Potential Conflicts of Interest

This work was supported by BUMED advanced medical development program 604771 N C165 001 A0812. Some of the authors are employees of the US Government. This work was prepared as part of their official duties. Title 17 U.S.C. §105 provides that “Copyright protection under this title is not available for any work of the United States Government.” Title 17 U.S.C §101 defined a US Government work as a work prepared by a military service member or employees of the US Government as part of that person’s official duties. The opinions or assertions contained in this chapter are the private views of the authors and are not to be construed as reflecting the views, policy, or positions of the Department of the Navy, Department of Defense, nor that of the US Government. The authors of this chapter have no additional conflicts of interest to disclose.

References

1. Strong AL, Bowles AC, MacCrimmon CP, Frazier TP, Lee SJ, Wu X et al (2015) Adipose stromal cells repair pressure ulcers in both young and elderly mice: potential role of Adipogenesis in skin repair. Stem Cells Transl Med 4(6):632–642
2. Semon JA, Zhang X, Pandey AC, Alandete SM, Maness C, Zhang S et al (2013) Administration of murine stromal vascular fraction ameliorates chronic experimental autoimmune encephalomyelitis. Stem Cells Transl Med 2(10):789–796
3. Lee HK, Lim SH, Chung IS, Park Y, Park MJ, Kim JY et al (2014) Preclinical efficacy and mechanisms of mesenchymal stem cells in animal models of autoimmune diseases. Immune Netw 14(2):81–88
4. Nicoletti GF, De Francesco F, D'Andrea F, Ferraro GA (2015) Methods and procedures in adipose stem cells: state of the art and perspective for translation medicine. J Cell Physiol 230(3):489–495
5. De Miguel MP, Fuentes-Julian S, Blazquez-Martinez A, Pascual CY, Aller MA, Arias J et al (2012) Immunosuppressive properties of mesenchymal stem cells: advances and applications. Curr Mol Med 12(5):574–591
6. Gimble JM, Bunnell BA, Frazier T, Rowan B, Shah F, Thomas-Porch C et al (2013) Adipose-derived stromal/stem cells: a primer. Organogenesis 9(1):3–10
7. Crop MJ, Baan CC, Korevaar SS, Ijzermans JN, Pescatori M, Stubbs AP et al (2010) Inflammatory conditions affect gene expression and function of human adipose tissue-derived mesenchymal stem cells. Clin Exp Immunol 162(3):474–486
8. Davis TA, Anam K, Lazdun Y, Gimble JM, Elster EA (2014) Adipose-derived stromal cells promote allograft tolerance induction. Stem Cells Transl Med 3(12):1444–1450
9. Bourin P, Bunnell BA, Casteilla L, Dominici M, Katz AJ, March KL et al (2013) Stromal cells from the adipose tissue-derived stromal vascular fraction and culture expanded adipose tissue-derived stromal/stem cells: a joint statement of the International Federation for Adipose Therapeutics and Science (IFATS) and the International Society for Cellular Therapy (ISCT). Cytotherapy 15(6):641–648
10. Kern S, Eichler H, Stoeve J, Kluter H, Bieback K (2006) Comparative analysis of mesenchymal stem cells from bone marrow, umbilical cord blood, or adipose tissue. Stem Cells 24(5):1294–1301
11. Yu G, Floyd ZE, Wu X, Halvorsen YD, Gimble JM (2011) Isolation of human adipose-derived stem cells from lipoaspirates. Methods Mol Biol 702:17–27

Chapter 8

High Definition Confocal Imaging Modalities for the Characterization of Tissue-Engineered Substitutes

Dominique Mayrand and Julie Fradette

Abstract

Optimal imaging methods are necessary in order to perform a detailed characterization of thick tissue samples from either native or engineered tissues. Tissue-engineered substitutes are featuring increasing complexity including multiple cell types and capillary-like networks. Therefore, technical approaches allowing the visualization of the inner structural organization and cellular composition of tissues are needed. This chapter describes an optical clearing technique which facilitates the detailed characterization of whole-mount samples from skin and adipose tissues (ex vivo tissues and in vitro tissue-engineered substitutes) when combined with spectral confocal microscopy and quantitative analysis on image renderings.

Key words Confocal microscopy, Imaging, Tissue structure, Skin, Adipose tissue, Tissue engineering, Apoptosis, Microvascular networks

1 Introduction

Histological analysis of thin tissue sections is commonly the initial step performed to assess how tissue-engineered substitutes structurally resemble the tissue of origin. In comparison, immunofluorescence imaging of whole-mount specimens (in toto labeling) using confocal laser scanning microscopy can reveal the three-dimensional tissue structures. For example, the extent and morphology of vascular networks within engineered substitutes or native tissues can be investigated. However, it can be challenging to assess the spatial distribution within thick tissues because light scattering interferes with image acquisition beyond approximately 100 μm [1].

Absorption and scattering of light will occur as the focal point progresses within the tissue, a phenomenon dependent on the thickness and density of the tissue [2], but also on its composition in water, lipids and hemoglobin content [1, 3]. This limitation of confocal microscopy in terms of image acquisition can be greatly improved using optical clearing protocols, which increase light penetrance [4].

Bruce A. Bunnell and Jeffrey M. Gimble (eds.), *Adipose-Derived Stem Cells: Methods and Protocols*, Methods in Molecular Biology, vol. 1773, https://doi.org/10.1007/978-1-4939-7799-4_8, © Springer Science+Business Media, LLC, part of Springer Nature 2018

Indeed, researchers in the fields of embryology and neurology have developed various useful protocols that can be adapted to circumvent this limitation. Optical clearing is a development process which, when used in combination with different imaging modalities, contributes to exceptional image quality of whole-mount specimens of appreciable size such as murine and zebrafish embryos or brains [1, 5, 6]. Optical clearing methods therefore represent a simple and inexpensive approach to help reveal the structural complexity of tissues or engineered constructs and assess their homogeneity and metabolic state (viable or apoptotic cells, etc.).

In the field of tissue engineering, the in vitro production of substitutes featuring higher complexity and multiple cell types is progressing rapidly. Adipose-derived stem/stromal cells (ASCs) are widely used for tissue engineering purposes and their multilineage differentiation potential makes them excellent candidates for the reconstruction of various tissues such as adipose, bone or skin substitutes [7]. For all tissue types that are engineered in vitro, an important challenge is to ensure a rapid vascularization of voluminous substitutes upon grafting. This prompted the development of various strategies to generate prevascularized tissues in vitro, on the basis that a preformed network of capillary-like structures would ensure a rapid connection to the host vasculature and improve graft survival [8–11]. Indeed, endothelial cells seeded into a substitute will proliferate, migrate and organize themselves to form a network of capillaries within the tissue. It is then important to assess the distribution (homogeneity) and morphology of the resulting structures. Few studies tackled the challenge of providing extensive structural imaging of 3D tissue-engineered constructs [12, 13]. A previous study usefully described the advantages of multiphoton microscopy and optical clearing to probe decellularized/recellularized engineered tissues with a focus on large blood vessels and lung tissue scaffolds [12]. Multiphoton-based imaging is less affected by light diffusion. Consequently, thicker specimens can be observed using multiphoton than confocal-based microscopy [14, 15], which on the other hand is more commonly available at imaging platforms.

In this chapter, we successfully used methyl salicylate-based optical clearing to highlight the possibilities and benefits of this technique when combined with regular laser scanning confocal imaging for detailed analyses of tissue-engineered constructs. Using human skin or adipose tissue samples as well as in vitro prevascularized adipose substitutes engineered from ASCs (Fig. 1), we provide a protocol enabling a comprehensive assessment of the spatial distribution and morphology of vascular networks within tissues (Figs. 2, 3 and 4). We also describe how to detect apoptotic cells and provide a tissue cell death index (Fig. 5). The assessment of cell viability can be particularly important for thick tissues engineered in vitro considering that cell death is most probable within

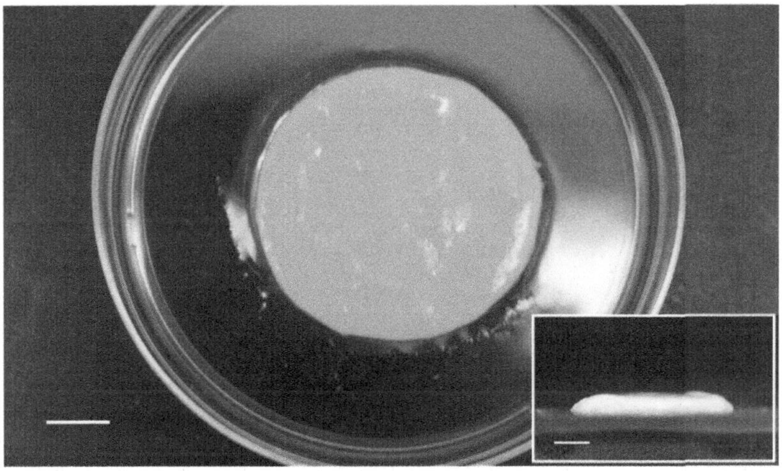

Fig. 1 Macroscopic appearance of a large tissue-engineered adipose substitute of 16 cm^2 and approximately 1 mm in thickness. Bar = 1 cm. The insert represents a side-view of the construct. Bar = 2 mm

the tissue's central regions, which are usually beyond standard imaging capacity. Both of these examples highlight the benefits of performing such imaging modalities not only to assess the substitute's structural features before grafting but also to investigate in vitro the impact of pharmacological compound treatment (such as proangiogenic or antiangiogenic molecules) in a quantitative manner from rather large tissue sample volumes.

2 Materials

2.1 Tissue Samples

1. Samples of resected tissues such as human skin and adipose tissue (*see* **Note 1**), or tissue-engineered substitutes produced in vitro (*see* **Note 2**).

2.2 Sample Treatment

1. 2.0 mL snaplock microcentrifuge clear propylene tubes.
2. PBS (phosphate buffer saline): (g/L) 0.9 g Na_2PO_4, 0.2 g KH_2PO_4, 0.2 g KCl, 8 g NaCl, pH 7.4.
3. Fixative: 4% formaldehyde in PBS.
4. Methanol gradients (50 and 100%): Mix one part pure methanol in one part of deionized water to generate a 50% methanol solution.
5. DMSO (dimethyl sulfoxide).
6. Modified Dent's fixative solution: Mix one part DMSO with three parts of methanol (*see* ref. 16).
7. Bleaching solution: Mix four parts of modified Dent's fixative in one part 30% hydrogen peroxide. Use fresh (*see* ref. 17).

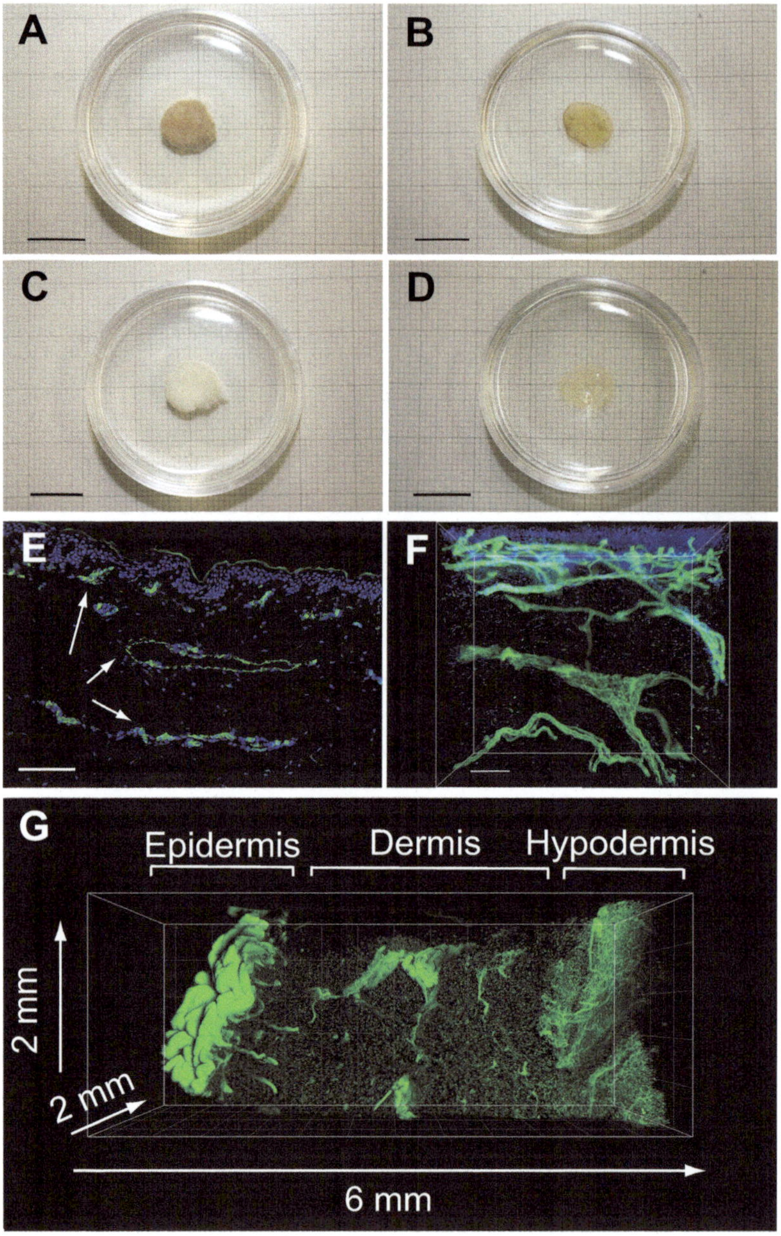

Fig. 2 Impact on skin's transparency of a bleaching treatment combined with optical clearing. Macroscopic aspect of (**a**) untreated skin sample, (**b**) optically cleared skin sample, (**c**) bleached skin before, and (**d**) after combination with optical clearing. Bars = 1 cm. Detection of CD31-expressing structures (in green, arrows) on (**e**) thin transverse cryosection of human skin or (**f**) whole-mount preparation of optically cleared skin sample, which reveals the morphology and spatial distribution of structures such as the epidermis and the vascular network. Bars = 100 μm. (**g**) Nuclei labeling of a large 6 mm × 2 mm × 2 mm skin sample reveals the entire skin's thickness from the epidermis to the hypodermis. Nuclei were labeled with (**e** and **f**) Hoechst or (**g**) SYBR® Green I

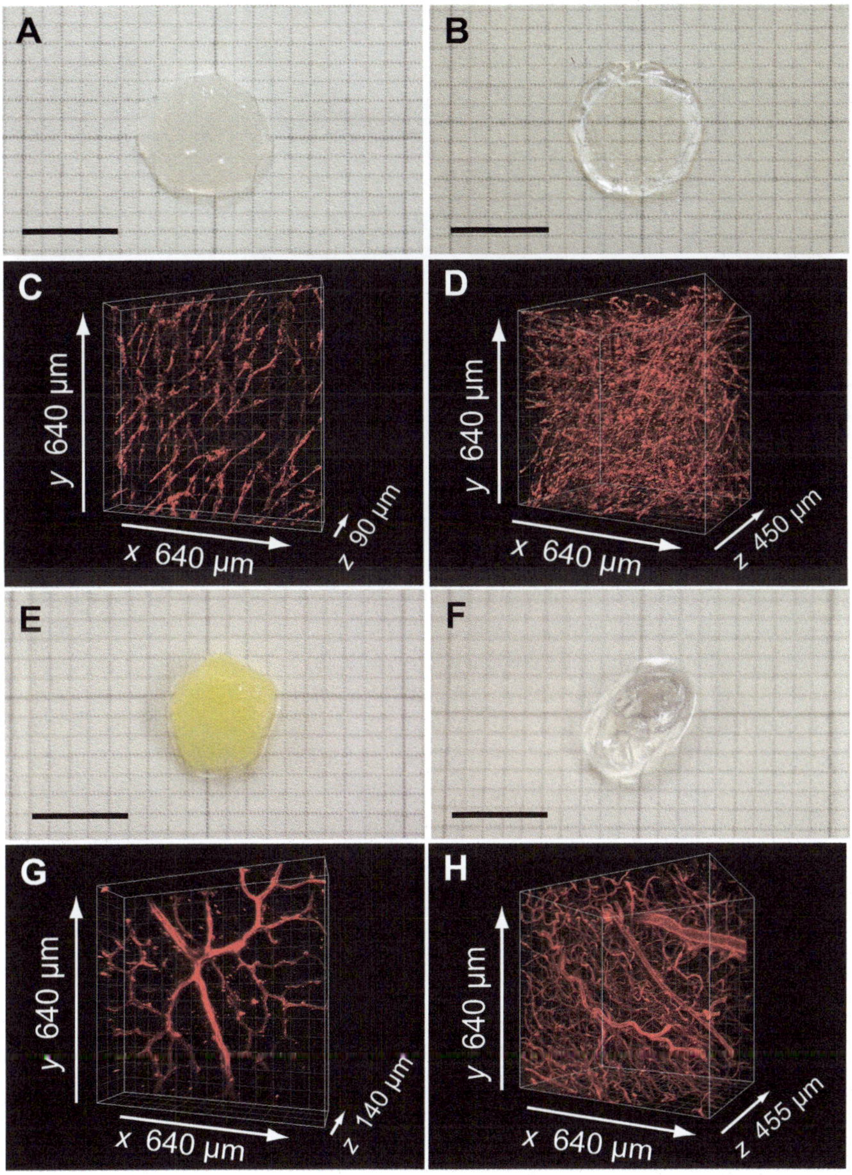

Fig. 3 Impact of optical clearing on adipose tissue samples. Tissues that were (**a**, **c**, **e**, **g**) untreated or (**b**, **d**, **f**, **h**) optically cleared illustrate the treatment's impact on the transparency of the tissue. Macroscopic appearance of samples from (**a**, **b**) an in vitro prevascularized reconstructed adipose tissue and (**e**, **f**) native adipose tissue. Bars = 5 mm. (**c** and **d**, **g** and **h**) Detection of CD31-labeled vascular structures and confocal imaging on optically cleared samples enabled to scan deeper into (**d**) reconstructed and (**h**) native adipose tissues compared to their respective untreated samples counterparts (**c**, **g**)

8. Methyl salicylate gradients (33, 66, and 100%): For a 33% solution, mix one part of methyl salicylate with two parts of methanol. For a 66% solution, mix two parts of methyl salicylate with one part of methanol.
9. Cover glasses: 24 mm × 55 mm × 0.17 mm.

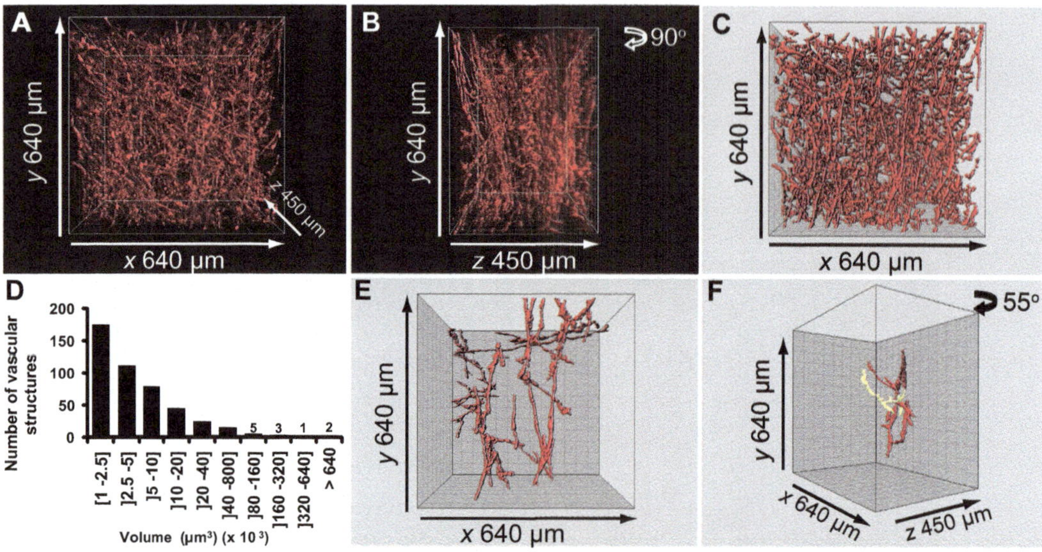

Fig. 4 Detailed analysis of the capillary network within engineered tissues. (**a**) Front view of a whole-mount preparation of an optically cleared sample from an in vitro prevascularized reconstructed adipose tissue labeled with CD31 and scanned with the confocal microscope. (**b**) Side view (thickness) after a 90° rotation. (**c**) Isosurface rendering of a section of the adipose substitute generated with the Imaris software. (**d**) Frequency distribution of the number of capillary-like structures comprised within specified volume intervals analyzed from the image shown in (**c**). Examples of detailed analysis: (**e**) Structures featuring a volume between 160,000 and 320,000 μm^3 have been selected from the image shown in (**c**). (**f**) Example of the selection of a single capillary-like structure (in yellow), shown after a 55° rotation

10. Silicone isolators: different thicknesses are available (0.5 mm, 1 mm, 1.5 mm and 2 mm, Grace Biolabs, Bend, OR, USA).
11. Clear nail polish.
12. Blocking buffer: 1% bovine serum albumin (BSA) in PBS.
13. Staurosporine: from *Streptomyces* sp. (Sigma-Aldrich, catalog number S5921).
14. Culture medium: Dulbecco's Modified Eagle's medium (D-MEM)–Ham's F12 medium 1:1, 1.2 g/L $NaHC0_3$, 3.574 g/L HEPES free acid, 312.5 μL/L 2 N HCl containing 10% of fetal bovine serum and antibiotics.
15. Platform shaker.

2.3 Antibodies and Dyes

1. Primary antibodies (Table 1) (*see* **Note 3**).
2. Fluorochrome-conjugated antibodies (Table 2) (*see* **Note 4**).
3. Nucleic dyes for double-stranded DNA: Hoechst 33258 (Sigma-Aldrich, #B2883) and SYBR® Green I (Invitrogen, #S7563).

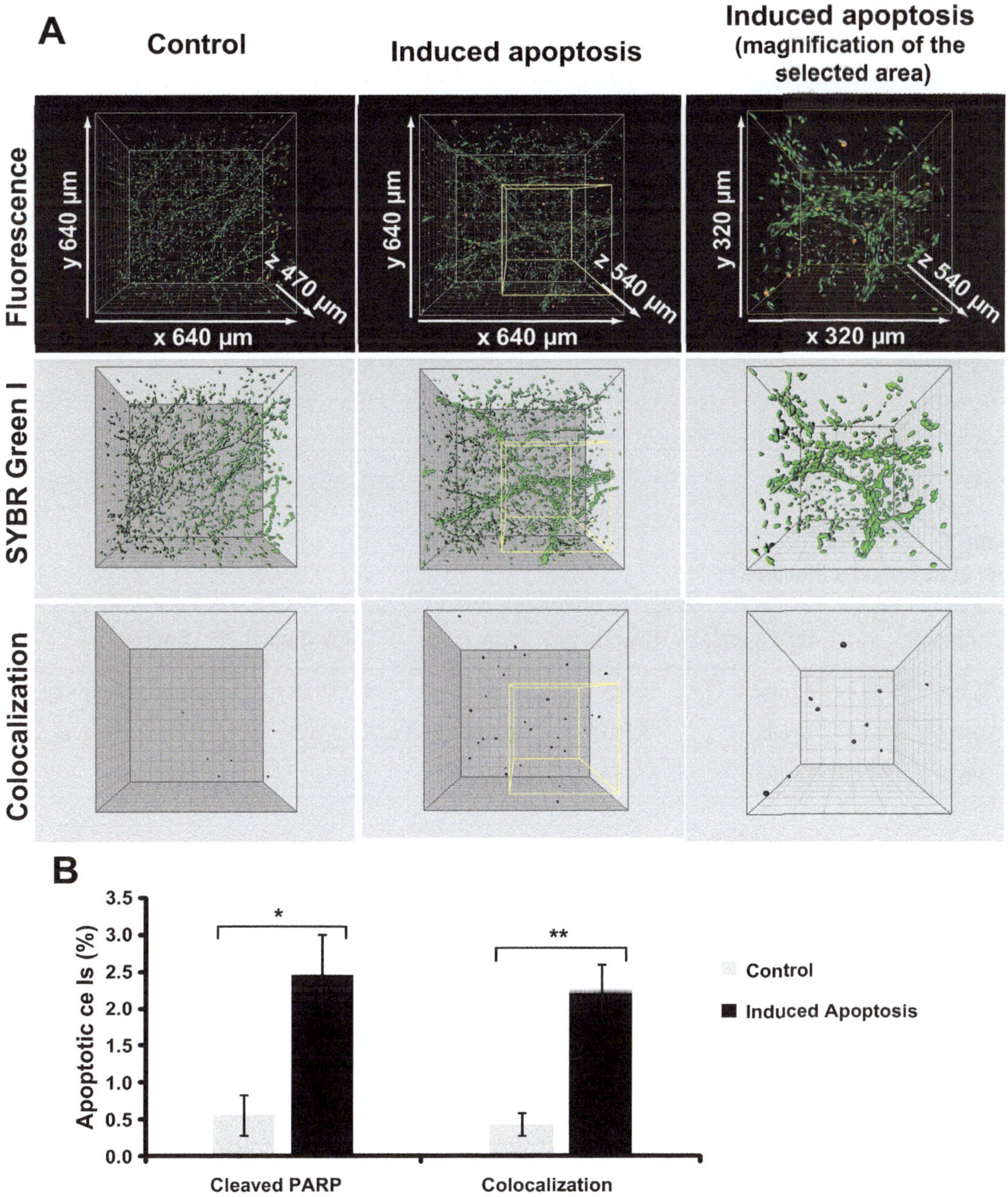

Fig. 5 Quantification of apoptotic cells by confocal microscopy on fixed tissue samples. (**a**) The quantification of apoptotic cells was performed for vehicle-treated adipose tissue explants (left column) and for staurosporine-treated samples (middle column). Higher magnifications of the selected areas are shown in the far right column. Cell detection was performed by double-labeling (top row) of cleaved PARP (apoptotic cells, in red) and SYBR® Green I (total nuclei, in green). The middle row represents Imaris-generated isosurface reconstructions from total nuclei images, while the bottom row indicates the colocalization of cleaved PARP expressing cells with SYBR® Green I positive nuclei. (**b**) Quantification of endogenous and induced apoptosis from the 3D image renderings ($n = 4$ per group, $^{*}P = 0.0037$, $^{**}P = 0.0007$, Student's t-tests)

Table 1
List of primary antibodies used

Antigen	Antigen localization	Host species	Type	Clonality	Catalog number	Source
Human CD31	Endothelial cell junctions	Sheep	IgG	Polyclonal	AF806	R&D systems
Human cleaved PARP [4B5BD2]	Apoptotic cell's nucleus	Mouse	IgG1	Monoclonal	AB110315	Abcam
Negative isotype control antibody	–	Sheep	IgG	Polyclonal	12-515	Millipore
Negative isotype control antibody [1E2.2]	–	Mouse	IgG1	Monoclonal	CBL610	Millipore

Table 2
List of detection antibodies used

Conjugate	Host species	Reactivity	Type	Catalog number	Source
Alexa Fluor 488	Donkey	Sheep	IgG (H + L)	A-11015	Life technologies
Alexa Fluor 633	Donkey	Sheep	IgG (H + L)	A-21100	Life technologies
CF633	Donkey	Mouse	IgG (H + L)	20124	Biotium

2.4 Image Acquisition

1. LSM 700 inverted confocal scanning microscope system controlled by Zen 2010 Software (Carl Zeiss) and equipped with plan-apochromat 10×/0.45 and plan-apochromat 20×/0.75 objectives.
2. Imaris Scientific 3D/4D processing and analysis software v7.0.0 (Bitplane, MA, USA).

3 Methods

Carry out all procedures at room temperature unless otherwise specified. Incubation times and volumes should first be optimized according to the type of tissue, the antibodies used and the size of the samples (*see* **Note 5**).

3.1 Sample Preparation

1. Fix the tissue samples in 4% formaldehyde for 2 h.
2. For the following steps, use clear 2.0 mL tubes and a 500 μL working volume of the solutions.
3. Perform a permeabilization step to promote the penetration of the antibodies. Dehydrate the samples in a methanol series (50% once and 100% twice, 30 min each step).

4. Incubate the samples for 2 h in the modified solution of Dent's fixative.
5. When appropriate, add a bleaching step to further enhance imaging (optional, *see* **Note 6**). Incubate the samples overnight at 4 °C in the bleaching solution.
6. Rehydrate the samples by washing 30 min in a 50% methanol solution, followed by three rinses of 30 min in PBS.

3.2 Labeling

The labeling protocol should be adapted and optimized for the detection of the structures of interest, especially when different antigens are targeted (*see* **Note 7**).

1. Saturate the samples with 500 μL of 1% PBS-BSA solution for 30 min.
2. Incubate the samples with 200 μL of the primary antibody diluted in 1% PBS-BSA for 24 h to 72 h at 4 °C with gentle agitation to ensure a uniform labeling.
3. Rinse three times with 500 μL of PBS, 30 min each time.
4. Incubate the samples with 200 μL of the conjugated antibody diluted in 1% PBS-BSA for 24 h to 72 h at 4 °C, in the dark, with gentle agitation to ensure a uniform labeling.
5. Rinse three times with 500 μL of PBS, 30 min each time.
6. If a nucleic acid labeling is required, it can be added to the conjugated antibody solution. Refer to the technical sheet for suggested dilutions.

3.3 Optical Clearing

1. Dehydrate the samples in a methanol series (50% once and 100% twice, 500 μL), 30 min each, in the dark, followed by a methyl salicylate series (33, 66, and 100% twice, 500 μL), 30 min each (*see* **Note 8**).
2. Mount the sample in 100% methyl salicylate between two cover glasses separated with a silicone isolator featuring a thickness proportionate to the size of the sample (*see* **Note 9**).
3. Seal with clear nail polish.
4. Keep in the dark at 4 °C until imaging.

3.4 Image Acquisition

1. Image the tissues using a confocal scanning microscope system.
2. If available on your system, use the auto *z* brightness correction function during image acquisition to reduce loss of light when penetrating deeper into the specimen (*see* **Note 10**).

3.5 Image Analysis

1. Produce isosurface reconstructions of the structures from the acquired images using the Imaris 7.0.0 software or another software compatible with your platform.
2. In order to quantify the vascular network, calculate the total volume of vascular structures from the vascular network surface

rendering and report it based on the total volume of sample analyzed. Each structure can be analyzed individually by using selected criteria such as the volume (Fig. 4d–f).

3. To determine a tissue cell death index, calculate the ratio of apoptotic nuclei positive for cleaved PARP (Poly (ADP-Ribose) Polymerase) reported on the total number of cells present in the sample analyzed (*see* **Note 11**) (Fig. 5).

4 Notes

1. For human tissues and substitutes engineered from human cells, donor consent and institutional ethical committee approval are required.
2. Labeling and imaging can be performed on human tissue-engineered constructs such as in vitro prevascularized adipose substitutes from which samples were harvested for the featured analyses (*see* Figs. 1, 3, and 4). These large tissue constructs, which are rich in endogenous matrix components, undifferentiated ASCs and in vitro differentiated adipocytes as well as human vascular endothelial cells, were produced according to modifications of the protocols described in [18, 19].
3. Preferentially choose antigens that will provide a strong signal. For example, immunodetection of the CD31 antigen, an adhesion molecule particularly abundant at the intercellular junctions of the endothelial cells, is useful to highlight the tissue's vascular and lymphatic networks [20].
4. Some proteins, such as matrix components, are strongly autofluorescent in the visible range. If combined with a weak labeling of the target protein, the expected positive signal is difficult to distinguish through background noise. A good option to circumvent this problem is to select probes with near-infrared wavelengths (633 and 647 nm) [21].
5. The proposed incubation times and volumes were adequate for samples in the range of 2 × 2 × 1 mm in size and should be increased for larger samples. The first consideration guiding the choice of sample size must be the limitation imposed by the working distance of the objective during acquisition. Sample size will also be mostly directed by the anisotropy of the tissue, which impacts the scattering of light. For example, scattering of light is greater in the epidermis than in the dermis [22].
6. Light is strongly absorbed by chromophores such as melanin and hemoglobin. If appropriate, an optional bleaching step can be performed before labeling to further enhance imaging [23, 24]. Human skin samples benefit from the bleaching and optical clearing treatments (*see* Fig. 2a–d).

7. The immunofluorescence procedure proposed here is optimal for the analysis presented in this chapter. The labeling protocol must be adapted based on the antigens and detection antibodies used. The length of the incubation times with the antibodies must be adjusted according to the size of the samples in order to favor antibody penetration.
8. Optical clearing significantly reduces the scattering of light and allows the acquisition of images to a significantly greater depth (*see* Fig. 3). For example, the optical clearing of reconstructed adipose tissue allowed the acquisition of an image from a specimen nearly five to ten times thicker when a 20× objective is used, which means that images representing a depth of 450–500 μm can be acquired at this magnification, which is the working distance limit of the objective. The key to a good clarification is to match the refraction index of both the tissue and the chosen clarifying agent [4, 25]. Other agents than methyl salicylate can be used to perform optical clearing, such as BABB (1:2 benzyl alcohol–benzyl benzoate solution), which is also an excellent clarifying agent with a slightly different refraction index than methyl salicylate. Benzyl benzoate, in combination with tetrahydrofuran in replacement of methanol, has been reported to improve the preservation of GFP expression in mouse brain specimen [26]. The Sca*le* method based on the mixed solution of urea, glycerol, and Triton has also been reported to improve the preservation of fluorescent specimen for brain and muscle studies [13, 27]. Our protocol proposes to use methanol for the dehydration steps. When applicable, it might be advantageous to replace it by ethanol to better preserve the fluorescence of reporter genes. In all cases, precaution must be taken to minimize the shrinkage of the specimen during dehydration. It is therefore recommended to proceed with a progressive alcohol gradient (30, 50, 70, 90, and 100%) rather than a steeper gradient (50 and 100%) [25].
9. Mounting the sample between two cover glasses is a delicate step, but it allows to scan the tissue on both sides. Precautions must be taken when using methyl salicylate and BABB clearing agents since they are corrosive and could damage the objectives.
10. The auto *z* brightness correction function allows an automatic linear interpolation of the acquisition settings including laser power and gain level during the scan in the *z* axis. The settings are then adjusted automatically during the acquisition for each given *z*-position.
11. Induction of apoptosis was performed by incubation of small pieces of fresh adipose tissue for 24 h at 37 °C in presence of 5 μM of staurosporine in culture medium. Negative controls were incubated in the same conditions in medium containing 1.25% DMSO. The benefits of this imaging approach include

the use of fixed tissue samples and the detection of low levels of apoptotic cells deeper into the tissue (*see* Fig. 5).

Acknowledgments

This work was supported by the Canadian Institutes of Health Research (CIHR) grant #111233. The confocal imaging system was obtained through the *Fonds des leaders* program from Canada Foundation for Innovation (CFI) to Julie Fradette. We thank former and current members of the laboratory who provided tissues and insights during the development of these protocols. We acknowledge V.J. Moulin for sharing staurosporine and SYBR® Green I reagents. J.F. is a Fonds de Recherche du Québec-Santé (FRQS) career award scholar.

References

1. Liu YA, Chen Y, Chiang AS, Peng SJ, Pasricha PJ, Tang SC (2011) Optical clearing improves the imaging depth and signal-to-noise ratio for digital analysis and three-dimensional projection of the human enteric nervous system. Neurogastroenterol Motil 23(10):e446–e457
2. Calve S, Ready A, Huppenbauer C, Main R, Neu CP (2015) Optical clearing in dense connective tissues to visualize cellular connectivity in situ. PLoS One 10(1):e0116662
3. Aswathy RG, Yoshida Y, Maekawa T, Kumar DS (2010) Near-infrared quantum dots for deep tissue imaging. Anal Bioanal Chem 397(4):1417–1435
4. Zucker RM, Hunter S, Rogers JM (1998) Confocal laser scanning microscopy of apoptosis in organogenesis-stage mouse embryos. Cytometry 33(3):348–354
5. Zucker RM, Hunter ES 3rd, Rogers JM (2000) Confocal laser scanning microscopy of morphology and apoptosis in organogenesis-stage mouse embryos. Methods Mol Biol 135:191–202
6. Metcalfe WK, Myers PZ, Trevarrow B, Bass MB, Kimmel CB (1990) Primary neurons that express the L2/HNK-1 carbohydrate during early development in the zebrafish. Development 110(2):491–504
7. Gimble JM, Katz AJ, Bunnell BA (2007) Adipose-derived stem cells for regenerative medicine. Circ Res 100(9):1249–1260
8. Rochon MH, Fradette J, Fortin V, Tomasetig F, Roberge CJ, Baker K, Berthod F, Auger FA, Germain L (2010) Normal human epithelial cells regulate the size and morphology of tissue-engineered capillaries. Tissue Eng Part A 16(5):1457–1468
9. Aubin K, Vincent C, Proulx M, Mayrand D, Fradette J (2015) Creating capillary networks within human engineered tissues: impact of adipocytes and their secretory products. Acta Biomater 11:333–345
10. Kang JH, Gimble JM, Kaplan DL (2009) In vitro 3D model for human vascularized adipose tissue. Tissue Eng Part A 15(8):2227–2236
11. Baiguera S, Ribatti D (2013) Endothelialization approaches for viable engineered tissues. Angiogenesis 16(1):1–14
12. Calle EA, Vesuna S, Dimitrievska S, Zhou K, Huang A, Zhao L, Niklason LE, Levene MJ (2014) The use of optical clearing and multiphoton microscopy for investigation of three-dimensional tissue-engineered constructs. Tissue Eng Part C Methods 20(7):570–577
13. Decroix L, Van Muylder V, Desender L, Sampaolesi M, Thorrez L (2015) Tissue clearing for confocal imaging of native and bio-artificial skeletal muscle. Biotech Histochem:1–8
14. Conchello JA, Lichtman JW (2005) Optical sectioning microscopy. Nat Methods 2(12):920–931
15. Dufour P, Dufour S, Castonguay A, McCarthy N, De Koninck Y (2006) Microscopie à deux photons pour l'imagerie cellulaire fonctionnelle: avantages et enjeux: ou Un photon c'est bien… mais deux c'est mieux! Med Sci 22(10):837–844

16. Dent JA, Polson AG, Klymkowsky MW (1989) A whole-mount immunocytochemical analysis of the expression of the intermediate filament protein vimentin in Xenopus. Development 105(1):61–74
17. Dickie R, Bachoo RM, Rupnick MA, Dallabrida SM, Deloid GM, Lai J, Depinho RA, Rogers RA (2006) Three-dimensional visualization of microvessel architecture of whole-mount tissue by confocal microscopy. Microvasc Res 72(1–2):20–26
18. Vermette M, Trottier V, Menard V, Saint-Pierre L, Roy A, Fradette J (2007) Production of a new tissue-engineered adipose substitute from human adipose-derived stromal cells. Biomaterials 28(18):2850–2860
19. Proulx M, Aubin K, Lagueux J, Audet P, Auger M, Fortin MA, Fradette J (2015) Magnetic resonance imaging of human tissue-engineered adipose substitutes. Tissue Eng Part C Methods 21(7):693–704
20. Wang XN, McGovern N, Gunawan M, Richardson C, Windebank M, Siah TW, Lim HY, Fink K, Li JL, Ng LG, Ginhoux F, Angeli V, Collin M, Haniffa M (2014) A three-dimensional atlas of human dermal leukocytes, lymphatics, and blood vessels. J Invest Dermatol 134(4):965–974
21. Hilderbrand SA, Weissleder R (2010) Near-infrared fluorescence: application to in vivo molecular imaging. Curr Opin Chem Biol 14(1):71–79
22. Anderson RR, Parrish JA (1981) The optics of human skin. J Invest Dermatol 77(1):13–19
23. Zonios G, Bykowski J, Kollias N (2001) Skin melanin, hemoglobin, and light scattering properties can be quantitatively assessed in vivo using diffuse reflectance spectroscopy. J Invest Dermatol 117(6):1452–1457
24. d'Esposito AN, Daniil, Desjardins A, Walker-Samuel S, Lythgoe MF (2015) Optical clearing of the mouse brain and light attenuation quantitation. In: Duco Jansen E (ed) Optical interactions with tissue and cells XXVI, vol 9321. SPIE, Bellingham, p 248
25. Bucher D, Scholz M, Stetter M, Obermayer K, Pfluger HJ (2000) Correction methods for three-dimensional reconstructions from confocal images: I. Tissue shrinking and axial scaling. J Neurosci Methods 100(1–2):135–143
26. Becker K, Jahrling N, Saghafi S, Weiler R, Dodt HU (2012) Chemical clearing and dehydration of GFP expressing mouse brains. PLoS One 7(3):e33916
27. Hama H, Kurokawa H, Kawano H, Ando R, Shimogori T, Noda H, Fukami K, Sakaue-Sawano A, Miyawaki A (2011) Scale: a chemical approach for fluorescence imaging and reconstruction of transparent mouse brain. Nat Neurosci 14(11):1481–1488

Chapter 9

Adipose-Derived Stromal Vascular Fraction Cells and Platelet-Rich Plasma: Basic and Clinical Implications for Tissue Engineering Therapies in Regenerative Surgery

Pietro Gentile and Valerio Cervelli

Abstract

Cell-based therapy and regenerative medicine offer a paradigm shift in regard to various diseases causing loss of substance or volume and tissue or organ damage. Recently, many authors have focused their attention on mesenchymal stem cells for their capacity to differentiate into many cell lineages. The most widely studied types are bone marrow mesenchymal stem cells and adipose derived stem cells (ADSCs), which display similar results. Based on the literature, we believe that the ADSCs offer advantages because of lower morbidity during the harvesting procedure. Additionally, platelet-rich plasma can be used in this field for its ability to stimulate tissue regeneration. The aim of this chapter is to describe ADSC preparation and isolation procedures, preparation of platelet-rich plasma, and the application of ADSCs in regenerative plastic surgery. We also discuss the mechanisms and future role of ADSCs in cell-based therapy and tissue engineering.

Key words Adipose stromal vascular cells, Platelet-rich plasma, Platelet rich lipotransfert, Adipose-derived stem cells

1 Introduction

In recent years, many scientists have shown the existence of cells in the adult body that are capable of repairing and regenerating damaged tissues. Adipose tissue is a multifunctional organ that contains various cellular types, such as mature adipocytes and the stromal vascular fraction (SVF). The adipose SVF provides a rich source [1] of adipose-derived stem cells (ADSCs) and can easily be isolated from human adipose tissue, representing a viable alternative to bone marrow mesenchymal stem cells (BM-MSCs) [2]. Moreover, these cells can be added onto a scaffold, such as purified adipose tissue or alloplastic material, which stimulates the long-term cell retention and subsequent colonization. These techniques can be performed to treat soft tissue defects, scars, and burn injury and to regenerate various damaged tissues.

Bruce A. Bunnell and Jeffrey M. Gimble (eds.), *Adipose-Derived Stem Cells: Methods and Protocols*, Methods in Molecular Biology, vol. 1773, https://doi.org/10.1007/978-1-4939-7799-4_9, © Springer Science+Business Media, LLC, part of Springer Nature 2018

Platelet-rich plasma is a concentration of autologous human platelets in a small volume of plasma, containing at least seven major growth factors (including platelet-derived growth factors [PDGFs], basic fibroblast growth factor [bFGF], vascular endothelial growth factor [VEGF], insulin-like growth factor-1 [IGF-1], and transforming growth factor-β [TGF-β] released by actively degranulated platelets and known to facilitate the wound-healing process. Consequently, platelet-rich plasma (PRP) is currently used in the field of the regenerative medicine because of its ability to stimulate the tissue regeneration.

The aim of this chapter is (a) to describe the isolation procedures, molecular characterization, and differentiation capacity of ADSCs; (b) to describe the preparation of PRP and methodological changes to the preparation of PRP in the literature; and (c) to discuss the clinical application of ADSCs and PRP in regenerative plastic surgery.

Freshly isolated SVF cells are a heterogeneous mixture of endothelial cells, smooth muscle cells, pericytes, fibroblasts, mast cells, and preadipocytes.

Culture of these cells under standard conditions eventually (within the first few passages) results in the appearance of a relatively homogeneous population of mesodermal or mesenchymal cells [2] that were first named processed lipoaspirate cells but are now more commonly called ADSCs (or ASCs, which can be confused with ASCs for adult stem cells).

The Mesenchymal and Tissue Stem Cell Committee of the International Society for Cellular Therapy (ISCT) has recently proposed a minimal set of four criteria to define human mesenchymal stem cells (MSCs) [3]:

1. MSCs are plastic-adherent when maintained under standard culture conditions.
2. MSCs have the capacity for osteogenic, adipogenic, and chondrogenic differentiation.
3. MSCs express CD73, CD90, and CD105.
4. MSCs lack expression of the hematopoietic lineage markers c-kit, CD14, CD11b, CD34, CD45, CD19, CD79, and human leukocyte antigen-DR.

The most commonly studied MSCs are BM-MSCs and ADSCs. The ADSCs meet the majority of the ISCT's criteria for MSCs; however, Lin et al. [4] have found that ADSCs exist as CD34+ CD31− CD104b− smooth muscle actin—cells in the capillary and in the adventitia of larger vessels in vivo. In the capillary, these cells coexist with pericytes and endothelial cells, both of which are possibly related to ADSCs. In the larger vessels, these ADSCs exist as specialized fibroblasts (having stem cell properties) in the adventitia.

Corselli et al. [5] previously demonstrated that human pericytes, which encircle capillaries and microvessels, give rise in culture to genuine MSCs. This raised the question whether all MSCs are derived from pericytes. Pericytes and other cells defined on differential expression of CD34, CD31, and CD146 were sorted from the stromal vascular fraction of human white adipose tissue. Besides pericytes, CD34+ CD31− CD146− CD45− cells, which reside in the outmost layer of blood vessels, the tunica adventitia, natively express MSC markers and give rise in culture to clonogenic multipotent progenitors identical to standard bone marrow-derived MSCs. Despite common MSC features and developmental properties, adventitial cells and pericytes retain distinct phenotypes and genotypes through culture.

However, in the presence of growth factors involved in vascular remodeling, adventitial cells acquire a pericyte-like phenotype.

In conclusion, Corselli et al. [5] demonstrated the coexistence of two separate perivascular MSC progenitors: pericytes in capillaries and microvessels and adventitial cells around larger vessels.

Factors such as donor age, adipose tissue type (white or brown) and anatomical location (subcutaneous or visceral adipose tissue), type of surgical procedure, culturing conditions, exposure to plastic, plating density, and medium formulations might influence both the proliferation rate and the differentiation capacity of these cells. For example, peroxisome proliferator-activated receptor-γ (PPAR-γ) (an indicator of adipogenic potential and commitment) is expressed more by the adipose tissue samples taken from the arm [6].

In addition, there is evidence that the ability of adipose precursor cells to grow and differentiate varies among fat depots and changes with age. It has also been demonstrated that a PPAR-γ agonist aids in the induction of differentiation in cells from all depots and ages. Younger patients have increased PPAR-γ-2 expression in all depots, whereas older patients have consistent elevated expression only in the arm and thigh depots [6]. In fact, the growth rate of ADSCs from SVF was higher in a group of patients 25–30 years old than in older patients. Superficial abdomen adipose tissue appears to be less susceptible to apoptotic stimuli [6].

Levi et al. [7] reported that subcutaneous fat depots retain markedly different osteogenic and adipogenic potentials. Osteogenesis is most robust in adipose-derived stromal cells from the flank and thigh, as compared with those from the arm and abdomen ($p < 0.05$ by all markers examined). This is accompanied by elevations in bone morphogenetic protein 4 and bone morphogenetic protein receptor 1B ($p < 0.05$ by all markers examined).

The osteogenic advantage of cells from the flank and thigh is also observed when the paracrine effects of these cells are analyzed.

Conversely, those cells isolated from the flank have a lesser ability to undergo adipogenic differentiation. Adipose-associated HOX genes are less expressed in flank-derived adipose-derived stromal cells. Variations exist between fat depots in terms of adipose-derived stromal cell osteogenic and adipogenic differentiation.

Differences in HOX expression and bone morphogenetic protein signaling may underlie these observations. This study indicates that the choice of fat depot derivation of adipose derived stromal cells may be an important one for future efforts in tissue engineering.

Neither the type of surgical procedure nor the anatomical site of the adipose tissue affects the total number of viable cells that can be obtained from the SVF [8]. However, since different anatomical localizations of fat tissues have their own metabolic characteristics, such as lipolytic activity, fatty acid composition, and gene expression profile, the source of subcutaneous adipose tissue grafts (abdominal-subcutaneous vs. peripheral-subcutaneous) might influence the long-term characteristics of the fat graft.

In a study by Aksu et al. [9] in which human ADSCs were isolated from superficial and deep adipose layers of the abdominoplasty specimens obtained from patients (male and female) undergoing elective surgeries, there was no significant difference in the degree of osteogenic differentiation between the ADSCs from both depots in the female. In the male, the superficial depot ADSCs differentiated faster and more efficiently than those of the deep depot. Male ADSCs from both depots differentiated more effectively than female ADSCs from both depots.

The frequency of proliferating SVF cells and the population doubling time are dependent on the surgical procedure, with some advantages for resection and tumescent liposuction compared with ultrasound-assisted liposuction [10]. In one study comparing BM-MSCs and lipoaspirate-derived ADSCs [10] from the same patient, no significant differences were observed regarding the yield of adherent stromal cells, growth kinetics, cell senescence, multilineage differentiation capacity, or gene transduction efficiency. Metabolic characteristics and fat cell viability seem not to differ when comparing standard liposuction with syringe aspiration, and no unique combination of preparation or harvesting techniques has appeared superior to date [8]. Although attachment and proliferation capacity are more pronounced in ADSCs derived from younger donors compared with older donors, the differentiation capacity is maintained with aging [8]. ADSCs have the same differentiation potential as described for BM-MSCs. However, some characteristics, such as the colony frequency and the maintenance of proliferating ability in culture, seem even to be superior in ADSCs compared with BM-MSCs [8].

The proliferation of ADSCs can be stimulated by fibroblast growth factor 2 (FGF-2) via the FGF-receptor-2 [11], by sphingosylphosphorylcholine via activation of c-jun N-terminal kinase (JNK) [11], by platelet-derived growth factor via activation of JNK [12], and by oncostatin M via activation of the microtubule-associated protein kinase kinase (MEK)/extracellular signal-regulated kinase (ERK) and the JAK3/STAT1 pathway [13]. ADSCs do express an autocrine FGF-2 loop that maintains their self-renewal ability in vitro [14]. Since inhibition of MEK1 reduces the clonogenic potential of SVF without affecting their differentiation potential, the ERK1/2 signaling pathway seems to be involved in the FGF-2-mediated self-renewal [14]. In addition, the longevity of human ADSCs can be extended by overexpression of the catalytic subunit of the human telomerase gene [15]. ADSCs are known to secrete potent growth factors, such as VEGF, hepatocyte growth factor (HGF), and IGF-1 [16]. Tumor necrosis factor can significantly increase the secretion of VEGF, HGF, and IGF-1 from ADSCs by a p38 mitogen-activated protein kinasedependent mechanism.

2 Materials

2.1 Manual SVF Extraction

1. The buffer (500 ml or more of phosphate-buffered saline [PBS] or KRB) and freshly prepared collagenase solution.
2. PBS solution with 1% bovine serum albumin.

   Adipose tissue in sterile 250-ml plastic bottles (each bottle could accommodate ~100 ml of tissue).
3. 0.1% collagenase type I (C130; Sigma-Aldrich, Milan, Italy, http://www.sigmaaldrich.com)
4. Erythrocyte lysis buffer (155 mM NH_4Cl, 10 mM $KHCO_3$, and 0.1 mM EDTA).
5. 100-mm Falcon strainer (Becton, Dickinson and Company, Sunnyvale, CA, http://www.bd.com).

   PRP preparation made by authors Gentile and Cervelli:
6. Cascade Medical Enterprises, Plymouth, UK.
7. PRL—Platelet Rich Lipotransfert-Corios, San Giuliano Milanese.

   Such devices include:
8. The CATS (Fresenius, Wilmington, DE, http://www.fmcna.com).
9. Sequestra (Medtronic, Minneapolis, http://www.medtronic.com).
10. Hemonetics Cell Saver 5 (Haemonetics Corp., Braintree, MA, http://www.haemonetics.com), and others [17–19].

Many surgical procedures require the use of relatively small volumes of platelet rich plasma [20]. Consequently, small, compact office systems have been developed that produce approximately 6 ml of platelet-rich plasma from 45–60 ml of blood [20–22]. There are many such systems, including:

1. The GPS (Biomet, Warsaw, IN, http://www.biomet.com).
2. PCCS (Implant Innovations, Inc., Palm Beach Gardens, FL, http://www.3i-online.com).
3. Symphony II (DePuy, Warsaw, IN, http://www.depuy.com).
4. SmartPReP (Harvest Technologies Corp.).
5. Norwell, MA (http://www.harvesttech.com).
6. Magellan (Medtronic) [17, 19–22].

Although all operate on a small volume of drawn blood (45–60 ml) and on the principle of centrifugation, these systems differ widely in their ability to collect and concentrate platelets, with approximately 30–85% of the available platelets collected and a less-than-twofold to approximately eightfold increase in the platelet concentration over baseline [17, 18].

There are several devices for the PRP preparation, such as:

1. Fibrinet (Cascade Medical Enterprises, Plymouth, UK, http://www.cascademedical.com).
2. Regen (Regen Lab, Le Mont-sur-Lausanne, Switzerland, http://www.regenkit.com).
3. Plateltex (Plateltex S.R.O.,Bratislava, Slovakia, http://www.plateltex.com).
4. and Vivostat (Vivostat A/S, Borupvang, Denmark, http://www.vivostat.com).

Generally, we prepared PRP according to the Cascade method or PRL procedure, and in all casesweworked under a protocol approved by our institution's transfusion service.

In general, most systems, whether large or small volume, do not concentrate the plasma proteins of the coagulation cascade [17, 19]. The concentration of plasma protein levels above baseline can be achieved through secondary ultrafiltration, as is done with the UltraConcentrator (Interpore Cross, Irvine, CA) and the Access System (Interpore Cross), in which the buffy coat collected from a centrifugation stage is passed through hollow fibers with an effective pore size of 30 kDa. This system removes by filtration up to two-thirds of the aqueous phase; thus, the concentrations of the retained plasma proteins and formed elements are increased substantially [23].

In an interesting work, Mazzucco et al. described the different growth factor concentrations that are obtained through different devices (Fibrinet, Plateltex, and Regen) and a homemade method

[24]. The PDGF-BB, TGF-!, and IGF-1 were detected in lower concentrations with the use of Fibrinet. In contrast, the Regen method showed high concentrations of TGF-B, bFGF, and IGF-1, whereas the Plateltex method showed a high level of endothelial growth factor [24].

3 Methods

According to Yu's protocol [25] for manual SVF extraction, we performed the following steps:

1. (a) We warmed up the buffer (500 ml or more of phosphate-buffered saline [PBS] or KRB) and freshly prepared collagenase solution. (b) We prepared PBS solution with 1% bovine serum albumin, filtered the solution, and warmed it at 37 °C.
2. We dispensed a volume of adipose tissue in sterile 250-ml plastic bottles (each bottle could accommodate ~100 ml of tissue). We then added an equal volume of warm PBS, agitated it to wash the tissue for 3–5 min, and aspirated the infranatant solution. The wash was repeated three times.
3. We then added an equal volume (60–70 ml) of warm 0.1% collagenase type I (C130; Sigma-Aldrich, Milan, Italy, http://www.sigmaaldrich.com) into the bottles containing the washed adipose tissue sample and placed them in a shaking water bath at 37 °C at ~75 rpm for 60 min.
4. Samples were then centrifuged for 10 min at 600 × *g* at room temperature.
5. We removed the supernatant, containing mature adipocytes and oil, and the underlying layer of collagenase solution.
6. The pelleted cells were centrifuged at 300 × *g* for 5 min at room temperature in erythrocyte lysis buffer (155 mM NH_4Cl, 10 mM $KHCO_3$, and 0.1 mM EDTA) and incubated for 5 min at room temperature.
7. After centrifugation at 350 × *g* for 5 min, the pellet was resuspended in a few microliters of growth medium and passed through a 100-mm Falcon strainer (Becton, Dickinson and Company, Sunnyvale, CA, http://www.bd.com).
8. Counting the cellular population using a hemocytometer, we obtained ~250,000 ± 34,782 nucleated cells per milliliter of adipose tissue (12 donors were used to develop this value) [26].

For automatic-enzymatic SVF extraction, the Celution system (Cytori Therapeutics, Inc., San Diego, CA, http://www.cytori.com) was used [4].

1. Adipose tissue was introduced into the cell-processing device and subsequently washed to remove red blood cells and debris, and then enzymatically digested with collagenase type I.
2. Upon digestion of adipose tissue and release of mononuclear cells from the adipose tissue matrix, the released cells were transferred into the centrifuge processing vessel.
3. The cells within the suspension were concentrated by short centrifugation and wash cycles.
4. The cycles were repeated until the entire volume of input cell suspension has been processed and the cell population had been localized into the output chamber.
5. The cells were washed one final time and then suspended for use in 5 ml of Ringer's lactate solution.
6. The SVF pellet was resuspended in erythrocyte lysis buffer, and after centrifugation at 350 × *g* for 5 min, the cellular population was counted using hemocytometer. The cells yield was ~50,000 ± 6956 nucleated cells per milliliter of adipose tissue (ten donors were used to develop this value) [26].

For not-enzymatic SVF extraction, by the PRL—Platelet rich Lipotransfert (Corios, San Giuliano milanese, Italy) was used.

1. Fat (80 ml) was subjected to automatic filtration and centrifugation cycles at 800 × *g* per 10 min after which 40 ml of the suspension was extracted from the bag.
2. The suspension was further filtered through 0.2 μm filter and 20 ml of the enhanced SVF (e-SVF) suspension was obtained.
3. Tthe e-SVF suspension was added and mixed with the centrifuged fat graft. Using specific microcannulas for implantation, the SVF-enhanced fat graft was transferred into 10-ml syringes and aseptically reinjected into the soft tissue defect.

We hypothesize that the mechanism of regeneration of the tissue is as follows: targeting of damaged areas and release of angiogenic and antiapoptotic factors, followed by formation of new vessels and oxygenation. The potential benefit of SVF supplementation could be explained by the ability of cells that exist within the SVF population [27] to secrete various growth factors that improve survival and increase vascularization [28, 29], leading to an increased survival of the graft as shown by a study on a rodent [30]. Our results documented that the SVF cell yield from the manual system was much greater than that of the automatic system.

3.1 PRP Preparation from a Small Volume of Blood

1. PRP was prepared from a small volume of blood (18 cc) according to the method of Cascade-Esforax system, 2,3,6 with modifications.

2. Briefly, to prepare PRP, blood was taken from a peripheral vein using sodium citrate as an anticoagulant.
3. The our traditional preparation of PRP consisted of a 1100 × *g* for 10 min of centrifugation, which allows the platelets to remain suspended in the plasma, while the leukocytes and erythrocytes are displaced to the bottom of the tube.
4. A rapid centrifugation can cause mechanical forces and can raise the temperature, thus inducing changes in the ultrastructure of platelets that, in turn, can initiate a partial activation, with a consequent loss of its content.
5. Our PRP protocol uses Ca2+ to induce platelet activation and exocytosis of the a-granules. Calcium acts as a necessary cofactor for platelet aggregation. When Ca2+ is used to induce platelet activation, the secretion of the growth factors contained in the granules is slow.

3.2 PRP Preparation from a Large Volume of Blood by PRL Procedure

1. PRP was prepared from a large volume of blood (60–120 cc) according to the method of PRL with modifications.
2. Briefly, to prepare PRP, blood was taken from a peripheral vein using sodium citrate as an anticoagulant 3 cc for each 20 cc syringe.
3. Our traditional preparation of PRP consisted of a 400 × *g* for 10 min of centrifugation, which allows the platelets to remain suspended in the plasma, while the leukocytes and erythrocytes are displaced to the bottom of the tube.
4. OurPRP protocol does not use Ca2+ to induce platelet activation and exocytosis of the a-granules.
5. We obtained 20–45 ml average of PRP.

3.3 PRL Procedure According by Gentile and Cervelli Methods

1. PRP was prepared from a large volume of blood (60–120 cc) according to the method of PRL with modifications.
2. Briefly, to prepare PRP, blood was taken from a peripheral vein using sodium citrate as an anticoagulant 3 cc for each 20 cc syringe.
3. The our traditional preparation of PRP consisted of a 400 × *g* for 10 min of centrifugation, which allows the platelets to remain suspended in the plasma, while the leukocytes and erythrocytes are displaced to the bottom of the tube.
4. Our PRP protocol does not use Ca2+ to induce platelet activation and exocytosis of the a-granules.
5. We obtained 20–45 ml average of PRP.
6. Fat (60 ml) was subjected to automatic filtration and centrifugation cycles at 470 × *g* per 10 min after which 40 ml of the suspension was extracted from the bag.

7. The suspension was further filtered through 120 μm filter and 20 ml of the enhanced SVF (e-SVF) suspension was obtained.
8. The e-SVF suspension was added and mixed with the centrifuged fat graft.
9. The authors add 20% of SVF (0.2 ml) obtained to each milliliter of fat graft centrifuged; at the same time the authors add 20% of PRP (0.2 ml) obtained to each milliliter of fat graft centrifuged.
10. Using specific microcannulas for implantation, the SVF and PRP-enhanced fat graft was transferred into 10-ml syringes and aseptically reinjected into the soft tissue defect with "Gently" injection.

The types of PRP preparation may be divided into two categories: *manual methods* and methods that rely on use of *semiautomatic or automatic devices.* Briefly, the process of preparing PRP consists of four phases: blood collection, centrifugation for platelet concentration, induction of gelation (if the PRP is to be used in gel form), and activation.

1. The *manual preparation of PRP* consisted of a slow centrifugation:
2. The platelets were to remain suspended in the plasma.
3. The leukocytes and erythrocytes were displaced to the bottom of the tube. A rapid centrifugation can cause mechanical forces and elevate the temperature, thus inducing changes in the ultrastructure of platelets that, in turn, can initiate a partial activation, with a subsequent loss of their granular content [31]. Current systems for the preparation of platelet concentrations routinely report the use of various centrifugation rates (we used 1100 × *g*). After centrifugation, the buffy coat layer, consisting of platelets and white blood cells, was sequestered in a volume of 9 ml of plasma.
4. Anitua et al. [32] reported the use of two centrifugation rates.
5. Blood was collected into 3.8% (wt/vol) sodium citrate and centrifuged at 4500 × *g* for 12 min at 4 °C to obtain platelet-poor plasma (PPP) or at 460 × *g* for 8 min to obtain PRP.
6. Calcium chloride was added to PPP and PRP at a final concentration of 22.8 mM.
7. The secretion of growth factor begins with platelet activation.
8. Standard cell separators and salvage devices can be used to produce platelet-rich plasma. These devices operate on a unit of blood and typically use continuous-flow centrifuge bowl or continuous-flow disk separation technology and both a hard (fast) and a soft (slow) spin, yielding platelet concentrations of 2–4× baseline [17, 18].

There is increasing evidence of the ability of the SVF to differentiate into cells of nonmesodermal origin, such as neurons, endocrine pancreatic cells, hepatocytes, endothelial cells, and cardiomyocytes.

Accordingly, we have elected to describe this process using the term cross-differentiation (cross-germ plasticity).

The transcriptional and molecular events triggering the lineage-specific mesodermal differentiation into adipocytes [33–35], myocytes [36–38], osteocytes [39, 40], and chondrocytes [39] are well known, and several reviews focus on that point. In the case of adipocyte differentiation, although several transcriptional key events regulating the differentiation of preadipocytes into mature adipocytes have been identified in the last decade, master genes committing the multipotent mesenchymal stem cells to adipoblasts are still awaiting discovery. Recently, a transcriptional coactivator with PDZ-binding motif tafazzin gene (TAZ) was identified as an early "molecular rheostat" modulating mesenchymal stem cell differentiation [41, 42]. Whereas runx-2, the key osteogenic transcription factor, triggers MSC to enter an osteogenic differentiation program, adipogenic differentiation is mainly promoted by PPAR. It is mainly of interest how these two transcription factors are regulated to determine these alternative cell fates. Hong and Yaffe [42] demonstrated that TAZ coactivates runx-2-dependent gene transcription and inhibits PPARdependent gene transcription. As a net result, osteogenic differentiation is favored. By modulating TAZ expression in cell lines, mouse embryonic fibroblasts, primary MSCs in culture, and in zebrafish in vivo, Hong and Yaffe [42] were successful in triggering osteogenic versus adipogenic differentiation. These results indicate that TAZ functions as a real molecular rheostat that allocates MSCs to either osteogenic or adipogenic differentiation.

In this context, catenin signaling and Wnt3a are important mediators in reducing the osteogenic differentiation in ADSCs [43]. However, TAZ has been shown specifically only in BMMSCs, although a recent article by Cho et al. [44] suggests that TAZ is expressed in ADSCs.

4 Notes

1. As reported, PRP induced an increase of ADSC number without any morphological changes compared with control. There was a statistically significant increase, by approximately fourfold, at 4 and 6 days, when cells were preconfluent ($p < 0.02$). After 8 days, at confluence, there was a threefold increase of ADSC numbers in PRP cultures compared with controls. Oil Red O staining did not reveal any significant difference in intracytoplasmic lipid accumulation compared with PRP-treated and control ADSCs [31].

2. The present findings demonstrated that different concentrations of PRP have effects on ADSC proliferation. In the range from 0.2 to 0.4 ml, we observed an increase in the rate of proliferation and, more precisely, determined in vitro that the concentration of 0.4 ml was optimal; higher concentrations did not further affect the proliferation rate.
3. We expect that there will be new findings reported in the literature about the selection of the most appropriate regenerative methods. Indeed, there are already many publications regarding the use of PRP with/without fat graft in plastic and reconstructive surgery.
4. We have reported results obtained using PRP mixed with fat grafting in the treatment of chronic lower-extremity ulcers [45–47], loss of substance on the lower limbs [47], and application of the enhanced stromal vascular fraction in posttraumatic lower extremity ulcers [26].
5. In a study by Crovetti et al. [48], 24 patients with chronic cutaneous ulcers were treated with a series of PRP gel treatments. Nine patients demonstrated complete wound healing. There were no adverse effects encountered, and all patients noted decreased pain [48]. Another wound study by McAleer et al. [49] involved 24 patients with 33 chronic nonhealing lowerextremity wounds. The wounds were injected with PRP every 2 weeks. Successful wound closure and epitheliazation were obtained in 20 wounds. The mean time for closure was 11.15 weeks [49]. These findings were particularly significant because all patients had failed previously available treatment methods. Rozman and Bolta [50] reported the use of only PRP in skin and soft-tissue losses on the basis of its bactericidal and cell proliferation-promoting properties.
6. Cervelli et al. described the evaluation of the healing rate of chronic skin ulcer [45] according to the method of Kazakos et al. [51] with minor modifications. Kazakos et al. [51] performed a study to assess the benefits of using autologous PRP gel in the treatment of 59 patients affected by acute limb soft tissue wounds. The clinical end points were the healing rate and/or the time required to bring about adequate tissue regeneration to undergo reconstructive plastic surgery. That study showed that PRP gel treatment can be a valuable and effective aid in the management of acute trauma wounds.
7. Powell et. al. [52] describe anti-inflammatory properties with reduced edema and ecchymosis associated with the autologous platelet gel in eight women after deep-plane rhytidectomy (face lifting). PRP was also shown to be effective in stopping capillary bleeding in the surgical flaps of a series of 20 patients undergoing various types of cosmetic surgery (face lift, breast

size changes, or neck lifts) reported by Man et al. [20]. Guerrerosantos [53] reported the use of fat tissue without PRP in patients affected by Romberg syndrome disease and facial defects. In the facial tissue atrophy types 1 and 2, Guerrerosantos [53] reported that treatment with lipoinjection alone, based on the use of fat tissue without PRP and/or ADSC, was sufficient. In addition, Guerrerosantos et al. [54] reported some cases treated with rhytidoplasty combined with pursing plication suspension sutures and lipoinjection. They showed interesting results in the facial and cervical flaccidity and concluded that this combination of procedures provides a three-dimensional esthetic improvement in contour and volume, a reduced time of convalescence and recovery, and decreased risk for complications, especially in the facial nerves [54].

8. Recently, the use of fat graft with platelet-rich plasma [55] and with only lipostructure technique [56] has been reported in patients affected by Romberg syndrome. Platelet-rich plasma mixed with centrifuged fat tissue followed by the Coleman technique [43, 57] (for reconstructing the three-dimensional projection of the facial contour and restoring the superficial density of the facial tissues) yielded results that support the efficacy of this combination treatment. This was further confirmed on the basis of patient satisfaction.
9. On the basis of both in vitro experiments and preclinical studies, ADSCs have been applied to various clinical fields. In the first clinical case, autologous ADSCs were used for the regenerative treatment of widespread traumatic calvarial bone defects [29]. A 7-year-old girl with posttraumatic calvarial defects was treated with autologous cancellous iliac bone combined with autologous ADSCs, fibrin glue, and a biodegradable scaffold. Postoperative computed tomography showed new bone formation, and almost complete calvarial continuity was obtained.
10. The transfer of ADSCs combined with free fat has been reported to play an important role in maintaining the volume of the injected fat tissue. Injection of free fat together with ADSCs isolated from the equivalent liposuction aspirates, termed cell-assisted lipotransfer, could become an alternative to soft tissue augmentation surgery, including cosmetic breast augmentation [28].
11. A new aim could be the use of fresh ADSCs isolated from half of the fat tissue mixed with platelet-rich plasma and recombined with the other half. Autologous ADSC therapy could also be used to treat fistulas in patients with Crohn's disease. In a pilot study of five patients with Crohn's disease, the external openings in six of eight fistulas were closed by inoculation of the fistulas with autologous ADSCs [33]. Since that report was published,

ADSCs have been also used to repair tracheomediastinal fistulas caused by cancer ablation. The therapeutic potential of ADSCs for wound healing can be anticipated for the treatment of chronic ulcers caused by radiation therapy [29]. Finally, clinical trials of ADSCs for the treatment of both chronic heart failure and acute myocardial infarction have begun in Europe.

12. Since their isolation and characterization nearly a decade ago, adipose-derived stem cells have become one of the most popular adult stem cell populations for research in soft tissue engineering and regenerative medicine applications.
13. Compared with other stem cell sources, ADSCs offer several advantages, including an abundant autologous source, minor invasive harvesting (liposuction), significant proliferative capacity in culture, and multilineage potential. Numerous preclinical studies have been pursued, with early clinical data appearing in the literature. This review focuses on the published clinical and preclinical data to date using ADSCs and PRP for soft tissue reconstruction, with particular attention to experimental models and methodologies.
14. Our opinion on the future directions for making tissue regenerative therapies more effective is that ADSCs and growth factors will be used in all pathologies for autologous reconstruction of our tissue and organs. Although standardization of ADSC harvesting and processing techniques, as well as long-term results of existing clinical studies remain to be addressed, the known biological properties of ADSCs suggest a potential role in enhancing fat graft retention and facilitating minimally invasive reconstructive treatments. Although clinical applications are being reported, well-controlled clinical studies are needed to demonstrate safety and efficacy.

References

1. Katz AJ, Tholpady A, Tholpady SS et al (2005) Cell surface and transcriptional characterization of human adipose-derived adherent stromal (hA-DAS) cells. Stem Cells 23:412–423
2. Zuk PA, Zhu M, Mizuno H et al (2001) Multilineage cells from human adipose tissue: implications for cell-based therapies. Tissue Eng 7:211–228
3. Dominici M, Le Blanc K, Mueller I et al (2006) Minimal criteria for defining multipotent mesenchymal stromal cells. The international society for cellular therapy position statement. Cytotherapy 8:315–317
4. Lin K, Matsubara Y, Masuda Y et al (2008) Characterization of adipose tissue-derived cells isolated with the Celution™ system. Cytotherapy 10:417–426
5. Corselli M, Chen CW, Sun B et al (2012) The tunica adventitia of human arteries and veins as a source of mesenchymal stem cells. Stem Cells Dev 21(8):1299–1308
6. Schipper BM, Marra KG, Zhang W et al (2008) Regional anatomic and age effects on cell function of human adipose-derived stem cells. Ann Plast Surg 60:538–544
7. Levi B, James AW, Glotzbach JP et al (2010) Depot-specific variation in the osteogenic and adipogenic potential of human adipose-derived stromal cells. Plast Reconstr Surg 126:822–834

8. Xu Y, Malladi P, Wagner DR et al (2005) Adiposederived mesenchymal cells as a potential cell source for skeletal regeneration. Curr Opin Mol Ther 7:300–305
9. Aksu AE, Rubin JP, Dudas JR et al (2008) Role of gender and anatomical region on induction of osteogenic differentiation of human adiposederived stem cells. Ann Plast Surg 60: 306–322
10. De Ugarte DA, Morizono K, Elbarbary A et al (2003) Comparison of multi-lineage cells from human adipose tissue and bone marrow. Cells Tissues Organs 174:101–109
11. Jeon ES, Song HY, Kim MR et al (2006) Sphingosylphosphorylcholine induces proliferation of human adipose tissue-derived mesenchymal stem cells via activation of JNK. J Lipid Res 47:653–664
12. Kang YJ, Jeon ES, Song HY et al (2005) Role of c-Jun N-terminal kinase in the PDGF-induced proliferation and migration of human adipose tissue-derived mesenchymal stem cells. J Cell Biochem 95:1135–1145
13. Song HY, Jeon ES, Jung JS et al (2005) Oncostatin M induces proliferation of human adipose tissue-derived mesenchymal stem cells. Int J Biochem Cell Biol 37:2357–2365
14. Zaragosi LE, Ailhaud G, Dani C (2006) Autocrine fibroblast growth factor 2 signaling is critical for self-renewal of human multipotent adipose-derived stem cells. Stem Cells 24:2412–2419
15. Jun ES, Lee TH, Cho HH et al (2004) Expression of telomerase extends longevity and enhances differentiation in human adipose tissue-derived stromal cells. Cell Physiol Biochem 14:261–268
16. Wang M, Crisostomo P, Herring C et al (2006) Human progenitor cells from bone marrow or adipose tissue produce VEGF, Hgf, and IGF-1 in response to Tnf by a p38 mitogen activated protein kinase dependent mechanism. Am J Physiol Regul Integr Comp Physiol 291:R880–R884
17. Kevy SV, Jacobson MS (2004) Comparison of methods for point of care preparation of autologous platelet gel. J Extra Corpor Technol 36:28–35
18. Siebrecht MA, De Rooij PP, Arm DM et al (2002) Platelet concentrate increases bone ingrowth into porous hydroxyapatite. Orthopedics 25:169–172
19. Waters JH, Roberts KC (2004) Database review of possible factors influencing point-of-care platelet gel manufacture. J Extra Corpor Technol 36:250
20. Man D, Plosker H, Winland-Brown JE (2001) The use of autologous platelet-rich plasma (platelet gel) and autologous platelet-poor plasma (fibrin glue) in cosmetic surgery. Plast Reconstr Surg 107:229
21. Marlovits S, Mousavi M, Gabler C et al (2004) A new simplified technique for producing platelet-rich plasma: a short technical note. Eur Spine J 13(suppl 1):S102–S106
22. Lozada JL, Caplanis N, Proussaefs P et al (2001) Platelet-rich plasma application in sinus graft surgery: part I. Background and processing techniques. J Oral Implantol 27:38
23. Hood AG, Arm DM. Topical application of autogenous tissue growth factors for augmentation of structural bone graft fusion. Paper presented at: American Society of Extra-Corporeal Technology 11th annual symposium on new advances in blood management, 20–23 Apr 2004, Las Vegas, NV
24. Mazzucco L, Balbo V, Cattana E et al (2009) Not every PRP-gel is born equal. Evaluation of growth factor availability for tissues through four PRP-gel preparations: Fibrinet, RegenPRPKit, Plateltex and one manual procedure. Vox Sang 97:110–118
25. Yu G, Floyd E, Wu X et al (2011) Isolation of human adipose-derived stem cells from lipoaspirates. Methods Mol Biol 702:17–27
26. Cervelli V, Gentile P, De Angelis B et al (2011) Application of enhanced stromal vascular fraction and fat grafting mixed with PRP in post-traumatic lower extremity ulcers. Stem Cell Res 6:103–111
27. Rigotti G, Marchi A, Galie M et al (2007) Clinical treatment of radiotherapy tissue damage by lipoaspirate transplant: a healing process mediated by adipose-derived adult stem cells. Plast Reconstr Surg 119:1409–1422
28. Yoshimura K, Sato K, Aoi N et al (2008) Cellassisted lipotransfer for cosmetic breast augmentation: supportive use of adipose-derived stem/stromal cells. Aesthet Plast Surg 32:48–55
29. Lendeckel S, Jodicke A, Christophis P et al (2004) Autologous stem cells (adipose) and fibrin glue used to treat widespread traumatic calvarial defects: case report. J Craniomaxillofac Surg 32:370–373
30. García-Olmo D, García-Arranz M, García LG et al (2005) A phase I clinical trial of the treatment of Crohn's fistula by adipose mesenchymal stem cell transplantation. Dis Colon Rectum 48:1416–1423
31. Cervelli V, Gentile P, Scioli MG et al (2009) Application of platelet-rich plasma to fat grafting during plastic surgical procedures: clinical and in vitro evaluation. Tissue Eng Part C 15:625–634
32. Anitua E, Sa'nchez M, Nurden AT et al (2006) Autologous fibrin matrices: a potential source

of biological mediators that modulate tendon cell activities. J Biomed Mater Res A 77:285–293
33. Rosen ED (2002) The molecular control of adipogenesis, with special reference to lymphatic pathology. Ann N Y Acad Sci 979:143–158. discussion 188–196
34. Schaffler A, Muller-Ladner U, Scholmerich J et al (2006) Role of adipose tissue as an inflammatory organ in human diseases. Endocr Rev 27:449–467
35. Lane MD, Tang QQ (2005) From multipotent stem cell to adipocyte. Birth Defects Res A Clin Mol Teratol 73:476–477
36. Charge SB, Rudnicki MA (2004) Cellular and molecular regulation of muscle regeneration. Physiol Rev 84:209–238
37. Brand-Saberi B (2005) Genetic and epigenetic control of skeletal muscle development. Ann Anat 187:199–207
38. Tajbakhsh S (2005) Skeletal muscle stem and progenitor cells: reconciling genetics and lineage. Exp Cell Res 306:364–372
39. Otto WR, Rao J (2004) Tomorrow's skeleton staff: mesenchymal stem cells and the repair of bone and cartilage. Cell Prolif 37:97–110
40. Dani C (1999) Embryonic stem cell-derived adipogenesis. Cells Tissues Organs 165:173–180
41. Hong JH, Hwang ES, McManus MT et al (2005) TAZ, a transcriptional modulator of mesenchymal stem cell differentiation. Science 309:1074–1078
42. Hong JH, Yaffe MB (2006) TAZ: a beta-cateninlike molecule that regulates mesenchymal stem cell differentiation. Cell Cycle 5:176–179
43. Coleman SR (1997) Facial recontouring with lipostructure. Clin Plast Surg 24:347
44. Cho HH, Shin KK, Kim YJ et al (2010) NF-kappaB activation stimulates osteogenic differentiation of mesenchymal stem cells derived from human adipose tissue by increasing TAZ expression. J Cell Physiol 223:168–177
45. Cervelli V, Gentile P, Grimaldi M (2009) Regenerative surgery: use of fat grafting combined with platelet-rich plasma for chronic lower-extremity ulcers. Aesthet Plast Surg 33:340–345
46. Cervelli V, Gentile P, Casciani CU (2009) Use of platelet gel in chronic lower extremity ulcers. Plast Reconstr Surg 123:122e–123e
47. Cervelli V, De Angelis B, Lucarini L et al (2010) Tissue regeneration in loss of substance on the lower limbs through use of platelet-rich plasma, stem cells from adipose tissue, and hyaluronic acid. Adv Skin Wound Care 23:262–272
48. Crovetti G, Martinelli G, Issi M et al (2004) Platelet gel for healing cutaneous chronic wounds. Transfus Apher Sci 30:145–151
49. McAleer JP, Kaplan E, Persich G (2006) Efficacy of concentrated autologous platelet-derived growth factors in chronic lower extremity wounds. J Am Podiatr Med Assoc 96:482–488
50. Rozman P, Use BZ (2007) of platelet growth factors in treating wounds and soft tissue injuries (review). Acta Dermatovenerol Alp Panonica Adriat 16:156–165
51. Kazakos K, Lyras DN, Verettas D et al (2009) The use of autologous PRP gel as an aid in the management of acute trauma wounds. Injury 40:801–805
52. Powell DM, Chang E, Farrior EH (2001) Recovery from deep-plane rhytidectomy following unilateral wound treatment with autologous platelet gel: a pilot study. Arch Facial Plast Surg 3:245–250
53. Guerrerosantos J (2008) Evolution of technique: face and neck lifting and fat injections. Clin Plast Surg 35:663–676
54. Guerrerosantos J, Guerrerosantos F, Orozco J (2007) Classification and treatment of facial tissue atrophy in Parry-Romberg disease. Aesthet Plast Surg 31:424–434
55. Cervelli V, Gentile P (2009) Use of cell fat mixed with platelet gel in progressive hemifacial atrophy. Aesthet Plast Surg 33:22–27
56. Grimaldi M, Gentile P, Labardi L et al (2008) Lipostructure technique in Romberg syndrome. J Craniofac Surg 19:1089–1091
57. Coleman SR (1995) Long-term survival of fat transplants: controlled demonstrations. Aesthet Plast Surg 19:421–425

Chapter 10

Myogenic Differentiation of ASCs Using Biochemical and Biophysical Induction

Pinar Yilgor Huri, Justin Morrissette-McAlmon, and Warren L. Grayson

Abstract

Adipose-derived stem/stromal cells (ASCs) constitute a very promising source for cell therapy and tissue engineering approaches as they can be easily obtained in large quantities with comparatively minimal patient discomfort. Moreover, ASCs have multilineage differentiation capacity. Among these, differentiation capacity along the myogenic lineage is of particular interest since myogenic precursors are scarce and obtaining a large number of cells from skeletal muscle biopsies is difficult. Here, we describe a method to effectively induce ASC myogenesis through the combination of biochemical (cocktail including 5-azacytidine and horse serum) and biophysical (dynamic culture via uniaxial cyclic strain) stimulation. This method results in multinucleated cells that are positive in myogenic markers including Pax 3/7, desmin, myoD, and myosin heavy chain.

Key words Adipose-derived stem/stromal cell, Cyclic strain, Dynamic culture, Myogenesis, Skeletal muscle, 5-azacytidine

1 Introduction

The global demand for contractile muscle grafts is steadily increasing. Therefore, engineering functional, contractile, implantable muscle grafts using autologous cells would provide a radical therapeutic advancement [1]. One major barrier to engineering functional muscle for therapeutic purposes is the identification of appropriate cell sources. Various cell types, including primary myogenic cells [2], satellite cells [3], and muscle-derived stem cells [4], have been used to engineer skeletal muscle grafts. However, several drawbacks are associated with these approaches. For example, satellite cells, which are the resident stem cells of the skeletal muscle, are scarce and cannot be expanded in cell culture to high passages [5]. Likewise, skeletal muscle tissue from which primary myocytes can be isolated is limited. Therefore, autologous stem cells with myogenic differentiation capacity that can be easily isolated in large numbers are of great importance for muscle regenerative therapies.

Bruce A. Bunnell and Jeffrey M. Gimble (eds.), *Adipose-Derived Stem Cells: Methods and Protocols*, Methods in Molecular Biology, vol. 1773, https://doi.org/10.1007/978-1-4939-7799-4_10,

Mesenchymal stem cells can be isolated from various sources within the body and hold great promise for clinical applications due to high proliferative and differentiation capacities. Among these, bone marrow derived mesenchymal stem cells (MSCs) were first shown to possess myogenic differentiation capacity upon induction with DNA demethylating agent 5-azacytidine (5-Aza) [6]. In non-myogenic cell cultures, 5-Aza induces genomic demethylation that overcomes an epigenetic barrier and multinucleation of cells leading to myogenesis [7, 8]. Similar to prior observations with MSCs, the myogenic differentiation capacity of adipose derived stem/stromal cells (ASCs) were explored and the first evidence of myogenesis was shown in ASC cultures including horse serum [9] and later 5-Aza [10]. ASCs are a very promising cell source for clinical applications including muscle regenerative therapies, as they are abundant, easily procurable and have been successfully differentiated into various cell types [11–13]. Along with induction with soluble factors such as horse serum and 5-Aza, substrate stiffness and anisotropy were shown to effectively induce ASC myogenesis. ASCs differentiate into the myogenic lineage when cultured on substrates mimicking muscle stifness [14] and alignment [15]. Our previous studies have shown that ASC myogenesis could be further enhanced by mimicking the loading profile of the natural muscle tissue. ASCs align, form multinucleated fibers and express muscle specific transcription factors and proteins (Pax3/7, myoD, desmin, and myosin heavy chain) upon biochemical (5-Aza and horse serum) and biophysical (11% uniaxial cyclic strain) stimulation [16, 17]. Here we describe the methods to obtain ASC myogenesis by using biochemical and biophysical induction.

2 Materials

2.1 Equipment

1. Laminar flow cabinet.
2. CO_2 incubator.
3. Inverted light microscope.
4. Hemocytometer.
5. Centrifuge.
6. Orbital shaker.
7. Autoclave.
8. Flexcell system (FX-5000™) (Flexcell International Corp, Hillsborough, NC).

2.2 General Supplies

1. Absorbent bench protectors (sterilized under UV in the hood).
2. Sterile 250 mL bottle.
3. Sterile 15 and 50 mL tubes.

4. Serological pipette and tips.
5. Aspirator tips (sterile).
6. Cell culture flasks (T175).
7. Cryovials.
8. UniFlex culture plate (Collagen I coated) (Flexcell).

2.3 Cell Culture Supplies

2.3.1 ASC Isolation

1. Collagenase type I solution:

   Stock solution: Prepare 10× stock solution by dissolving 1 g of collagenase Type I (Worthington Biochemical Corp.) and 10 g of bovine serum albumin (BSA, Sigma) within 100 mL of PBS with Ca^{2+} on a stirrer plate. Sterile filter the 10× stock, make 5 mL aliquots and store at −20 °C for several months.

   Working solution: Thaw stock solution, warm in water bath, and use within 1 h after thawing (also warm a bottle of PBS with Ca^{2+}). Dilute the stock 1:10 in PBS with Ca^{2+}.
2. Stromal medium:

   High-glucose DMEM (GIBCO Invitrogen), 10% fetal bovine serum (FBS, Atlanta Biologicals), 1% penicillin–streptomycin (GIBCO Invitrogen).
3. PBS (with and without Ca^{2+}) (Mediatech, Cellgro).
4. Red blood cell lysing buffer.

   Stock solution: Prepare 10× stock solution 1.55 M NH_4Cl, 0.1 M $KHCO_3$, 1 mM EDTA within 100 mL Millipore H_2O. Dissolve stock components completely on a stirrer plate. Sterile filter and store at 4 °C for several months.

   Working solution: Thaw stock solution and warm in water bath. Dilute the stock 1:10 in distilled water.

2.3.2 ASC Culture

1. Expansion medium: High-glucose DMEM (GIBCO Invitrogen), 10% FBS (Atlanta Biologicals), 1% penicillin–streptomycin (GIBCO Invitrogen), 1 ng/mL bFGF (PeproTech).
2. Culture medium (CM): Low-glucose DMEM (GIBCO Invitrogen), 10% FBS (Atlanta Biologicals), 1% penicillin–streptomycin (GIBCO Invitrogen), 1 ng/mL bFGF (PeproTech).
3. Myogenic induction medium: Low-glucose DMEM, 1% FBS, 5% horse serum (GIBCO Invitrogen), 1% penicillin–Streptomycin, 10 μM 5-Azacytidine (Sigma).
4. Trypsin (0.25%) with ethylenediaminetetraacetic acid (EDTA) (1 mM) (Invitrogen).
5. Freezing medium: 10% DMSO (Sigma), 50% high-glucose DMEM, 40% FBS.

2.3.3 ASC Characterization Supplies

1. DPBS without Ca^{2+}/Mg^{2+} (Cellgro) with 2% FBS (Atlanta Biological).
2. FITC-conjugated CD90 and CD105 (BD Biosciences).
3. PE-conjugated CD73 (BD Biosciences).
4. FITC-IgG1 (BD Biosciences).
5. PE-IgG1 (BD Biosciences).
6. Accuri C6 Flow Cytometer.
7. FloJo Analysis Software.
8. Accutase (Sigma-Aldrich).

2.4 Immunocytochemistry Supplies

1. 3.7% paraformaldehyde.
2. NaN_3 (optional, for long-term sample storage).
3. Triton X-100.
4. Normal serum or BSA.
5. Primary antibody, fluorescence-labeled secondary antibody.
6. 15 cm petri dish.
7. Parafilm.
8. Kimwipes.
9. DAPI.
10. Glycerol (50% in PBS) or anti-fade mounting medium.
11. Clear nail polish.
12. Orbital shaker.
13. Fine-tipped forceps.

3 Methods

3.1 ASC Isolation

1. ASCs are isolated from human subcutaneous adipose lipoaspirate.
2. Line surface of biological hood with absorbent bench protectors. *See* Fig. 1a for hood setup.
3. Wash ~100 mL lipoaspirate with equal volume warmed PBS in a sterile 250 mL bottle then allow 3–5 min for phase separation. Top phase contains ASCs within the stromal vascular fraction; bottom phase contains blood and other aqueous components.
4. Pipette out the lower liquid phase and discard it, then repeat washing step 2–3 times until lower phase becomes clear. Multiple washes are necessary since blood and other aqueous compartments generally can not be washed out with a single washing step.

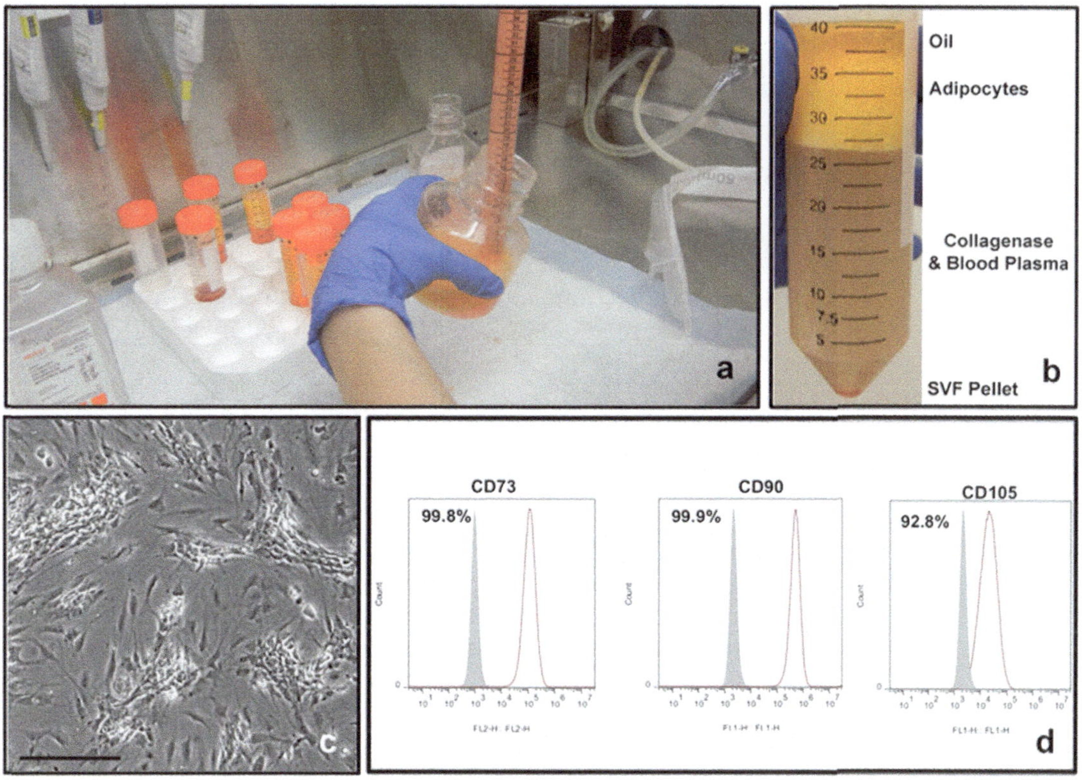

Fig. 1 Steps during ASC isolation and characterization. (**a**) Cell culture hood setup during isolation with hood surface lined with absorbent bench protectors. (**b**) Layers of tissue after centrifugation. (**c**) Representative image of P0 ASCs at ~80% confluence. Scale bar: 250 μm. (**d**) Representative image of flow cytometry analyses

5. Pour fat into multiple 50 mL tubes and add an equal volume of warmed working Collagenase Type I Solution to each tube. Shake to mix then lie tubes sideways on an orbital shaker for 1 h at ~140 rpm (or until the tissue appears smooth) at 37 °C.
6. Centrifuge tubes at 300 × *g* for 5 min, shake to dislodge and break up the pellet, then centrifuge again at 300 × *g* for 5 min. This helps reduce adipocyte contamination within the pellet. Layers should resemble those in Fig. 1b.
7. Carefully pipette out the top layer of oil/fatty tissue as well as middle liquid layer and discard it. Leave ~10 mL of liquid at the bottom to safeguard the pellet.
8. Resuspend pellet with 20 mL PBS.
9. Pipet the cell suspensions through a 100 μm cell strainer. Collect in fresh 50 mL tubes.
10. Bring volume up to 50 mL with PBS.
11. Centrifuge at 300 × *g* for 5 min.
12. Pipette out and discard all oil and PBS.

13. Resuspend pellet in 10 mL 1× red blood cell lysis buffer. Incubate 7 min at room temperature.
14. Add PBS to dilute the buffer, centrifuge at 300 × *g* for 5 min.
15. Resuspend pellet in 10 mL of Stromal Medium. Count cells.
16. Split cells into appropriate number of flasks (~33 mL of lipoaspirate tissue or 10–15 × 10^6 cells per T175 flask).
17. Bring total Stromal Medium volume to 30 mL per flask.
18. Change medium after 48 h continuing with Stromal Medium. Change medium every 2–3 days until cells reach 80–90% confluence (Fig. 1c).
19. Trypsinize cells with 0.25% trypsin–EDTA and cryopreserve as passage 0 (P0) ASC in Freezing Medium.

3.1.1 ASC Expansion

1. Thaw ASCs and plate cells in T175 flasks at 5000 cell/cm^2 (~1 × 10^6 cells per T175 flask).
2. Culture cells within Expansion Medium until 70–80% confluent. Passing at higher confluencies can affect proliferative ability, morphology, and differentiation potential.

3.1.2 ASC Characterization

1. Remove media from flask and wash with PBS without Ca^{2+}/Mg^{2+}.
2. Add 3–5 mL of Accutase that has been previously warmed to the flask. Make sure that the flask is coated entirely with solution and incubate flask for 3–5 min at 37 °C.
3. Observe under the microscope to ensure the cells have detached from the flask. Neutralize with ASC Expansion medium. Cells should be counted on hemocytometer.
4. 1 × 10^5 cells should be placed in 1.5 mL microcentirfuge tubes and spun down at 300 × *g* for 5 min.
5. Neutralized media is removed and replaced with 100 μL cold PBS without Ca^{2+}/Mg^{2+} containing 2% FBS for a concentration of 1 × 10^6 cells. Place the cell suspension on ice for 10 min for blocking.
6. After blocking, add the appropriate test amount of antibody and isoptype control to each of the samples. Samples are incubated on ice for 30 min. All antibodies (*see* Subheading 2.3.3) and isotype controls are fluorophore conjungated and purchased from BD Biosciences, USA.
7. Wash cell suspension with 1 mL of PBS solution without Ca^{2+}/Mg^{2+} containing 2% FBS. Spin cell suspension at 300 × *g* for 5 min.
8. Resuspend cells at a concentration of 1 × 10^6 cells/mL with PBS solution without Ca^{2+}/Mg^{2+} containing 2% FBS.
9. Run samples in the Accuri C6 Flow Cytometer.
10. Gate and Analyze results accordingly. *See* Fig. 1d for sample results.

3.2 Seeding on UniFlex® Culture Plate

1. Aspirate media from the T175 flask and wash with 10 mL of 1× PBS without Ca^{2+}/Mg^{2+}.
2. Add ~2 mL of 0.25% trypsin–EDTA that has been briefly warmed to activate the enzyme. Ensure that all regions of the bottom of the flask have been coated with trypsin. Incubate at 37 °C for 3–5 min.
3. Knock the side of the flask to disturb any nonadherent cells and add Expansion Medium at least double the volume of the trypsin.
4. Collect cell suspension in a 15 mL conical tube. Take 20 μL sample and count the cells.
5. Centrifuge the tube at 300 × *g* for 5 min.
6. Resuspend the pellet in culture media (CM) at an appropriate concentration for seeding at 5000 cells/cm².
7. Pipet 50,000 cells into each well of the UniFlex® plate (Fig. 2a). Culture cells within 3 mL of CM.

3.3 Myogenic Induction

1. At 24 h following seeding, begin preparation of myogenic induction media (MIM) by adding 1% penicillin–streptomycin, 1% FBS, and 5% horse serum to low glucose DMEM. Place this mixture in the water bath to warm (*see* **Note 1**).

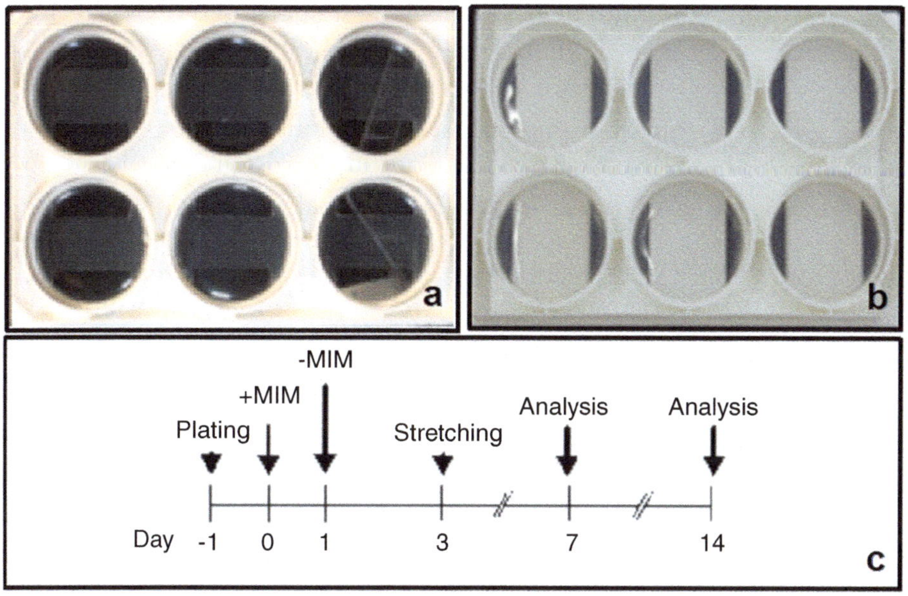

Fig. 2 Flexcell® tension system setup. UniFlex® plate before (**a**) and after (**b**) being placed on the post. (**c**) Timeline of ASC culture showing plating, application of myogenic induction medium (MIM), the application of strain, various assessment time points

2. Weigh out the appropriate amount of 5-Aza to make a 5 mg/mL stock solution in dI H_2O–acetic acid (1:1, v/v). Immediately sterile filter and add an appropriate volume of the stock solution of 5-Aza to the prewarmed MIM to a final concentration of 10 μM (*see* **Notes 2** and **3**).
3. Aspirate the CM from the UniFlex® culture wells and add 3 mL of the freshly prepared MIM to the induced group (*see* **Note 4**).
4. At 24 h following application of MIM to the induced group, aspirate the MIM from the wells and wash the wells with warmed 1× PBS without Ca^{2+}/Mg^{2+}. Add 3 mL of fresh CM to each well (*see* **Note 4**). For the remainder of the study feed every-other-day with CM.

3.4 Dynamic Culture

1. Apply silicone lubricant to each post of the BioFlex® baseplate and place UniFlex® plates on the posts (*see* Fig. 2b).
2. Place each of the plates on the posts by first removing the rubber gasket and placing the plate securely in the gasket and then placing the plate and gasket combination into the tray. Make sure that all sides of the gasket are securely in the tray and the plate is secure in the gasket (*see* Fig. 2c for final baseplate setup). Without this, the vacuum pressure will not work to deform the plate's membranes.
3. Begin dynamic culture 48 h following the removal of MIM from the induced group. *See* Fig. 2d for timeline.
4. Turn on the computer, pump, and FlexCell® FlexLink®.
5. Once all four plates are secure in the tray, place the tray in the incubator and secure the vacuum valves.
6. Close the incubator and return to the computer. Open the FlexSoft® FX-5000™ software program.
7. When the program opens, create a new strain regimen selecting the correct plate type and assigning the correct time, amplitude, frequency, and waveform for strain application, in this case 1 h, 11%, 1 Hz, and sine wave respectively (*see* **Note 5**).
8. Assign the regimen that was created and set the pump to the auto position. Do not start the regimen until the pump has had time to pressurize completely. This usually only takes a couple of minutes. Start the program that has been assigned (*see* **Note 6**).
9. After an hour has passed the machine will automatically stop and the pump will switch from the auto setting to off, meaning the pump will need to be reset for each subsequent group (*see* **Note 7**).
10. ASCs should be strained one hour per day for up to 18 days for a total of 21 days of culture following application of MIM (*see* **Note 8**).

3.5 Immunocytochemistry

3.5.1 Fixation

1. Carefully cut out the membrane out of the UniFlex® plate and place in a 6 well plate with cell seeded part facing upwards.
2. Fix cells with 3.7% formaldehyde for 20 min at room temperature.
3. Wash 3 × 5 min with PBS.
4. Either continue staining protocol or store at 4 °C (if storing for >1 week store in 0.1% NaN_3 to reduce bacterial growth).

3.5.2 Blocking and Permeabilization

1. All steps are in a well plate on an orbital shaker at 60 rpm, excepting the antibody incubations.
2. Permeabilize cells with 0.2% Triton-X 100 in PBS for 10 min at room temperature.
3. Wash 5 min with PBS.
4. Block cells with 10% normal serum (use serum from the animal in which the secondary antibody was raised) or 1% BSA in PBS for 30 min at room temp.
5. Follow one of the following staining procedures (Subheadings 3.5.3, 3.5.4, or 3.5.5).

3.5.3 Staining for Muscle Specific Markers

1. Incubate samples on a droplet of primary antibody solution (rabbit anti-desmin (1:50, Santa Cruz Biotechnology), mouse anti-MyoD1 (1:125, Sigma), goat anti-Pax 3/7 (1:50, Santa Cruz Biotechnology), mouse anti-skeletal myosin (MHC, Sigma, 1:400) diluted in 1% normal serum or 0.1% BSA overnight at 4 °C.

   It is best to do the antibody staining in a staining chamber. To prepare the chamber, line the bottom of a 15 cm petri dish with Parafilm. Draw and label a corresponding grid on a piece of paper and tape to the underside of the dish under the Parafilm. This will allow each drop of incubating sample-antibody to be labeled. Soak four Kimwipes in water and place one at each corner of the petri dish to keep it humidified during incubation. Pipette droplets of antibody solution throughout the grid. Volume should be optimized but is ~100 μL for 18 mm coverslips and 70 μL for 15 mm coverslips.

   Lift membranes out of well plate using fine-tipped forceps, paying careful attention to which side has the cell monolayer. Gently touch the membrane edge to a Kimwipes to wick away excess PBS, being careful not to touch the cell monolayer.

   Place the membrane *cell-side-down* onto the antibody droplet. Incubate at 4 °C overnight.
2. Transfer the membrane back to the well plate with the cell-side-up.
3. Wash 3 × 5 min in PBS.

4. Using a new piece of Parafilm in the staining chamber and following the procedure above, incubate samples on a droplet of fluorescent secondary antibody diluted in 1% normal serum or 0.1% BSA for 1 h at room temperature in the dark. Typical dilution of secondary antibody is 1:200 or 1:400.
   Note: From this point on, protect samples from exposure to light.
5. Transfer the membranes back to the well plate with the cell-side-up.
6. Wash 3 × 5 min in PBS.
7. Stain with DAPI diluted in PBS at 1:2000 for 5 min.
8. Wash 3 × 5 min in PBS.

3.5.4 Phalloidin Staining

1. Following blocking and permeabilization steps, incubate samples for 1 h with TRITC-conjugated phalloidin (Sigma) for actin filament staining (1:40 in 1% normal serum or 0.1% BSA).
2. Wash 3 × 5 min in PBS.
3. Stain with DAPI diluted in PBS at 1:2000 for 5 min.
4. Wash 3 × 5 min in PBS.

3.5.5 BrdU Staining

1. Perform Bromodeoxyuridine (BrdU, Sigma) incorporation and immunostaining in order to identify the proliferating cells in culture.
2. After plating cells, add 10 μm BrdU to the culture medium for 18 h.
3. Rinse the wells with PBS with Ca^{2+}/Mg^{2+}.
4. Fix the samples as described in Subheading 3.5.1.
5. Incubate samples in 2 N HCl in 0.5% Triton-X-100 in dI water for 30 min in order to denature the DNA.
6. Wash extensively with PBS.
7. Block the samples with 5% serum in PBS for 30 min.
8. Prepare the staining chamber as described in Subheading 3.5.3 and incubate samples with mouse anti-BrdU conjugated to Alexa Fluor®647 (1:50, Invitrogen) at 4 °C overnight.
9. Counterstain nuclei with DAPI diluted in PBS at 1:1000 for 10 min. DAPI concentration and staining time is doubled here as it helps increase the signal with denatured DNA.

3.5.6 Mounting Samples

1. Pipette 3 μL of 50% glycerol on a clean microscope slide.
2. Lift membrane with fine-tipped forceps and gently dab edge on Kimwipes to remove excess PBS, taking care not to disturb the cell monolayer.
3. Lay membrane on a glycerol droplet with cell-side-down.
4. Seal edges with nail polish and store in the dark.

3.5.7 Assessment of Multinucleation

1. Observe samples costained with Phalloidin-TRITC and DAPI under fluorescence microscope and take multiple representative images.
2. Observe areas where multiple nuclei are observed within a continuous cytoskeleton. *See* Fig. 3a for a sample image.
3. Count nuclei within the continuous cytoskeleton. Count at least six samples and six different areas from each sample and calculate the mean nuclei/cell number.

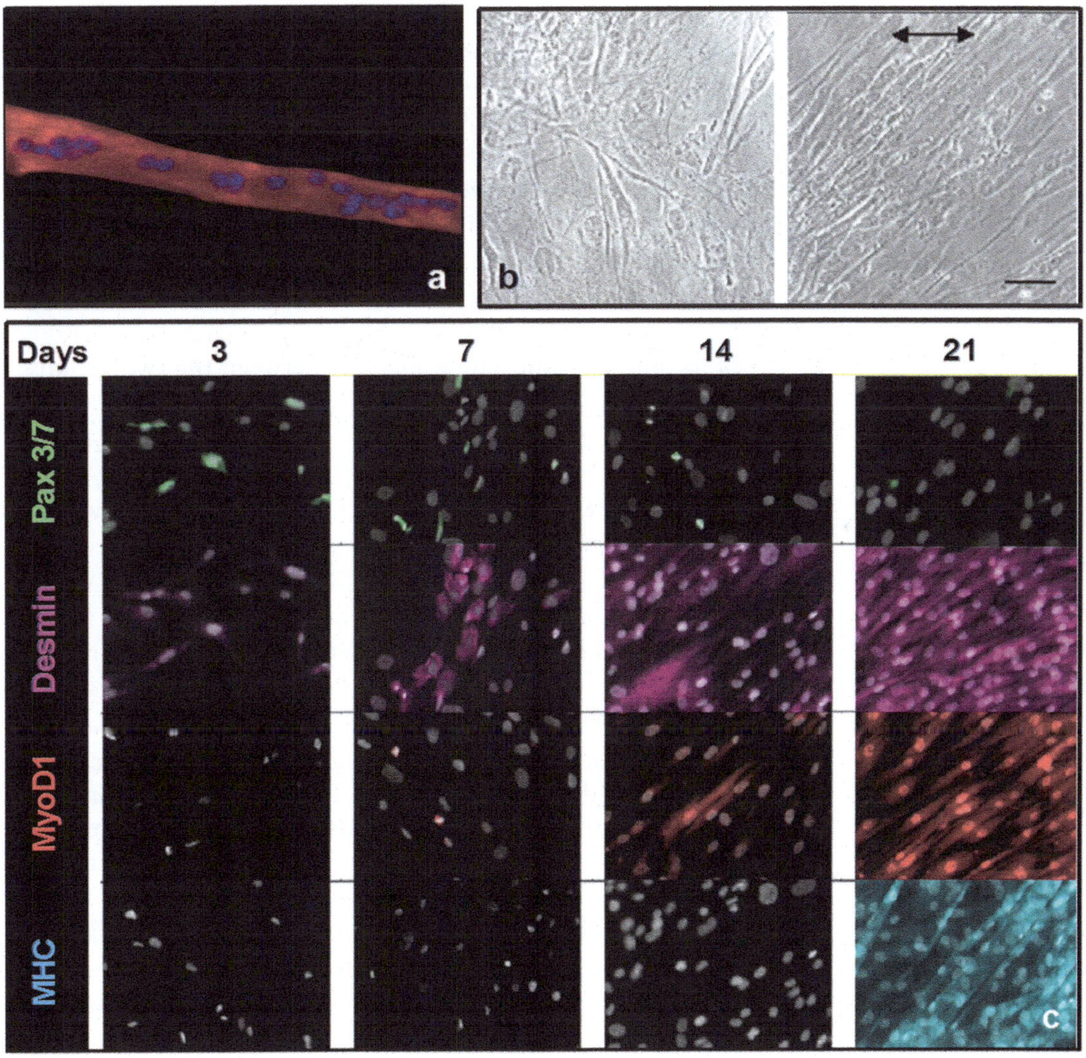

Fig. 3 Representative images of ASC myogenesis. (**a**) Multinucleation is assessed by counting the nuclei within the continuous cytoskeleton from samples costained with Phalloidin-TRITC and DAPI. (**b**) ASCs align ~45° with respect to the direction of strain (right image) (bidirectional arrow indicate the direction of strain) as compared to random cytoskeletal organization in case of static culture (left image) on Day 21. Scale bar: 50 μm. (**c**) Representative images of immunofluorescence staining for Pax 3/7, desmin, myoD, and myosin heavy chain on days 3, 7, 14, and 21 for dynamic culture conditions. ASCs were positive for desmin, myoD, and myosin heavy chain on day 21

4 Notes

1. The MIM is made initially without 5-Aza and warmed so that immediately following the introduction of the 5-Aza to the media, it can be applied to the induced group. This is necessary due to the quick degredation of 5-Aza in solution.
2. A slight excess of 5 mg/mL 5-Aza solution should be made to ensure enough solution remains following the sterile filtering step. Sterile filtering is necessary if 5-Aza stock is not sterile and cannot be weighed in a sterile manner. All of the steps following the creation of the 5-Aza stock solution should be completed as quickly as possible to maximize efficacy of the induction treatment since 5-Aza degrades by ~10% every few hours.
3. 5-Aza is soluble in dI H_2O–acetic acid (1:1, v/v) up to 5 mg/mL, thus the solubilized 5-Aza can be prepared in any solution up to a 5 mg/mL concentration, but should be remade for each use due to the quick degredation rate in solution.
4. As a matter of consistency, it may be appropriate to change the CM in the wells that are not induced to maintain consistency of feeding and nutrient availability between the two groups.
5. The parameters used for the dynamic culture (11% uniaxial cyclic strain, 0.5 Hz, 1 h/day) were chosen to mimic the daily excercise a skeletal muscle may undergo in vivo. A strain amplitue of 11% is the maximum limit of the Flexcell system to provide uniaxial strain.
6. Once the strain regimen has begun it may be prudent to open the incubator briefly to inspect the plates and ensure that the vacuum seal has been successful and t*he membra*nes are being pulled down over the posts. This motion should be visible to the naked eye.
7. To maintain consistency throughout the experiment, the plates should be removed from the posts as *quickly a*fter the end of the hour of straining as possible as it may be that there is a slight static strain in the membrane caused from sitting on the post and the ASC growth and alignment may also be affected by plates remaining on the post during periods where cyclic strain is not being applied.
8. It has been noted that plastic deformation of the UniFlex® plate membranes occurs throughout the study as early as 1 week into the strain regimen and may affect the strain the ASCs are receiving as the study continues.
9. During BrdU staining, efficiency of DNA denaturation is dependent on fixation method. If cells were fixed with ethanol, 2 N HCl for ~20–30 min may be sufficient. While formaldehyde preserves the cellular morphology better, it also

makes denaturation less efficient. Therefore, higher acid concentration or higher temperature is required.

10. Representative results of ASC myogenesis are presented in Fig. 3.

References

1. Qazi TH, Mooney DJ, Pumberger M, Geissler S, Duda GN (2015) Biomaterials based strategies for skeletal muscle tissue engineering: existing technologies and future trends. Biomaterials 53:502–521
2. Juhas M, Bursac N (2014) Roles of adherent myogenic cells and dynamic culture in engineered muscle function and maintenance of satellite cells. Biomaterials 35:9438–9446
3. Hosseinzadeh S, Soleimani M, Rezayat SM, Ai J, Vasei M (2014) The activation of satellite cells by nanofibrous poly epsilon-caprolacton constructs. J Biomater Appl 28:801–812
4. Usas A, Maciulaitis J, Maciulaitis R, Jakuboniene N, Milasius A, Huard J (2011) Skeletal muscle-derived stem cells: implications for cell-mediated therapies. Medicina 47:469–479
5. Carlson ME, Conboy IM (2007) Loss of stem cell regenerative capacity within aged niches. Aging Cell 6:371–382
6. Wakitani S, Saito T, Caplan AI (1995) Myogenic cells derived from rat bone marrow mesenchymal stem cells exposed to 5-azacytidine. Muscle Nerve 18:1417–1426
7. Hupkes M, Jonsson MK, Scheenen WJ, van Rotterdam W, Sotoca AM, van Someren EP et al (2011) Epigenetics: DNA demethylation promotes skeletal myotube maturation. FASEB J 25:3861–3872
8. Kaur K, Yang J, Eisenberg CA, Eisenberg LM (2014) 5-Azacytidine promotes the transdifferentiation of cardiac cells to skeletal myocytes. Cell Reprogram 16:324–330
9. Mizuno H, Zuk PA, Zhu M, Lorenz HP, Benhaim P, Hedrick MH (2002) Myogenic differentiation by human processed lipoaspirate cells. Plast Reconstr Surg 109:199–209. discussion 10–11
10. Meligy FY, Shigemura K, Behnsawy HM, Fujisawa M, Kawabata M, Shirakawa T (2012) The efficiency of in vitro isolation and myogenic differentiation of MSCs derived from adipose connective tissue, bone marrow, and skeletal muscle tissue. In Vitro Cell Dev Biol Anim 48:203–215
11. Hutton DL, Logsdon EA, Moore EM, Mac Gabhann F, Gimble JM, Grayson WL (2012) Vascular morphogenesis of adipose-derived stem cells is mediated by heterotypic cell-cell interactions. Tissue Eng A 18:1729–1740
12. Hutton DL, Moore EM, Gimble J, Grayson WLPDGF (2013) Spatiotemporal cues induce development of vascularized bone tissue by adipose-derived stem cells. Tissue Eng A 19(17–18):2076–2086
13. Zuk PA, Zhu M, Ashjian P, De Ugarte DA, Huang JI, Mizuno H et al (2002) Human adipose tissue is a source of multipotent stem cells. Mol Biol Cell 13:4279–4295
14. Choi YS, Vincent LG, Lee AR, Dobke MK, Engler AJ (2012) Mechanical derivation of functional myotubes from adipose-derived stem cells. Biomaterials 33:2482–2491
15. Choi YS, Vincent LG, Lee AR, Kretchmer KC, Chirasatitsin S, Dobke MK et al (2012) The alignment and fusion assembly of adipose-derived stem cells on mechanically patterned matrices. Biomaterials 33:6943–6951
16. Yilgor Huri P, Cook CA, Hutton DL, Goh BC, Gimble JM, DiGirolamo DJ et al (2013) Biophysical cues enhance myogenesis of human adipose derived stem/stromal cells. Biochem Biophys Res Commun 438:180–185
17. Huri PY, Wang A, Spector AA, Grayson WL (2014) Multistage adipose-derived stem cell myogenesis: an experimental and modeling study. Cell Mol Bioeng 7:497–509

Chapter 11

Isolation of Murine Adipose-Derived Stromal/Stem Cells for Adipogenic Differentiation or Flow Cytometry-Based Analysis

Gail Kilroy, Marilyn Dietrich, Xiying Wu, Jeffrey M. Gimble, and Z. Elizabeth Floyd

Abstract

Murine models of obesity or reduced adiposity are a valuable resource for understanding the role of adipocyte dysfunction in metabolic disorders. Adipose tissue stromal vascular cells or primary adipocytes derived from murine adipose tissue and grown in culture are essential tools for studying the mechanisms underlying adipocyte development and function. Herein, we describe methods for the isolation, expansion, and long-term storage of murine adipose-derived stromal/stem cells along with protocols for inducing adipogenesis in this cell population or isolating the adipose stromal vascular fraction cells for flow cytometric analysis.

Key words Murine, Adipose-derived stem cells (ASCs), Collagenase, Isolation, Mesenchymal stem cells (MSCs), Hematopoietic stem cells (HSCs), Stromal vascular fraction (SVF), Adipogenesis, Flow cytometry

1 Introduction

Adipose tissue can be separated into mature adipocytes and a stromal vascular fraction containing a heterogeneous mesenchymal cell population that includes hematopoietic stem cells, endothelial cells, erythrocytes, fibroblasts, pericytes, myeloid and lymphoid cells, and adipose stromal cells [1, 2]. When the heterogeneous stromal vascular fraction (SVF) is cultured, the adherent cells form a more homogeneous cell population of mesenchymal stem/progenitor cells that are referred to as adipose-derived stromal/stem cells (ASCs). The murine ASCs are commonly used in basic research and have the capacity to undergo adipogenic, osteogenic, chondrogenic, myogenic, and neuronal differentiation in vitro [3–6]. Mesenchymal stem/progenitor cells (MSCs) were originally characterized from bone marrow, but harvesting MSCs from adipose tissue has several

Bruce A. Bunnell and Jeffrey M. Gimble (eds.), *Adipose-Derived Stem Cells: Methods and Protocols*, Methods in Molecular Biology, vol. 1773, https://doi.org/10.1007/978-1-4939-7799-4_11, © Springer Science+Business Media, LLC, part of Springer Nature 2018

advantages over acquiring these cells from the bone marrow, including better accessibility and the greater abundance of the MSCs in adipose tissue. Thus, the murine ASCs are a valuable resource for studying mechanisms controlling the differentiation potential of MSCs or studies specifically focused on adipogenesis and the role of mature adipocytes in metabolic disorders related to obesity.

Recent data clearly establishes that obesity is often accompanied by an influx of immune cells into the adipose tissue, leading to chronic inflammation of the adipose tissue (reviewed in [7]). In obesity, the adipose tissue stromal vascular fraction cells may primarily be composed of immune cells, including macrophages and T and B lymphocytes. Consequently, understanding the contribution of these adipose tissue immune cells to obesity-related health problems has become an area of increasing interest. As these cells are commonly analyzed via flow cytometric analysis of cell surface markers, steps for isolating murine stromal vascular cells for flow cytometry are included as part of a protocol describing the isolation and expansion of murine ASCs and the differentiation of the ASCs into adipocytes in vitro.

2 Materials

2.1 Tissue

Adipose tissue: gonadal (epididymal in males, parametrial in females), inguinal, retroperitoneal (along the dorsal abdominal wall, near the kidney, but distinct from perirenal fat), mesenteric (along the intestines, containing multiple lymph nodes), suprascapular brown adipose tissue.

2.2 Supplies

1. 50 ml polypropylene conical tubes, sterile.
2. 0.2 μm filter units.
3. 2 ml microtubes, autoclave to sterilize.
4. 35 mm tissue culture plates.
5. 100 μm cell strainer, sterile.
6. 15 ml polystyrene conical tubes, sterile.
7. Freezing apparatus (Nalgene Cyro freezing container, catalog number 5100-0001).
8. Hemocytometer.
9. Scissors and forceps, autoclave to sterilize.
10. Additional supplies, if isolating the adipose tissue SVF for flow cytometry applications.
    (a) 12-well tissue culture plate, sterile.
    (b) 5 ml polystyrene round bottom tubes.
    (c) 5 ml polystyrene round bottom tubes with 40 μm strainer.

2.3 Equipment

1. Inverted microscope.
2. Shaking water bath or shaking incubator at 37 °C.
3. Centrifuge, 4 °C and room temperature.
4. Biosafety hood.
5. 5% CO_2 humidified Incubator.

2.4 Buffers

All buffers are sterile filtered using a 0.2 μm filter unit.

1. *Phosphate Buffered Saline (PBS)*: 37 mM NaCl, 2.7 mM KCl, 10 mM Na_2HPO_4, 2 mM KH_2PO_4 pH 7.4. (Sterile solutions of PBS are available commercially.)
2. *Collagenase solution*: Weigh out 10 mg of type 1 collagenase and dissolve in 10 ml Hanks Buffered Saline Solution (HBSS) (Ca^{2+} and Mg^{2+} free). Add 3.3 units DNase 1 and 10 μl 1 M $MgCl_2$. After sterile filtration, warm the solution to 37 °C. This solution should be used within 1 h of its preparation.
3. *FACS staining buffer*: 0.1% fatty acid-free bovine serum albumin (BSA, fraction V) in PBS. Sterile filtered using a 0.2 μm filter.

2.5 Media Solutions

All solutions are sterile filtered using a 0.2 μm filter unit.

1. *Stromal medium*: DME/F:12 with 15% fetal bovine serum (FBS) and 2× antibiotic solution from 100× stock (penicillin/streptomycin) (*see* **Note 1**).
2. *Differentiation medium*: In advance, prepare and aliquot the following stock solutions and store frozen at −20 °C until required.
   (a) A 66 mM stock solution of biotin (2000-fold concentration) dissolved in 1 N sodium hydroxide.
   (b) A 34 mM stock solution D-pantothenate (2000-fold concentration) dissolved in water.
   (c) A 1 mM dexamethasone (1000-fold concentration) dissolved in water or ethanol depending on its formulation (*see* **Note 2**).
   (d) A 5 mM stock solution of rosiglitazone or equivalent PPARγ agonist dissolved in dimethyl sulfoxide (1000-fold concentration).

   Prepare fresh before use:

   (e) A 500 mM stock solution of 1-methyl 3-isobutyl xanthine (methylisobutylxanthine, MIX; 1000-fold concentration) dissolved in 50 mM NaOH.
   (f) A 200 μM stock solution of insulin (2000-fold concentration) dissolved in 10 mM HCl, made in plastic tube, *not glass*.

3. *Adipocyte maintenance medium*: This solution is prepared in an identical manner as differentiation medium except that it does not contain either the methylisobutylxanthine or rosiglitazone; these two stock solutions should be omitted. Use this solution within 2 weeks of its preparation.
4. *Freezing medium*: The freezing medium consists of 15% FBS, 75% DME/F:12, and 10% dimethyl sulfoxide. Use this solution within 2 weeks of its preparation.
5. *Oil Red O staining solution*: Weigh out 0.7 g Oil Red O. Dissolve in 200 ml isopropanol. Filter through a 0.2 μm filter. Store at room temperature as stock solution. To stain cells, make a 60% Oil Red O solution in distilled water by adding 30 ml of the Oil Red O stock to 20 ml distilled water. Mix and let stand at room temperature 30 min before use and filter again if particulates are present. Use this solution within 24 h of its preparation.

3 Methods

3.1 Obtaining the Adipose Tissue

1. Ensure the ASC isolation procedure is initiated as soon as possible after the tissue is extracted. This is especially important to consider if a large number of samples are being obtained.
2. Warm buffers along with the Collagenase solution and Stromal Media in the 37 °C water bath.
3. Place bench protector down in biosafety hood.
4. Sacrifice animal using anesthesia or CO_2 asphyxiation as approved by the AVMA Guidelines on Euthanasia. Saturate the fur with 70% ethanol and place the animal ventral side up.
5. Using sterile tissue forceps, pull up on the skin and cut the skin vertically toward the head.
6. Using sterile scissors and forceps, open the abdominal cavity and remove the desired fat pads.
7. Weight the tissue and place in a sterile 50 ml tube containing enough sterile PBS to cover the tissue. If collecting the tissue for flow cytometry, the 50 ml tube and PBS should be prechilled on ice.

3.2 Processing the Adipose Tissue

1. Working under a laminar flow hood, use sterile forceps to transfer the tissue to the bottom of a 35 mm tissue culture dish containing enough HBSS to rinse tissue well.
2. Transfer the rinsed tissue to the lid of the 35 mm tissue culture dish and mince the adipose tissue into small pieces using sterile scissors.
3. Place the minced tissue in sterile 50 ml centrifuge tubes. If digesting less than 2.0 g adipose tissue, use a minimum of

10 ml collagenase solution to ensure complete digestion. If digesting ≥2 g tissue, add 5 ml collagenase solution/gram adipose tissue to each centrifuge tube (*see* **Note 3**).

4. Cap and wipe the centrifuge tubes with 70% ethanol, dry, then wrap with parafilm before placing the tubes in a 37 °C shaking water bath or incubator. Shake at 100 rpm for 60 min until the tissue appears smooth on visual inspection. If a shaking water bath is not available, swirl the bottles every 5–10 min in a 37 °C water bath.

3.3 Isolate the Stromal Vascular Fraction (SVF)

1. After the collagenase digestion, filter solution through a sterile 100 μm filter into a sterile 50 ml conical tube. Centrifuge the tubes at 450 × *g* at room temperature for 5 min. The stromal vascular fraction will be apparent as the dark red cells pelleted on the bottom. Carefully aspirate the oil on top and the primary adipocytes, which will appear as a yellow layer of floating cells (*see* **Notes 4** and **5**). Leave a small amount of the brown collagenase solution so that the pelleted stromal vascular fraction is not disturbed.
2. Resuspend the cells in 2 ml Ammonium-Chloride-Potassium (ACK) red cell lysis buffer with gentle vortexing or gentle trituration. Incubate at room temperature for 3–5 min.
3. Add 10 ml Hanks Buffered Saline Solution (HBSS) (Ca^{2+} and Mg^{2+} free) or PBS to neutralize the ACK red cell lysis buffer.
4. Centrifuge the resuspended cells at 450 × *g* for 5 min and carefully aspirate the remaining collagenase solution. When aspirating, the tip of the pipette should aspirate from the top so that the oil is removed as much as possible. The stromal cell pellet should be tightly packed at the bottom of the tube.
5. *If isolating the stromal vascular fraction cells for primary cell culture*: Add 2–3 ml of stromal media per gram adipose tissue (weight when harvested) to each centrifuge tube and resuspend the cells again with gentle vortexing or gentle trituration.
6. Plate the SVF obtained from 0.5–1.0 g adipose tissue in a 35 mm tissue culture dish or a single well of a 6 well plate.
7. 24 h after plating, aspirate the media from the plate, rinse the cells with PBS prewarmed to 37 °C and aspirate. Check the cells using the inverted microscope; if excess unattached cells remain after the first PBS rinse, repeat the rinse and aspirate.
8. Add fresh stromal media.
9. The media is changed every 2–3 days until the cells are 80–90% confluent. The cells may begin to detach if the media is removed completely; leave a small amount of media on the

cells with each media change. At 80–90% confluence, the cells are either passaged, stored as frozen stocks as passage one (P1) or differentiated.

10. *If isolating the stromal vascular fraction cells for flow cytometry after step 4*: The cells must be kept ice cold from this point onward.
11. Add 1 ml PBS containing 0.1% fatty acid-free BSA (FASC staining buffer) and resuspend the cells with gentle vortexing or trituration.
12. Filter the resuspended cells into a 5 ml polystyrene round bottom tube with a 40 μm cell strainer cap.
13. Use an aliquot of the cells, diluted with Trypan Blue (a 1:2 or 1:4 dilution is usually convenient) to count the cells using a hemacytometer.
14. *See* Subheading 3.5 for FACS staining steps.

3.4 Adipocyte Differentiation of the ASCs

When the cells reach about 90% confluence, the murine ASCs are ready to differentiate (*see* **Note 6**). Exchange the stromal media for the differentiation media and leave on the cells for 3 days. On day 3 post-induction, replace the differentiation media with the adipocyte maintenance media. The adipocyte maintenance media is changed every 2–3 days and mature adipocytes should be formed by day 8–12 post-induction (Fig. 1).

3.5 FACS Staining of the SVF Cells

For a more detailed description of FACS analysis of adipose tissue immune cell populations, including antibody selection and gating strategies, *see* Grant et al. [8].

1. Keep everything ice cold. Prechill centrifuge to 4 °C.
2. Place 1–2 × 10^6 SVF cells in a 5 ml polypropylene tube.
3. Centrifuge the cells at 450 × *g* for 5 minutes at 4 °C.
4. Carefully remove the supernatant using a transfer pipet. The pellet will be clearly visible. Disperse the cell pellet by adding 50ul staining buffer followed by gentle trituration. Immediately add 2 μl Fc Receptor Block (anti-mouse CD16/CD32, purified; flow cytometry) and mix gently.
5. Make a master mix containing all of the antibodies that will be used. Add 2 μl of each antibody per sample to the master mix.
6. Add the master mix to each tube at a total volume equivalent to 2 μl of each antibody (i.e., add 10 μl master mix if using five antibodies). Mix the samples well. Place on ice and cover with foil to protect from light for 45 min.
7. Add 1 ml ice-cold PBS and centrifuge at 450 × *g* for 5 min at 4 °C. Remove supernatant with transfer pipet. Disperse the pellet by adding 2 ml ice-cold PBS followed by gentle vortexing or trituration.

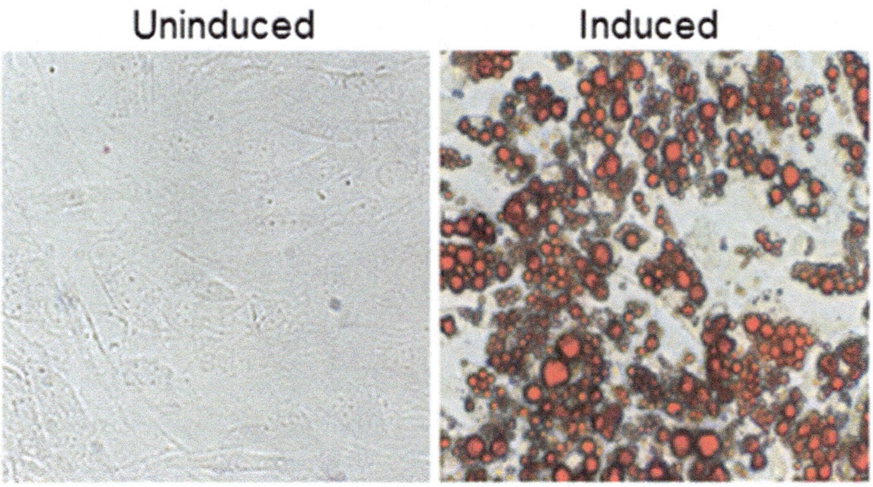

Fig. 1 Oil Red O staining of neutral lipid accumulation in murine adipose-derived stromal cells. ASCs were obtained from the inguinal adipose tissue of 4-week-old wild-type male mice. The ASCs were at passage number 2 and adipogenesis media was added after the cells reached 80–90% confluence. Neutral lipid staining was carried out in the uninduced and induced ASCs at day 8 post induction

8. Centrifuge at 450 × *g* for 5 min at 4 °C, then remove supernatant.
9. Add 500 μl 1% paraformaldehyde and gently vortex.
10. Samples may be stored in the dark at 4 °C until analyzed (Fig. 2).

3.6 Harvesting the Plated Cells

1. Remove media from tissue culture plate and rinse with 1.5–2 ml prewarmed sterile PBS if using a T75 flask (75cm^2 growth area). As a rule of thumb, use 0.025 ml PBS/cm^2 growth area of the plate.
2. Aspirate the PBS and add a minimum 0.025 ml/cm^2 growth area of 0.5% trypsin–EDTA solution. For example, if working in a 6-well plate format (9.5 cm^2 growth area), add 0.250 ml 0.5% trypsin–EDTA solution. If working in a 100 mm plate format (55cm^2 growth area), use at least 1.375 ml 0.5% trypsin–EDTA solution
3. Incubate in 37 °C incubator for 2–5 min. Do not allow the cells to remain in contact with the trypsin–EDTA solution longer than is necessary to detach the cells. When more than 90% of the cells have detached (verify using the inverted microscope), add a sufficient volume of stromal media to completely resuspend the cells.
4. Transfer the resuspended cells to a sterile 15 ml tube and centrifuge at 450 × *g* for 5 min at room temperature. Aspirate the media and resuspend the pelleted cells in 1–2 ml of stromal media. Use an aliquot of the cells and dilute with Trypan Blue (a 1:2 or 1:4 dilution is usually convenient) to count the cells using a hemacytometer.

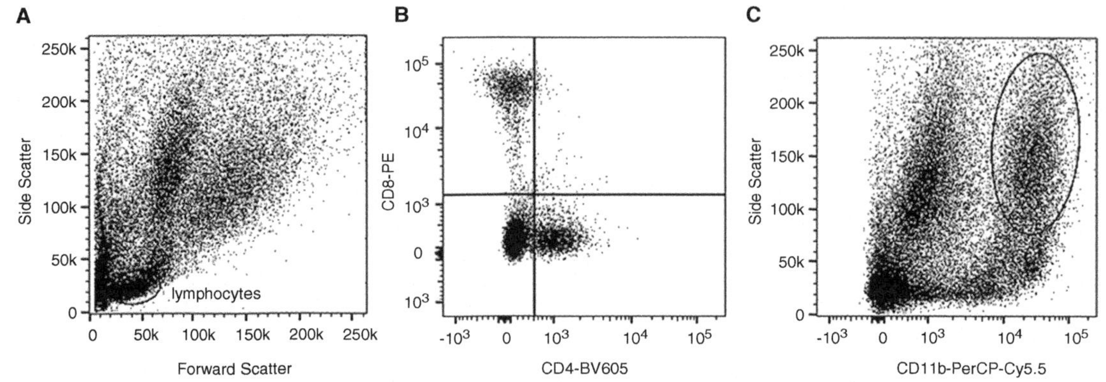

Fig. 2 Representative dot blots of murine stromal vascular fraction (SVF) cells. SVF cells were isolated from the epididymal adipose tissue of a 5-month-old wild-type mouse maintained on a high fat diet. (**a**) Total cell population with lymphocytes indicated in the lower left quadrant. (**b**) CD4+ T lymphocytes in the lower right quadrant and CD8+ T lymphocytes in the upper left quadrant were identified from the assigned lymphocyte region. (**c**) Macrophages were identified (circled region) from the CD11b+ population. Monoclonal antibodies used were anti-CD4-BV605, anti-CD8-PE, and anti-CD11b-PerCP-Cy5.5

3.7 Replating or Cryopreserving the Plated Cells

1. Replating the cells: Suspend the cells in stromal media and replate at 5000 cells/cm^2 for further passages or experiments (*see* **Note 7**).
2. Cryopreservation of the cells: Centrifuge the cells once more to obtain a cell pellet. Resuspend the pelleted cells in room temperature freezing media at a concentration of 1–2 × 10^6 cells per ml. Dispense 1 ml aliquots of the resuspended cells to sterile cryovials. Place the vials in the freezing apparatus (Nalgene Cyro freezing container) and place at −80 °C overnight. Transfer to long-term storage the next day.

3.8 Cell Fixation and Oil Red O Detection of Neutral Lipids

1. Cell fixation: To fix plated cells for Oil Red O staining, the media is removed and the cells are rinsed in PBS and the PBS is aspirated before adding an amount of 10% buffered formalin sufficient to completely cover the cells. Incubate the cells in the formalin for a minimum of 1 h at room temperature or overnight at 4 °C. The cells can remain in the formalin for a longer time, but this is not recommended unless it is necessary to collect multiple plates over several days before staining.
2. Oil Red O staining: After fixing the cells, add sufficient Oil Red O to each plate or well to completely cover the cells. Gently rock at room temperature for 1 h, and then rinse at least three times with distilled water. At this point, the cells can be imaged using a microscope/camera. If desired, the amount of Oil Red O staining can be quantified by removing the distilled water and adding 100% isopropanol. Add 0.375 ml 100% isopropanol/cm^2 (6-well plate is 9.5 cm^2/well; 12-well plate is 4.0 cm^2/well, 24-well plate is 2 cm^2/well). As a blank, add an equal amount of isopropanol to an empty well. Elution of the Oil

Red O should be complete within 10–15 min at room temperature. Determine the absorbance of the mixture at 500 nm (or 540 nm if using selected UV filters) and subtract the blank measurement from the other absorbances.

4 Notes

1. If fungal contamination is a concern, solutions of antibiotics combined with Amphotericin as an antimycotic agent are also commercially available.
2. Dexamethasone is available in a water-soluble and water-insoluble form from Sigma. For the water-insoluble form (Sigma D-4902), the 1 mM stock is made by adding 0.39 mg dexamethasome to 1 ml 100% ethanol. For the water-soluble form (Sigma D-2915), there is 69 mg dexamethasone/100 mg total weight. To obtain a 1 mM stock solution using the water-soluble form, add 6.0 mg dexamethasone mixture to 1 ml distilled water.
3. This amount of collagenase solution/gram tissue is a good starting point for determining how much collagenase solution is needed to obtain efficient digestion of the adipose tissue. Another approach is to add a volume of collagenase solution that is equal to the volume of minced adipose tissue.
4. The tissue can be filtered at this step to remove any remaining undigested pieces. Use a 100 μm filter (BD-Falcon) that is available for filtering small volumes. However, if it is desirable to collect the mature adipocytes at a later step, use a nylon filter (sterile) screen with a pore size of 250 μm.
5. As has been reported [9], it is also possible to separate small and large mature adipocytes for further analysis by using a 75 μm pore size filter at this step. Rather than aspirating the oil and yellow layer containing the adipocytes, carefully remove the oil and yellow layer and filter (BD-Falcon has filters of various pore sizes that work well for small volumes). The flow-through will contain the smaller adipocytes and the solution remaining on top of the filter will contain adipocytes greater than 75 μm in size.
6. ASCs obtained from young mice (3–6 weeks of age) or lean mice will typically differentiate more readily than ASCs obtained from older mice or obese animals. The yield of SVF cells or ASCs may also be affected by the age or body weight of the animal. With experience, this can be factored into the number of fat pads needed to obtain the desired number of cells for each application. In addition, ASCs from inguinal adipose tissue will differentiate more readily than ASCs from gonadal adipose tissue.
7. For reasons that are not quite clear, differentiation of the murine ASCs (and Human ASCs) occurs in a higher percentage

of the total number of cells when the cells are plated in smaller wells (i.e., the cell will better differentiate in a 12-well plate than a 6-well plate although both are originally seeded at the same concentration of cells per cm^2).

Acknowledgment

This work is supported by the National Institutes of Health (NIDDK, 1R01DK099625, Z.E.F.). This work used the Cell Biology and Bioimaging Core facility at Pennington Biomedical Research Center that is supported in part by COBRE (NIH 8P20-GM103528) and NORC (NIH 2P30-DK072476) center grants from the National Institutes of Health.

References

1. Cawthorn WP, Scheller EL, MacDougald OA (2012) Adipose tissue stem cells meet preadipocyte commitment: going back to the future. J Lipid Res 53(2):227–246. https://doi.org/10.1194/jlr.R021089
2. Bourin P, Bunnell BA, Casteilla L, Dominici M, Katz AJ, March KL, Redl H, Rubin JP, Yoshimura K, Gimble JM (2013) Stromal cells from the adipose tissue-derived stromal vascular fraction and culture expanded adipose tissue-derived stromal/stem cells: a joint statement of the International Federation for Adipose Therapeutics and Science (IFATS) and the International Society for Cellular Therapy (ISCT). Cytotherapy 15(6):641–648. https://doi.org/10.1016/j.jcyt.2013.02.006
3. Kelly KA, Tanaka S, Baron R, Gimble JM (1998) Murine bone marrow stromally derived BMS2 adipocytes support differentiation and function of osteoclast-like cells in vitro. Endocrinology 139(4):2092–2101
4. Thompson DL, Lum KD, Nygaard SC, Kuestner RE, Kelly KA, Gimble JM, Moore EE (1998) The derivation and characterization of stromal cell lines from the bone marrow of p53−/− mice: new insights into osteoblast and adipocyte differentiation. J Bone Miner Res 13(2):195–204
5. Safford KM, Hicok KC, Safford SD, Halvorsen YD, Wilkison WO, Gimble JM, Rice HE (2002) Neurogenic differentiation of murine and human adipose-derived stromal cells. Biochem Biophys Res Commun 294(2):371–379
6. Zheng B, Cao B, Li G, Huard J (2006) Mouse adipose-derived stem cells undergo multilineage differentiation in vitro but primarily osteogenic and chondrogenic differentiation in vivo. Tissue Eng 12(7):1891–1901. https://doi.org/10.1089/ten.2006.12.1891
7. Grant RW, Dixit VD (2015) Adipose tissue as an immunological organ. Obesity 23(3):512–518. https://doi.org/10.1002/oby.21003
8. Grant R, Youm YH, Ravussin A, Dixit VD (2013) Quantification of adipose tissue leukocytosis in obesity. Methods Mol Biol 1040:195–209. https://doi.org/10.1007/978-1-62703-523-1_15
9. Bluher M, Patti ME, Gesta S, Kahn BB, Kahn CR (2004) Intrinsic heterogeneity in adipose tissue of fat-specific insulin receptor knock-out mice is associated with differences in patterns of gene expression. J Biol Chem 279(30):31891–31901. https://doi.org/10.1074/jbc.M404569200. M404569200 [pii]

Chapter 12

Three-Dimensional Magnetic Levitation Culture System Simulating White Adipose Tissue

Hubert Tseng, Alexes C. Daquinag, Glauco R. Souza, and Mikhail G. Kolonin

Abstract

White adipose tissue (WAT) has attracted interest for tissue engineering and cell-based therapies as an abundant source of adipose stem/stromal cells (ASC). However, technical challenges in WAT cell culture have limited its applications in regenerative medicine. Traditional two-dimensional (2D) cell culture models, which are essentially monolayers of cells on glass or plastic substrates, inadequately represent tissue architecture, biochemical concentration gradients, substrate stiffness, and most importantly for WAT research, cell phenotypic heterogeneity. Physiological cell culture platforms for WAT modeling must recapitulate the native diversity of cell types and their coordination within the organ. For this purpose, we developed a three-dimensional (3D) model using magnetic levitation. Here, we describe our protocol that we successfully employed to build adipose tissue organoids (adipospheres) that preserve the heterogeneity of the constituent cell types in vitro. We demonstrate the capacity of assembling adipospheres from multiple cell types, including ASCs, endohtelial cells, and leukocytes that recreate tissue organization. These adipospheres mimicked WAT organogenesis in that they enabled the formation of vessel-like endothelial structures with lumens and differentiation of unilocular adipocytes. Altogether, magnetic levitation is a cell culture platform that recreates tissue structure, function, and heterogeneity in vitro, and serves as a foundation for high-throughput WAT tissue culture and analysis.

Key words White adipose tissue, Adipose stem cells, Stromal vascular fraction, Magnetic levitation, Adiposphere, Adipogenesis, 3D cell culture

1 Introduction

In the evolving field of tissue engineering, adipose stem/stromal cells (ASC) have grown in prominence due to their similarity to bone marrow mesenchymal stem cells (MSC) [1]. In particular, ASCs display a multipotency, proliferative capacity, and pericyte functionality in promoting endothelial proliferation and blood vessel formation that is commonly seen with bone marrow MSCs, but are more abundant [2–5]. As a result, there is a research interest in white adipose tissue (WAT), which contains ASCs, as well as differentiated adipocytes, endothelial cells, and infiltrating

Bruce A. Bunnell and Jeffrey M. Gimble (eds.), *Adipose-Derived Stem Cells: Methods and Protocols*, Methods in Molecular Biology, vol. 1773, https://doi.org/10.1007/978-1-4939-7799-4_12, © Springer Science+Business Media, LLC, part of Springer Nature 2018

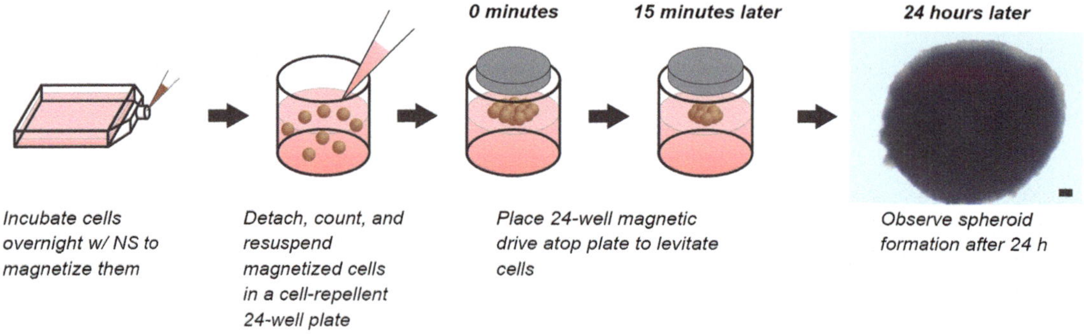

Fig. 1 Schematic of magnetic levitation of preadipocytes. Shown is a spheroid of 3T3-L1 preadipocytes in levitation. Scale bar = 100 μm

leukocytes [1]. This interest is rooted in not only the desire for a better understanding of ASC biology and growth, but also its promise as a deliverable therapeutic in it of itself as part of lipotransfer for cosmetic and functional tissue repair [6–8], as well as the numerous links between obesity and various diseases [9].

However important a research topic WAT has become, in vitro analysis of its many cellular components has proven difficult and limited its study. Traditionally, cells isolated from primary WAT would be seeded in two-dimensional (2D) monolayers on stiff plastic or glass substrates. These models do not represent native tissue environments, as they: lack a three-dimensional (3D) structure seen in vivo [10]; use rigid substrates that are much stiffer than native tissue [11–13]; have a uniform biochemical concentration gradient that is rarely seen in vivo [14]; and lack an extracellular matrix (ECM) and the accompanying cell-ECM interactions that drive tissue morphology and function [15–17]. More specifically to WAT, cells from the stromal vascular fraction (SVF), which includes ASCs, endothelial cells, and leukocytes, lose their heterogeneity when cultured together in 2D, as a result of its attachment to plastic and the domination of its mesenchymal fraction over time [18–20]. An ideal cell culture platform would replicate the 3D structure, phenotypic heterogeneity, and functionality of WAT.

Towards that end, we developed a tissue culture model of WAT using magnetic levitation (Fig. 1). The basis of magnetic levitation is the attachment of a magnetic nanoparticle assembly to cells to magnetize them and aggregate them using magnetic forces [21]. These nanoshuttles (NS), consisting of gold, iron oxide, and poly-L-lysine, work by electrostatically attaching to cell membranes, thereby magnetizing the cell. When resuspended in media, these cells can then be levitated off of any stiff substrate by placing a magnet above the culture vessel to aggregate cells towards the air liquid interface, where the cells will work with each other to build a larger 3D structure subsequently reinforced with ECM [21, 22].

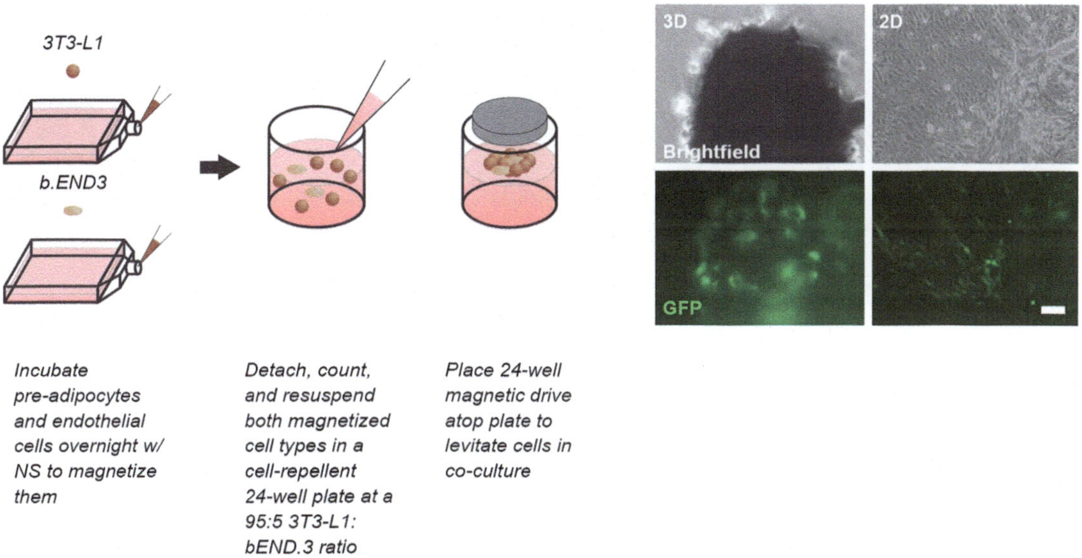

Fig. 2 Schematic of magnetic levitation of preadipocytes and endothelial cells in coculture. Shown is a spheroid coculture of GFP-tagged bEND.3 endothelial cells and 3T3-L1 preadipocytes. Formation of circular endothelial structures (bottom left) is indicative of vascularization. Note the nonuniform clustering of minimal organization of endothelial cells (green) in 2D, showing the loss of heterogeneity. Scale bar = 50 μm. Adapted from Daquinag et al. [32]

The advantage of this platform is that in levitating cells, these cells are maintained in a 3D environment that mimics native tissue structure, while being removed of any stiff substrate that in 2D could lead to the loss of phenotypic heterogeneity in SVF cells [21]. This platform has been used to create 3D models of many tissue types that recapitulate native structure and functionality, and cocultures with fine spatial organization [22–31]. We previously used this system to create adipospheres, which recapitulated native tissue structure, heterogeneity, and functionality (Figs. 2, 3, and 4) [32]. We describe here our protocol to create adipospheres using magnetic levitation and discuss our results gained from this model.

2 Materials

- Cells.
  - 3T3-L1 murine embryonic preadipocytes.
  - bEND.3 murine endothelial cells.
  - SVF harvested from primary murine WAT.
- Standard cell culture medium: Dulbecco's Modified Eagle Medium (DMEM), 10% fetal bovine serum (FBS), 100 U/mL penicillin, 100 U/mL streptomycin.

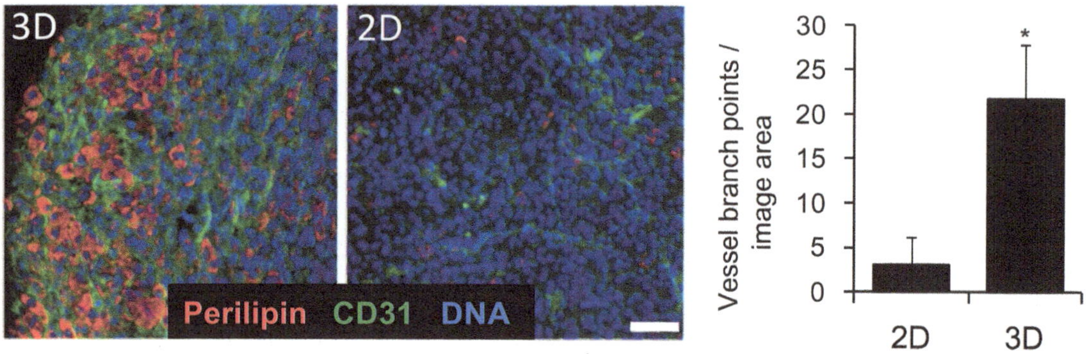

Fig. 3 Immunofluorescent stains of SVF cocultures in 3D (left) and in 2D (center) for CD31 (green) and perilipin (red). Nuclei are counterstained with DAPI (blue). SVF cells in 3D formed more vessel branch points than they did in 2D (right), supporting the idea that 3D environments promote vascularization. Scale bar = 50 μm and $* = p < 0.05$

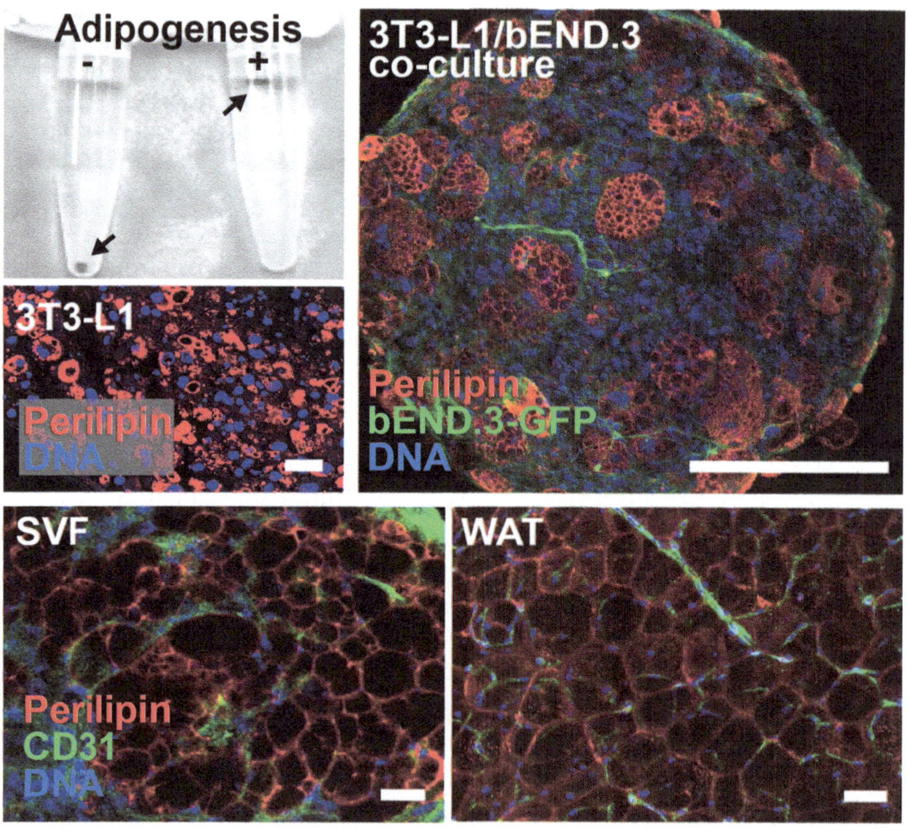

Fig. 4 Adipogenesis induction. Top left: Lipid droplet accumulation in differentiated 3T3-L1 preadipocytes is revealed from the buoyancy of the spheroid. Middle left: Positive immunofluorescent stains demonstrates lipid droplet accumulation. Top right: Adipospheres of 3T3-L1s and bEND.3 (top right) are also rich in perilipin and contain endothelial structures, demonstrating simultaneous lipogenesis and vascularization Bottom left: SVF adipospheres feature lipid accumulation and vascularization comparable to those of native mouse WAT (bottom right). Scale bar = 100 μm. Adapted from Daquinag et al. [32]

- SVF cell culture medium: DMEM, 10% FBS.
- Trypsin.
- Phosphate buffered saline (PBS).
- NanoShuttle (NS, Nano3D Biosciences, Houston, TX).
- 24-well plate with cell-repellent surface.
- 24-well magnetic drive (Nano3D Biosciences).
- 24-well plate lid insert (Nano3D Biosciences).
- Adipogenic induction medium: DMEM, 10% FBS, 0.5 mM isobutylmethyxanthine, 1 μM dexamethasone, 0.2 mM indomethacin, 1.7 μM insulin.
- Adipogenic maintenance medium: DMEM, 10% FBS, 1.7 μM insulin.

3 Methods

3.1 Preadipocyte Magnetic Levitation Spheroid Culture (Fig. 1)

1. Seed 3T3-L1s in a T75 flask and culture until 70–80% confluence.
2. Add 600 μL NS directly to the flask. Mix the NS throughout the media and leave flask in incubator overnight to magnetize cells (*see* **Notes 1** and **2**).
3. The next day, detach the cells with trypsin, count, resuspend cells at a concentration of 6×10^5 cells/mL, and distribute the cells with 400 μL/well in a 24-well cell-repellent plate (2.4×10^5 cells/spheroid) (*see* **Notes 3–5**).
4. Close the plate with the 24-well lid insert, 24-well magnetic drive, then plate lid (*see* **Notes 6** and **7**).
5. Move the plate of 3T3-L1s to an incubator, where the cells will aggregate under the magnetic field (*see* **Note 8**).
6. 3T3-L1s can be cultured for 14 days with periodic media exchange, typically every 2 days (*see* **Note 9**).

3.2 Predipocyte/Endothelium Magnetic Levitation Coculture (Fig. 2)

1. Seed 3T3-L1s and bEND.3s in T75 flasks and culture until 70–80% confluence.
2. Add 600 μL NS directly to each flask. Mix the NS throughout the media and leave flask in incubator overnight to magnetize cells (*see* **Notes 1** and **2**).
3. The next day, detach the cells with trypsin, count, and mix the cells with a ratio of 95:5 3T3-L1:bEND.3 (*see* **Notes 3** and **4**).
4. Resuspend cells at a concentration of 6.25×10^5 cells/mL, and distribute the cells with 400 μL/well in a 24-well cell-repellent plate (2.5×10^5 cells/spheroid) (*see* **Note 5**).
5. Close the plate, and move the plate of cocultures into an incubator (*see* **Notes 6–8**).

6. The cocultures are cultured for 14 days with periodic media exchange, typically every 2 days (*see* **Note 9**).

3.3 Primary WAT-Derived SVF Magnetic Levitation Culture (Fig. 3)

1. Isolate SVF from mixed subcutaneous or intraperitoneal WAT (*see* **Notes 10** and **11**).
2. Seed SVFs in a T75 flask and culture to 70–80% confluence.
3. Add 600 μL NS directly to the flask. Mix the NS throughout the media and leave flask in incubator overnight to magnetize cells (*see* **Notes 1** and **2**).
4. The next day, detach the cells with trypsin, count, and resuspend cells at a concentration of 8.75×10^5 cells/mL, and distribute the cells with 400 μL/well in a 24-well cell-repellent plate (3.5×10^5 cells/spheroid) (*see* **Notes 3–5**).
5. Close the plate, and move the plate of cocultures into an incubator (*see* **Notes 6–8**).
6. The spheroids are cultured for up to 21 days with periodic media xchange, typically every 2 days (*see* **Note 9**).

3.4 Adipocyte Differentiation of Cells in Magnetic Levitation Culture (Fig. 4)

1. Cells are magnetically levitated into spheroids and cultured for 1 day in standard cell culture media.
2. Spheroids are then induced for adipogenic differentiation by replacing the media with adipogenic induction media and leaving it for 3 days (*see* **Note 9**).
3. Adipogenic differentiation media is then replaced with adipogenic maintenance media, and cultured for up to 21 days with periodic media exchange, typically every 2 days (*see* **Notes 9** and **12**).

4 Notes

1. Ensure that the NS is homogenized before use, as it can separate into phases after a long period of time. Mix thoroughly until an even brown color is achieved.
2. The amount of NS can be optimized, the typical concentration that we use is 1 μL NS/1×10^4 cells, and we estimate that in a T75, the yield will be 6×10^6 cells.
3. Before detaching cells, view the cells under a microscope (brightfield) to verify magnetization. NS attaches via electrostatic force to the cells at small quantities (~50 pg/cell), and should appear as dark dots on the cell surface. Use brightfield rather than phase contrast to view the dark NS.
4. Magnetized cells will appear brown, which is a result of NS's brown color and not an indication of cytotoxicity.

5. Optimize the cell number for your cells and experiment. The size of the spheroid will vary based on the number of cells, time of culture, and extent of adipogenesis. Spheroids with a diameter of up to 0.5 cm can be formed and levitated. However, in spheroids that are too large, cell death and necrosis in the center of the spheroid will eventually occur due to poor oxygenation.
6. Do not autoclave magnets or insert. Magnets will demagnetize when exposed to high temperatures. Sterilize parts with 70% ethanol (v/v).
7. Take care to keep the plate flat. Tilting the plate too far could allow the cells to escape the media and attach to the plate lid.
8. The magnetic field will work fast to levitate cells. Typically cells will aggregate in the media within 15 min, and fully form a spheroid within 24 h.
9. Media can be easily exchanged by temporarily placing the magnetic drive underneath the 24-well plate to pull down and hold levitated cells while media is exchanged, then immediately replacing it atop the plate [28]. This method can also be used to stain cells inside the plate by using the magnet to hold samples while you add and remove solutions. For transfer, a MagPen (Nano3D Biosciences) consisting of a Teflon pen with magnetic rod can be used to pick up magnetically levitated cultures out of the plate and onto a coverslip for imaging.
10. Protocols of SVF isolation may vary [19], which may result in different relative representations of the distinct cell types in the adiposphere. Use your lab's specific protocol for SVF isolation.
11. SVF cells can also be magnetized directly in suspension without plating. Magnetization in suspension would require mixing cells with NS and shaking at room temperature for 30 min to 6 h. Measure efficiency of magnetization by separating magnetized and unmagnetized cells, and optimize this process for your specific application.
12. Adipogenesis can be assessed based on lipid buoyancy. To do this, remove the magnet temporarily. Adiposphere floatation will indicate lipid droplet formation.

References

1. Daquinag AC, Zhang Y, Kolonin MG (2011) Vascular targeting of adipose tissue as an anti-obesity approach. Trends Pharmacol Sci 32(5):300–307
2. Han J, Koh YJ, Moon HR et al (2010) Adipose tissue is an extramedullary reservoir for functional hematopoietic stem and progenitor cells. Blood 115:957–964
3. Bianco P (2014) "Mesenchymal" stem cells. Annu Rev Cell Dev Biol 30:677–704
4. Bianco P, Robey PG, Simmons PJ (2008) Mesenchymal stem cells: revisiting history, concepts, and assays. Cell Stem Cell 2:313–319
5. Kolonin MG, Simmons PJ (2009) Combinatorial stem cell mobilization. Nat Biotechnol 27(3):252–253

6. Choi JH, Gimble JM, Lee K et al (2010) Adipose tissue engineering for soft tissue regeneration. Tissue Eng Part B Rev 16:413–426
7. Scherberich A, Müller AM, Schäfer DJ et al (2010) Adipose tissue-derived progenitors for engineering osteogenic and vasculogenic grafts. J Cell Physiol 225(2):348–353
8. Güven S, Mehrkens A, Saxer F et al (2011) Engineering of large osteogenic grafts with rapid engraftment capacity using mesenchymal and endothelial progenitors from human adipose tissue. Biomaterials 32:5801–5809
9. Kolonin MG, Evans KW, Mani SA et al (2012) Alternative origins of stroma in normal organs and disease. Stem Cell Res 8(2):312–323
10. Schindler M, Nur-E-Kamal A, Ahmed I et al (2006) Living in three dimensions: 3D nanostructured environments for cell culture and regenerative medicine. Cell Biochem Biophys 45:215–227
11. Griffith LG, Swartz MA (2006) Capturing complex 3D tissue physiology in vitro. Nat Rev Mol Cell Biol 7:211–224
12. Peyton SR, Kim PD, Ghajar CM et al (2008) The effects of matrix stiffness and RhoA on the phenotypic plasticity of smooth muscle cells in a 3-D biosynthetic hydrogel system. Biomaterials 29:2597–2607
13. Pedersen JA, Swartz MA (2005) Mechanobiology in the third dimension. Ann Biomed Eng 33:1469–1490
14. Zhang S (2004) Beyond the Petri dish. Nat Biotechnol 22:151–152
15. Kleinman HK, Philp D, Hoffman MP (2003) Role of the extracellular matrix in morphogenesis. Curr Opin Biotechnol 14:526–532
16. Cukierman E, Pankov R, Stevens DR et al (2001) Taking cell-matrix adhesions to the third dimension. Science 294:1708–1712
17. Pampaloni F, Reynaud EG, Stelzer EHK (2007) The third dimension bridges the gap between cell culture and live tissue. Nat Rev Mol Cell Biol 8:839–845
18. Zuk PA, Zhu M, Mizuno H et al (2001) Multilineage cells from human adipose tissue: implications for cell-based therapies. Tissue Eng 7:211–228
19. Daquinag AC, Zhang Y, Amaya-Manzanares F et al (2011) An isoform of decorin is a resistin receptor on the surface of adipose progenitor cells. Cell Stem Cell 9:74–86
20. Traktuev DO, Merfeld-Clauss S, Li J et al (2008) A population of multipotent CD34-positive adipose stromal cells share pericyte and mesenchymal surface markers, reside in a periendothelial location, and stabilize endothelial networks. Circ Res 102:77–85
21. Souza GR, Molina JR, Raphael RM et al (2010) Three-dimensional tissue culture based on magnetic cell levitation. Nat Nanotechnol 5:291–296
22. Tseng H, Gage JA, Raphael RM et al (2013) Assembly of a three-dimensional multitype bronchiole coculture model using magnetic levitation. Tissue Eng Part C Methods 19:665–675
23. Molina JR, Hayashi Y, Stephens C et al (2010) Invasive glioblastoma cells acquire stemness and increased Akt activation. Neoplasia 12:453–463
24. Xu L, Gao G, Ren J et al (2012) Estrogen receptor β of host promotes the progression of lung cancer brain metastasis of an orthotopic mouse model. J Cancer Ther 3:352–358
25. Lee JS, Morrisett JD, Tung C-H (2012) Detection of hydroxyapatite in calcified cardiovascular tissues. Atherosclerosis 224:340–347
26. Castro-Chavez F, Vickers KC, Lee JS et al (2013) Effect of lyso-phosphatidylcholine and Schnurri-3 on osteogenic transdifferentiation of vascular smooth muscle cells to calcifying vascular cells in 3D culture. Biochim Biophys Acta 1830:3828–3834
27. Becker JL, Souza GR (2013) Using space-based investigations to inform cancer research on Earth. Nat Rev Cancer 13:315–327
28. Haisler WL, Timm DM, Gage JA et al (2013) Three-dimensional cell culturing by magnetic levitation. Nat Protoc 8:1940–1949
29. Timm DM, Chen J, Sing D et al (2013) A high-throughput three-dimensional cell migration assay for toxicity screening with mobile device-based macroscopic image analysis. Sci Rep 3:3000
30. Tseng H, Balaoing LR, Grigoryan B et al (2014) A three-dimensional co-culture model of the aortic valve using magnetic levitation. Acta Biomater 10:173–182
31. Jaganathan H, Gage J, Leonard F et al (2014) Three-dimensional in vitro co-culture model of breast tumor using magnetic levitation. Sci Rep 4:6468
32. Daquinag AC, Souza GR, Kolonin MG (2013) Adipose tissue engineering in three-dimensional levitation tissue culture system based on magnetic nanoparticles. Tissue Eng Part C Methods 19:336–344

Chapter 13

Isolation of Human Adipose-Derived Stem Cells from Lipoaspirates

Jie Li, J. Lowry Curley, Z. Elizabeth Floyd, Xiying Wu, Yuan Di C. Halvorsen, and Jeffrey M. Gimble

Abstract

Adipose tissue is as an abundant and accessible source of stem cells with multipotent properties suitable for tissue engineering and regenerative medical applications. Here, we describe methods from our own laboratory and the literature for the isolation and expansion of adipose-derived stem cells (ASCs). We present a large scale procedure suitable for processing >100 mL volumes of lipoaspirate tissue specimens by collagenase digestion and a related procedure suitable for processing adipose tissue aspirates without digestion.

Key words Adipose-derived stem cells (ASCs), Biopsy, Collagenase, Expansion, Human, Isolation, Lipoaspirate, Mesenchymal stem cells (MSCs), Stromal vascular fraction (SVF)

1 Introduction

Mesenchymal stromal/stem cells (MSC) were initially described in bone marrow and have been found subsequently in multiple tissues, including subcutaneous adipose tissue [1–4]. Although adipose-derived stromal cells had been termed "preadipocytes" [2, 3], multiple independent investigators have demonstrated that they are multipotent, with adipogenic, chondrogenic, neuronal-like, and osteogenic differentiation capability [5–8]. Consequently, they have now been identified as adipose-derived stromal/stem cells or ASCs [9]. This protocol, which includes information presented in earlier volumes of this series [10], describes the isolation of primary in vitro cultures of ASCs from human adipose tissue.

2 Materials

2.1 Tissue

Subcutaneous adipose tissue samples obtained from liposuction aspirates (*see* **Note 1**).

Bruce A. Bunnell and Jeffrey M. Gimble (eds.), *Adipose-Derived Stem Cells: Methods and Protocols*, Methods in Molecular Biology, vol. 1773, https://doi.org/10.1007/978-1-4939-7799-4_13, © Springer Science+Business Media, LLC, part of Springer Nature 2018

2.2 Supplies

1. 200 mL plastic centrifugation bottles (Nalgene).
2. 0.2 μm filter units.
3. 50 mL conical tubes.
4. 2 mL tubes.
5. Hemocytometer.
6. Freezing apparatus (alcohol container).
7. Fluorochrome conjugated monoclonal antibodies against stromal, hematopoietic, endothelial, pericytic, and related cell surface antigens.

2.3 Equipment

1. Inverted microscope—Nikon Eclipse TS100 with epi-fluorescence attachment (Mercury Lamp Illuminator model name: C-SHG) (Nikon Instruments Incorporation, Melville, NY) and equipped with a camera photometric cool-snap (Nikon).
2. MetaMorph Imaging Software (Universal Imaging Corporation).
3. Shaking water bath.
4. Centrifuge.
5. Biosafety hood.
6. CO_2 incubator.

2.4 Media Stock Solution (See Note 2)

All the media solutions are filtered through a 0.2 μm filter unit.

1. *0.1% Collagenase solution*: Weigh out 0.1 g of type I collagenase and 1 g of powdered bovine serum albumin (fraction V). Dissolve these in 100 mL of phosphate buffered saline (PBS) supplemented with 2 mM calcium chloride ($CaCl_2$). The volume of the collagenase solution used will be equal to the volume of the lipoaspirates (1:1). After sterile filtration, warm the solution to 37 °C. This solution should be used within 1 h of preparation.
2. Erythrocyte lysis buffer: Using sterile, distilled water, prepare 155 mM ammonium chloride (NH_4Cl), 10 mM potassium carbonate (K_2CO_3), and 0.1 mM EDTA. This solution should be used within 24 h of its preparation.
3. Stromal medium: To 500 mL of DMEM/Ham's F-12 medium, add 55 mL of fetal bovine serum (10%) and 5.6 mL of antibiotic (penicillin/streptomycin)–antimycotic (amphotericin) 100× stock solution. This solution should be used within 4 weeks of its preparation. All fetal bovine serum should be prescreened prior to purchase for its ability to support both cell proliferation and adipocyte differentiation.
4. Differentiation medium: In advance, prepare and aliquot the following stock solutions and store frozen at −20 °C until required.

(a) A 66 mM stock solution of biotin (2000-fold concentration) dissolved in 1 N sodium hydroxide.

(b) A 34 mM stock solution D-pantothenate (2000-fold concentration) dissolved in water.

(c) A 1 mM dexamethasone (1000-fold concentration) dissolved in water or ethanol depending on its formulation.

(d) A 250 mM stock solution of methylisobutylxanthine (1000-fold concentration) dissolved in dimethyl sulfoxide.

(e) A 200 μM stock solution of human insulin (2000-fold concentration) dissolved in PBS.

(f) A 5 mM stock solution of rosiglitazone (1000-fold concentration) or equivalent PPARγ agonist dissolved in dimethyl sulfoxide. Use this solution within 2 weeks of its preparation.

Prepare the differentiation medium containing the following final concentrations in DMEM/Ham's F-12: 3% fetal bovine serum, 0.25 mM IBMX, 66 μM biotin, 34 μM D-pantothenate, 5 μM rosiglitazone (or equivalent PPARγ2 ligand), 1 μM dexamethasone, 200 nM human insulin. Use this solution within 2 weeks of its preparation.

5. Adipocyte maintenance medium: This solution is prepared in an identical manner as differentiation medium except that it does not contain either the isobutylmethylxanthine or the PPARγ agonist; these two stock solutions should be omitted. Use this solution within 2 weeks of its preparation.

6. *Oil Red O staining solution*: Weigh out 0.5 g Oil Red O. Dissolve in 100 mL isopropanol. Filter through a 0.2 μm filter. Store at room temperature as stock solution. At the time of use, take 6 mL of Oil Red O stock solution. Add 4 mL of distilled water. Stand 1 h at room temperature before use. Use this solution within 24 h of its preparation.

7. Freezing medium: The freezing medium consists of 80% fetal bovine serum or 80% calf serum, 10% DMEM/Ham's F-12, and 10% dimethyl sulfoxide. Use this solution within 2 weeks of its preparation.

8. *Medium Sterility Test Procedure*: Prior to use, it is wise to test the sterility of the medium by removing a single milliliter from each bottle, placing it in a single well of a 24 well plate, and incubating it for 48 h in a humidified, 37 °C, CO_2 incubator. After this period, examine the plate using a phase contrast microscope for any evidence of contamination. If contaminated, immediately inactivate all bottles and test plates with 15% bleach solution and discard.

3 Methods

After transportation to the laboratory, the liposuction sample can be kept at room temperature for no more than 24 h prior to use. Before performing the experiment, warm up the water bath to 37 °C.

All the following procedures are performed in biosafety hoods. Investigators should be trained in the handling of human tissues and human pathogens prior to initiation of any studies (*see* **Note 3**).

3.1 Processing Large Volumes of Tissue (Liposuction Aspirates ≥100 mL)

1. Warm up buffer (500 mL or more of PBS or KRB). Line the surface of the biosafety hood with a disposable bench protector.
2. Warm up freshly prepared collagenase solution in the 37 °C water bath.
3. Prepare PBS (or KRB) solution with 1% BSA, filter the solution and warm it in the 37 °C water bath.
4. Prepare stromal medium: cf Media stock solution. This should have been done in advance of the procedure.
5. To maintain optimal sterile conditions, open the surgical container used for liposuction procedure under the biosafety hood (*see* **Note 4**). Dispense a volume of adipose tissue in sterile plastic bottles (Nalgene): for each 175 cm^2 Flasks (0.16 mL tissue/cm^2) it is recommended that you distribute about 33 mL of tissue; each bottle can accommodate ~100 mL of tissue. We routinely process a total of 200 mL of tissue to be plated in six 175 cm^2 Flasks. Add an equal volume of warm PBS. Agitate to wash the tissue and then allow phase separation for 3–5 min. Alternatively, you can spin the samples at 300 × *g* in an appropriate centrifuge for 1 min at room temperature. Suction off the infranatant solution (lower liquid phase). The wash is repeated several times until a clear infranatant solution is obtained (usually 3–4 times).
6. Add an equal volume (60–70 mL) of warm collagenase solution into the Nalgene bottles containing the clean adipose tissue sample. Wrap the bottles with Parafilm and place them in a 37 °C shaking water bath at ~200 rpm for 60–80 min until the tissue appears smooth on visual inspection (*see* **Note 5**).
7. Isolation stromal vascular fraction (SVF):
   (a) After digestion, spin the samples at 300 × *g* in an appropriate centrifuge for 5 min at room temperature. Take the samples out of the centrifuge and shake them vigorously to thoroughly disrupt the pellet and to mix the cells. This is to complete the separation of the stromal cells from the primary adipocytes. Repeat the centrifugation step.

8. After spinning, the stromal vascular fraction will form a pellet at the bottom of the bottle or tube (this will usually include a layer of dark red cells). Carefully remove the top layer of oil and fat, the primary adipocytes (a yellow layer of floating cells), and the underlying layer of collagenase solution. Leave behind a small volume of collagenase solution above the pellet so that the cells are not disturbed.

   (a) Resuspend the cells in 10 mL of warm PBS (or KRB) solution and transfer the solution containing the cells into a 50 mL conical tube. Centrifuge the cells at 300 × *g* in an appropriate centrifuge for 5 min at room temperature.

   Aspirate the remaining collagenase solution. When aspirating, the tip of the pipette should aspirate from the top so that the oil is removed as thoroughly as possible. The cell pellet should be at the bottom of the tubes.

9. Resuspend the cells in 15 mL of stromal medium (*see* **Note 6**).

   Divide the cells according to the number of flasks. The cells are plated at a density equivalent to approximately 0.18 mL of liposuction tissue aspirate per cm^2 of surface area (volume of ~33 mL of tissue for a 175 cm^2 flask).

   Divide the cells according to the number of flasks. In this protocol, we use about 200 mL of liposuction tissue. Thus, to each 175 cm^2 flask (×6), we add 2.5 mL of cell suspension and 32.5 mL of stromal medium (*see* **Note 7**).

10. Forty-eight hours[1] after plating, aspirate the medium from the flask. Wash the cells with prewarmed PBS (*see* **Note 8**). Add 35 mL of fresh stromal medium.

    The medium is then changed every 2–3 days until the cells achieve 80–90% confluence.

11. Remove medium from flasks and save the sterile "conditioned media" in a sterile tube for future cell culture application (this should be sterile filtered prior to such use). Add 10 mL of sterile warm PBS to the flasks and allow PBS to remain on cells for 2 min while flasks are in a horizontal position. Replace the PBS with 5 mL of Trypsin/EDTA solution (0.5%) (*see* **Note 9**). Incubate in incubator for 5–10 min. Verify under microscope that more than 90% of the cells have detached and then add 5–10 mL of stromal medium to allow the serum contained in the solution to neutralize the trypsin reaction.

    Proceed to cell counting by taking an aliquot of cells diluted in Trypan Blue (for a 1:1 dilution: add 10 μL of suspended cells to 10 μL of Trypan Blue). Count cells using the hemocytometer.

---

[1] This period can vary from 24 to 72 h depending on the number of cells attached to the plastic surface (observed under microscope).

12. After counting, you have several options:
    (a) Cryopreservation

        Suspend the cell pellet in room temperature freezing medium at a concentration of 1–2 × 10^6 cells/mL. Dispense 1 mL of aliquots of the cell suspension to sterile cryovials. Place cryovials in appropriate freezing apparatus (Mr. Frosty™ Freezing Container). Freeze cells to −80 °C. The next day, transfer the cells on dry ice or other frozen material to a liquid nitrogen storage container.

    (b) Use of cells for flow cytometry

        Harvest ~1.5 × 10^6 ASCs and centrifuge 300 × *g* for 5 min at room temperature in a 50 mL tube. Wash the ASCs twice with 10 mL cold PBS (Ca and Mg free) and resuspend cells in 500 μL cold PBS. Aliquot 50 μL of cells into ten 1.5-mL microcentrifuge tubes, add 50 μL volume of PBS containing a fluorochrome-conjugated monoclonal antibody or isotype control antibody (usually 5–10 μL) to each tube. Mix well. Incubate samples for 20–30 min at room temperature according to the manufacturer's instructions. Critical Note: Keep the tubes protected from light exposure to avoid bleaching of the fluorochrome. Wash cells with 1 mL PBS with 1% BSA and pellet cells @ 300 × *g* for 3 min at room temperature for three times. Resuspend cells in 500 μL of 1% formaldehyde in PBS to fix cells. Keep the tubes at 4 °C protected from light exposure until they can be analyzed on a flow cytometer within a 48 h period.

    (c) Replating the cells

        After cell count, suspend the cell pellet in stromal medium following the different concentrations listed in Table 1 to achieve a confluent culture within 24 h of replating.

    (d) Adipocyte differentiation

        When the cells reach between 80% and 90% confluence (before or after harvesting the cells), the ASCs are induced to differentiate. Aspirate the medium, add a small volume

Table 1
Table for plating

Plate	Area per plate	Cells per plate	Cells per well	Media per well
6 well plate	60 cm^2	1.8 × 10^6	30 × 10^4	2.5 mL
24 well plate	48 cm^2	1.44 × 10^6	6 × 10^4	1 mL
96 well plate	31 cm^2	0.93 × 10^6	10^4	200 μL

(about 1.5 mL for a 6 well plate) of prewarmed PBS + 1% antibiotic to wash the cells, and then remove the PBS by aspiration (*see* **Note 10**). Next, add the differentiation medium.

The cells will be maintained in the differentiation medium for 3 days.

Day +3 differentiation

Aspirate the differentiation medium and wash the cells with prewarmed PBS + 1% antibiotic (*see* **Note 11**). Then add a volume (2.5–3 mL for a 6 well plate) of adipocyte medium.

The adipocyte medium will be changed every 3 days until mature adipocytes are obtained (Day +9 to +12 differentiation) (*see* **Note 12**).

(e) Fixation of cells

After 12 days of differentiation, the cells can be fixed using 10% formalin solution, 4% paraformaldehyde, or 70% ethanol[2]. After removing the medium and washing the cells with PBS, immerse the cells in the fixative solution: 10% formalin, 4% paraformaldehyde, or 70% ethanol for 30 min, 10 min, or 1 h, respectively. Remove the fixative before staining (fixed cells can be stored at 4 °C for as long as several months although shorter times are recommended).

(f) Cell staining

Add 50 μL Oil Red O to each well for 15 min at room temperature. Rinse 3× or more with 50 μL distilled water. The rinse should become completely clear (no red coloring). Do not rinse with a volume larger than 50 μL. It will raise the level of the solution in the plastic well and cause staining of the wall of the well, resulting in high background. Elute the stain from the cells by adding 50 μL isopropanol per well. Elution is immediate. Read the OD_{540} using a plate reader. Subtract the background staining determined in blank wells (no cells) from the experimental points. Determine the relative staining intensity of the differentiated wells compared to the preadipocyte controls.

3.2 Processing Lipoaspirate Fluid (from [11])

1. Aspirate lipoaspirate fluid fraction using pipet, separating it from the lipoaspirate tissue fraction.
2. Transfer the lipoaspirate fluid (LAF) to a sterile plastic bottle (Nalgene).
3. Centrifuge for 10 min at 400 × *g* at room temperature.

[2] Using ethanol, there is a risk that the lipids will be eluted from the cells.

4. Resuspend LAF cell pellet in erythrocyte lysis buffer. Stand at room temperature for 5 min.
5. Filter LAF cell suspension through 100 μm filter.
6. Load cells onto Ficoll density gradient. Centrifuge for 20 min at 800 × *g* at room temperature.
7. Aspirate cells from gradient interface. Wash with PBS pre-warmed to 37 °C. Filter LAF cell suspension through 100 μm filter.
8. Count nucleated cell number using Trypan Blue staining and hemocytometer.
9. Suspend LAF cells in Stromal Medium. Plate at density of 6.4×10^4 cells/cm^2 on gelatin coated T175 cm^2. Maintain in culture and proceed as described under Subheading 3.1, **steps 10–12**.

4 Notes

1. Prior to the implementation of this protocol, all personnel involved in the processing of human lipoaspirate material or primary ASC cultures should complete safety training for the use of blood borne pathogens. Regardless of whether tissue donors have been screened for evidence of infection by hepatitis, HIV, or other transmissible agents, these precautions are mandatory. In addition, no glass containers or pipets should be used and the use of sharp objects (scissors, needles) during the processing steps should be minimized. All procedures involving the human tissue or cells should be conducted in a biological safety cabinet and with appropriate personnel protection gear.
2. Before purchase, the fetal bovine serum should be assayed to test for its ability to support adipogenesis.
3. Contamination—Microbial. In order to keep optimal sterile conditions, it is recommended that you open and close the container properly to avoid any potential contaminations. It is vital that the culture be examined regularly to confirm the absence of microbial contamination. To avoid this problem, 5% iodine solution can be added to the initial wash solution. The disinfectant is then washed away in subsequent washes. In order to avoid this problem in small adipose tissue samples, it is recommended to add in the PBS solution 1% of antibiotic solution and wash thoroughly the cells with this solution.
4. Contamination-Cellular. The blood cells could be a source of contamination and may reduce or prevent adhesion of stro-

mal cells; it is then important to wash thoroughly the cells with PBS (1% of antibiotic solution can be added). When removing the PBS from the cells, aspirate the solution up and down until the cells appear clean and free of red cells. An erythrocyte lysis buffer (155 mM NH_4Cl, 5.7 mM K_2HPO_4, 0.1 mM EDTA, pH 7.3) can serve to remove red blood cells. Endothelial cells (EC) could be another source of contamination. Intra-abdominal depots are more subject to this type of contamination when compared to subcutaneous adipose tissue, which is essentially free of EC. Therefore a filtration procedure can be performed by using a nylon mesh filter with a small pore size (25 μm).

5. After 1 h digestion, if pieces of undigested tissue are still observed in the tube, make sure that the collagenase solution is fresh and has not been maintained at room temperature for an extended period. This is necessary to maximize the enzyme efficiency. The collagenase solution can also be stored at −20 °C for a few days, with a minor loss of enzyme activity. Prior to use, the frozen solution is slowly thawed at room temperature and prewarmed to 37 °C. However, it is all right to not complete the digestion if a small amount of tissue fragments is observed in the solution.
6. A filtration procedure can be performed by using a nylon mesh filter with a small pore size (250–350 μm). The suspension is then centrifuged at 300 × *g* at room temperature to allow separation of the stromal vascular fraction from the mature adipocytes. The filtration can be performed after removing the floating mature adipocytes cells. This step will remove tissue fragments and is used by some investigators. However, the filtration procedure is not recommended for small amounts of adipose tissue.
7. To accelerate cell adhesion, the culture dishes can be precoated with extracellular matrix proteins, such as gelatin or Matrigel.
8. If the cells do not grow very well and do not appear healthy, there are a few options. The percent of preadipocytes obtained from the stromal vascular fraction after digestion is patient-dependent. Supplementation of the culture medium with 20–30% conditioned medium (saved from previous cultures) should facilitate the growth of the cells. If you do not have conditioned medium, another alternative is to increase the FBS contained in the stromal medium to 15–20%; however, this may promote premature adipogenesis. Signs of deterioration such as granularity around the nucleus, cytoplasmic vacuolations, and/or detachment of the cells from the plastic surface may indicate inadequate or toxic medium, microbial contamination or senescence of the primary cells.

9. It is important to not overexpose the cells to the trypsin/EDTA solution. This could decrease the cell viability.
10. If the adipocytes are detaching, do not dry the well when changing the medium since adipocytes tend to float when new medium is added.
11. According to the protocol, the medium is changed every 2–3 days. However, as the adipocytes mature, you may observe a yellowing of the culture medium: a drop in pH may account for this. As the pH falls from 7 to 6.5, cell growth will decline and cell viability falls at pH between 6.5 and 6. You can observe this change of pH by looking at the medium color change, going from red (pH 7) through yellow (pH ≤ 6), indicating the need for an immediate change of the medium.
12. If the cells do not differentiate very well, consider that the differentiation process may be patient dependent. The age of the donor can be a factor, since some studies suggest that the differentiation capacity is higher in culture from younger subjects compared to older people. To further enhance adipogenesis, the following alternatives are proposed:
    - You may try different PPARγ agonists (troglitazone, pioglitazone among others).
    - The addition of 5% rabbit serum (RS) can be added to the differentiation medium enhance differentiation (the ethyl acetate contain in the RS has been found 35-fold more abundant than in FBS [12].
    - Another alternative would be to perform the addition of the differentiation medium multiple times after a 3 day rest period; i.e., 3 days on in the presence of the differentiation medium and 3 days off in the presence of the adipocyte medium. Repeat this cycle until mature adipocytes are obtained.

5 Final Comments

Recently, there has been increased appreciation for the use of primary cell culture models in the study of human adipocyte differentiation in vitro. This protocol on the cultivation of human adipocyte precursor cells can be used by laboratories with access to human tissues in a scalable manner. The protocol can be adapted to the use of intact human adipose tissue which can be minced using dissecting scissors and then processed in a manner similar to the lipoaspirates.

References

1. Rangwala SM, Lazar MA (2000) Transcriptional control of adipogenesis. Annu Rev Nutr 20:535–559
2. Deslex S, Negrel R, Vannier C, Etienne J, Ailhaud G (1987) Differentiation of human adipocyte precursors in a chemically defined serum-free medium. Int J Obes (Lond) 11(1):19–27
3. Hauner H, Entenmann G, Wabitsch M, Gaillard D, Ailhaud G, Negrel R, Pfeiffer EF (1989) Promoting effect of glucocorticoids on the differentiation of human adipocyte precursor cells cultured in a chemically defined medium. J Clin Invest 84(5):1663–1670
4. Halvorsen YD, Bond A, Sen A, Franklin DM, Lea-Currie YR, Sujkowski D, Ellis PN, Wilkison WO, Gimble JM (2001) Thiazolidinediones and glucocorticoids synergistically induce differentiation of human adipose tissue stromal cells: biochemical, cellular, and molecular analysis. Metabolism 50(4):407–413
5. Zuk PA, Zhu M, Ashjian P, De Ugarte DA, Huang JI, Mizuno H, Alfonso ZC, Fraser JK, Benhaim P, Hedrick MH (2002) Human adipose tissue is a source of multipotent stem cells. Mol Biol Cell 13(12):4279–4295
6. Zuk PA, Zhu M, Mizuno H, Huang J, Futrell JW, Katz AJ, Benhaim P, Lorenz HP, Hedrick MH (2001) Multilineage cells from human adipose tissue: implications for cell-based therapies. Tissue Eng 7(2):211–228
7. Guilak F, Lott KE, Awad HA, Cao Q, Hicok KC, Fermor B, Gimble JM (2006) Clonal analysis of the differentiation potential of human adipose-derived adult stem cells. J Cell Physiol 206(1):229–237
8. Gimble JM, Katz AJ, Bunnell BA (2007) Adipose-derived stem cells for regenerative medicine. Circ Res 100(9):1249–1260
9. Mitchell JB, McIntosh K, Zvonic S, Garrett S, Floyd ZE, Kloster A, Halvorsen YD, Storms RW, Goh B, Kilroy GS, Wu X, Gimble JM (2006) The immunophenotype of human adipose derived cells: temporal changes in stromal- and stem cell-associated markers. Stem Cells 24:376–385
10. Dubois SG, Floyd EZ, Zvonic S, Kilroy G, Wu X, Carling S, Halvorsen YD, Ravussin E, Gimble JM (2008) Isolation of human adipose-derived stem cells from biopsies and liposuction specimens. Methods Mol Biol 449: 69–79
11. Yoshimura K, Shigeura T, Matsumoto D, Sato T, Takaki Y, Aiba-Kojima E, Sato K, Inoue K, Nagase T, Koshima I, Gonda K (2006) Characterization of freshly isolated and cultured cells derived from the fatty and fluid portions of liposuction aspirates. J Cell Physiol 208(1):64–76
12. Diascro DD Jr, Vogel RL, Johnson TE, Witherup KM, Pitzenberger SM, Rutledge SJ, Prescott DJ, Rodan GA, Schmidt A (1998) High fatty acid content in rabbit serum is responsible for the differentiation of osteoblasts into adipocyte-like cells. J Bone Miner Res 13(1):96–106

Chapter 14

Isolation of Murine Adipose-Derived Stromal/Stem Cells Using an Explant Culture Method

Jie Li, Hui Li, and Weidong Tian

Abstract

Adipose tissue provides valuable cell source for tissue engineering, regenerative medicine, and adipose tissue biology studies. The most widely used adipose-derived stromal/stem cells (ASCs) isolation protocol involves enzymatic digestion with collagenase. However, the yield of the method is poor even impossible to collect enough stromal vascular fraction (SVF) for expansion when the sample size is small, for instance only new born mice is available for cell culture. Here, we describe an efficient protocol for the isolation and expansion of ASCs using explant culture.

Key words Murine, Adipose-derived stromal/stem cells (ASCs), Collagenase, Explant, Mesenchymal stem cells (MSCs)

1 Introduction

Two methods have been widely adopted to isolate stem/stromal cells from mammal tissue: enzymatic dissociation of tissues and explant culture. ASCs with similar biological properties could be acquired from either methods using adipose tissue, which has proven to be an abundant source of stem cells [1]. The majority of methods isolating ASCs from mammalian adipose tissue use an approximately 1-h type I collagenase digestion. However, the digestion process damages the natural microenvironment of adipose tissue. Additionally, digestion parameters such as the length of the digestion time or concentration of enzyme activity may affect cell yield, viability, phenotype, and differentiation potential [2]. Compared to explant culture method, digestion techniques can be more labor-intensive and less cost-effective. Explant culture can be defined as an in vitro technique that allows small fragments of tissue to adhere to the growth surface, which will usually give rise to an outgrowth of cells. In particular, explant culture may be preferable when only small fragments of tissue are available. The process of explant culture is more moderate for ASCs to transfer from an

Bruce A. Bunnell and Jeffrey M. Gimble (eds.), *Adipose-Derived Stem Cells: Methods and Protocols*, Methods in Molecular Biology, vol. 1773, https://doi.org/10.1007/978-1-4939-7799-4_14, © Springer Science+Business Media, LLC, part of Springer Nature 2018

in vivo to an in vitro environment than using digestion culture and may simplify the process of satisfying the minimal manipulation criterion in US Food and Drug Administration (FDA) guideline. While the current protocol focuses on murine adipose tissue(s), similar isolation methods can be adopted in human or other mammalian adipose tissues as well [3–5].

A number of other non-enzymatic methods for isolating ASCs have been documented. One study found that directly plating the lipoaspirate can yield similar ASCs population compared to collagenase digestion methodology [6]. Research has also shown that the fluid portion of the liposuction aspirate contains a substantial amount of ASCs [7]. Shaking the lipoaspirates in ammonium-chloride-potassium lysing buffer following centrifuging is able to generate a mesenchymal stromal cell (MSC) population [8]. Vigorously washing the floating lipoaspirate can release a fraction of ASCs, which may be useful when processing large volumes of lipoaspirate [9]. These results suggest that mechanical disruption is able to release ASCs from the adipose tissue. Toward this end, a commercially available mechanical device called the "Lipogems device" is gaining acceptance using a self-contained kit to extract, manipulate and isolate ASCs [10]. Thus, plenty of nonenzymatic and enzymatic methods have been documented, which provide techniques to isolate and extract stem/stromal cells quickly and efficiently from adipose tissue.

2 Materials

2.1 Tissue

Subcutaneous inguinal adipose tissue from mouse or rat.

2.2 Supplies

1. 60 mm diameter culture dish or 25-cm^2 tissue culture (T25) flasks.
2. Scissors and forceps, autoclave to sterilize.
3. Hemocytometer.
4. Freezing apparatus.

2.3 Equipment

1. Phase-contrast microscope—Nikon Eclipse TS100 or equivalent.
2. Imaging software.
3. Biosafety hood.
4. CO_2 incubator.

2.4 Reagents and Buffers

1. 70% ethanol.
2. α-minimal essential medium (α-MEM) or DMEM/Ham's F-12 medium.

3. Fetal bovine serum (FBS).
4. Penicillin–streptomycin solution (10,000 U/mL).
5. 0.5% trypsin–EDTA solution.
6. Phosphate-buffered saline (PBS).

3 Methods

3.1 Isolation of the Adipose Tissue

All procedures are performed in a biosafety hood or other aseptic environment.

1. Ensure ASC isolation procedure is initiated within 20 min of tissue extraction, especially if a large number of animals are being euthanized for harvest on a single day.
2. Warm PBS in water bath at 37 °C.
3. Place bench protector down in hood.
4. Sacrifice animal using anesthesia or CO_2 asphyxiation as approved by the AVMA Guidelines on Euthanasia. Saturate the fur with 70% ethanol and place the animal ventral side up. A dissection board may be used if desired.
5. Using sterile tissue forceps, pull up on the skin and cut the skin vertically toward the head along the midline at the abdominal level.
6. Using sterile scissors and forceps, remove the desired fat pads from subcutaneous inguinal depots.
7. Place the tissue in a culture dish containing PBS and antibiotic.

3.2 Plate the Adipose Tissue Explants

1. Wash the adipose tissue twice with PBS and absorb excess PBS.
2. Transfer the adipose tissue to another dry container then mince into 1–2 mm^3 pieces using sterile scissors under aseptic conditions (*see* **Note 1**).
3. Place the tissue explants in culture dish or flask with 5 mm space between adjacent fragments (*see* **Note 2** and Fig. 1).
4. After adhesion (5–10 min, depends on the liquid content in the minced tissue), add fresh stromal media (α-MEM supplemented with 10% FBS, 100 U/mL penicillin–streptomycin) gently to the petri dish/flask without disturbing the explants (*see* **Note 3**).
5. Change the media every 2–3 days (*see* **Note 4**).
6. Remove the tissue explants at day 5 or day 6 (*see* **Note 5**).
7. At 80–90% confluence, the cells are either passaged, or stored as frozen stocks as passage zero (P0), or differentiated using an experimental protocol (*see* **Note 6**).

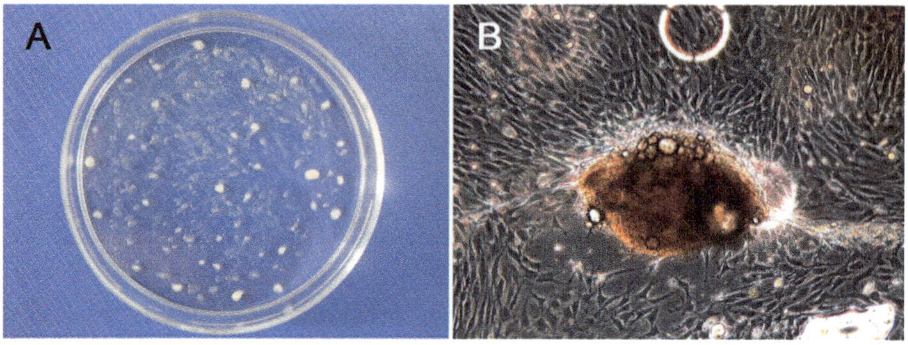

Fig. 1 The explant culture of the ASCc. (**a**) The adipose tissue fragments attached to the dish after 5–10 min. (**b**) ASC outgrowth from the tissue explants after 4 days

4 Notes

1. We prefer smaller tissue fragment than other groups [3] because central necrosis will happen due to low nutrient and oxygen exchange.
2. For reasons that are not quite clear, seeding explants at a too high density in the culture dish or flask will inhibit the growth of cells from the explant. Based on our experience, two fat pads from an adult mouse for five to six 60 mm diameter culture dishes is optimal. As reviewed in the literature, numerous substrates such as collagen, fibronectin, and basement membrane proteins are able to facilitate the attachment of the explants [11]. These substrates may enhance the outgrowth of ASCs from the explants.
3. It is critical that the tissue fragments tightly adhere to the culture dish or plate to obtain ASCs consistently and efficiently. As described by other independent studies and investigators, ASCs can only migrate from the adherent tissue fragments and seed as adherent cells from explants that are on the plastic but not from floating tissue fragments [12]. Stainless steel mesh has been used to keep the tissue fragments from floating [12], which might be helpful for ASCs isolation as well.
4. Cell outgrowth from the explants usually occurs during days 1–3. If the outgrowth of cells is not observed after 4–5 days, the explant culture failed and should be tried again.
5. After about 4–7 days, adipogenic differentiation of the ASCs will occur around the explants. Certain adipogenic factors secreted by the explants induced the differentiation process [13].
6. The explant culture method yield about 2.4×10^4 cells/mg tissue after 7 days culture [1]. The methods for characterization of the cells that grow out by explant culture are the same as other chapters in this book.

References

1. Jing W et al (2011) Explant culture: an efficient method to isolate adipose-derived stromal cells for tissue engineering. Artif Organs 35(2):105–112
2. Williams SK, McKenney S, Jarrell BE (1995) Collagenase lot selection and purification for adipose tissue digestion. Cell Transplant 4(3):281–289
3. Priya N et al (2014) Explant culture: a simple, reproducible, efficient and economic technique for isolation of mesenchymal stromal cells from human adipose tissue and lipoaspirate. J Tissue Eng Regen Med 8(9):706–716
4. Zeng G et al (2013) A rapid and efficient method for primary culture of human adipose-derived stem cells. Organogenesis 9(4):287–295
5. Gittel C et al (2013) Isolation of equine multipotent mesenchymal stromal cells by enzymatic tissue digestion or explant technique: comparison of cellular properties. BMC Vet Res 9:221
6. Busser H et al (2014) Isolation of adipose-derived stromal cells without enzymatic treatment: expansion, phenotypical, and functional characterization. Stem Cells Dev 23(19): 2390–2400
7. Yoshimura K et al (2006) Characterization of freshly isolated and cultured cells derived from the fatty and fluid portions of liposuction aspirates. J Cell Physiol 208(1):64–76
8. Baptista LS et al (2009) An alternative method for the isolation of mesenchymal stromal cells derived from lipoaspirate samples. Cytotherapy 11(6):706–715
9. Shah FS et al (2013) A non-enzymatic method for isolating human adipose tissue-derived stromal stem cells. Cytotherapy 15(8):979–985
10. Bianchi F et al (2013) A new nonenzymatic method and device to obtain a fat tissue derivative highly enriched in pericyte-like elements by mild mechanical forces from human lipoaspirates. Cell Transplant 22(11):2063–2077
11. Resau JH et al (1991) Explant organ culture: a review. Cytotechnology 7(3):137–149
12. Mori Y et al (2015) Improved explant method to isolate umbilical cord-derived mesenchymal stem cells and their immunosuppressive properties. Tissue Eng Part C Methods 21(4):367–372
13. Li J et al (2014) Secretory factors from rat adipose tissue explants promote adipogenesis and angiogenesis. Artif Organs 38(2):E33–E45

Chapter 15

Feline Adult Adipose Tissue-Derived Multipotent Stromal Cell Isolation and Differentiation

Carmel Fargason, Nan Zhang, and Mandi J. Lopez

Abstract

The cat, as a species, is somewhat new to the field of adult multipotent stromal cells. Despite the relative phylogenetic distance between the domestic cat, *Felis silvestris catus*, and humans, they share some similar health challenges like diabetes, kidney disease and asthma. There is a plethora of current investigative efforts focused on adult adipose tissue-derived multipotent stromal cell (ASC) therapies to address these and other conditions. Given the small size of domestic cats, particular attention to optimize cell isolation from relatively little tissue is a necessary condition of feline ASC studies and therapies. Additionally, there are some unique features of culture conditions to test and confirm feline ASC plasticity. This chapter contains a few of the novel aspects of feline ASC isolation and culture.

Key words Feline, Cat, Adipose, Adult multipotent stromal cell, Stem cell

1 Introduction

Feline multipotent stromal cells (MSCs) isolated from adipose tissue exhibit similar attributes to those from other species including in vitro culture expansion, multipotentiality, and surface protein expression [1]. It is well established, however, that culture conditions, including those for lineage induction, vary among species [2]. Hence, species-specific isolation and culture conditions as well as lineage differentiation are required for cell collection and characterization.

Potential therapeutic targets for feline adipose tissue-derived multipotent stromal cells (ASCs) include diabetes, chronic kidney disease and asthma (Table 1) [3–5]. In general, reported therapeutic doses of MSCs range from 10^5 to 10^8 cells/kg [2, 3, 6]. To derive autologous cell treatment doses, the number of ASCs per unit of harvested tissue as well as cell expansion rates must be determined. From this information, it is possible to predict the amount of tissue to be harvested for ASC isolation and expansion within an acceptable time period and passage number [7]. To

Bruce A. Bunnell and Jeffrey M. Gimble (eds.), *Adipose-Derived Stem Cells: Methods and Protocols*, Methods in Molecular Biology, vol. 1773, https://doi.org/10.1007/978-1-4939-7799-4_15, © Springer Science+Business Media, LLC, part of Springer Nature 2018

Table 1
Feline ASC trials

Treatment condition	Cell source	ASC dose	Administration	Treatment effect
Chronic Kidney Disease (CKD) [9]	Autologous	$0.1–4 \times 10^6$	1 intrarenal injection	Modest improvement in glomerular filtration rate and mild decrease in serum creatinine
CKD [3]	Allogeneic	2 or 4×10^6 cryopreserved 4×10^6 from cryopreserved adipose tissue	3 total intravenous injections administered every 2 weeks	Low dose of ASCs was well tolerated and serum creatinine improved. Adverse side effects associated with higher doses.
Chronic allergic feline asthma [5]	Allogeneic	$3.6–25 \times 10^6$ cryopreserved adipose tissue	6 total intravenous infusions administered biomonthly	Improved airway structure evaluated with computed tomography 8 months after treatment

achieve therapeutic ASC numbers within an ideal time frame, isolation conditions should be optimized for the greatest number of available ASCs without compromising expansion potential or plasticity.

A recent report surrounds the isolation of therapeutic doses of feline ASCs from epididymal adipose tissue [1]. According to the 2015–2016 National Pet Owners Survey from the American Pet Products Association, there are approximately 85.8 million privately owned cats in the USA, and, according to a 2012–2013 Humane Society survey, approximately 91% are surgically sterilized. Adipose tissue removed during elective sterilization procedures provides a unique opportunity to collect ASCs for allogeneic or autologous therapies. However, the amount of adipose tissue is often limited. The information in this chapter centers on isolation and characterization of ASCs from feline white adipose tissue collected during elective procedures.

2 Materials

2.1 Adipose Tissue Collection and ASC Isolation from Minimal Adipose Tissue Surrounding Excised Reproductive Organs

1. 15-mL conical tubes.
2. Phosphate-buffered saline (PBS).
3. 1% chlorhexidine in PBS.
4. Scalpel handle (#3) and blades (#10).
5. Tissue forceps.

6. 3 cm sterile petri dish.
7. 1 oz/30 mL graduated, glass bottle beakers.
8. Stir bars (1 cm) and stir plate.
9. 0.3% collagenase solution: 0.3% type I collagenase and 1% bovine serum albumin (BSA) dissolved in Krebs–Ringer bicarbonate buffer. Filter-sterilize with 0.22 μm filter. Warm to 37 °C. Make fresh the day of use.
10. 70 and 100 μm cell strainers.
11. Red cell lysis buffer: 155 mM NH_4Cl, 10 mM $KHCO_3$, and 0.1 mM EDTA, pH 7.3.
12. Complete stromal medium: Dulbecco's modified Eagle's medium F-12 (DMEM-F12) supplemented with 10% fetal bovine serum (FBS) and 1% antibiotic–antimycotic (AA) solution.
13. T25 culture flasks.

2.2 Cell Expansion

1. T75 culture flasks.
2. Phosphate-buffered saline (PBS).
3. 0.05% trypsin–ethylenediaminetetraacetic acid (EDTA).
4. 10% fetal bovine serum (FBS) in PBS.
5. 15-mL conical tubes.
6. Complete stromal medium: DMEM-F12 supplemented with 10% FBS and 1% AA solution.
7. Trypan Blue.
8. Hemocytometer.

2.3 Cryopreservation

1. T75 cell culture flasks.
2. PBS.
3. 0.05% trypsin–EDTA.
4. 10% FBS in PBS.
5. 15-mL conical tubes.
6. DMEM-F12.
7. Trypan Blue.
8. Hemocytometer.
9. Cryopreservation medium: 80% FBS, 10% DMEM-F12, 10% dimethylsulfoxide (DMSO).
10. 1.5 mL cryovials.
11. CoolCell® freezing container.

2.4 Flow Cytometry

2.4.1 Immunophenotype

1. T75 cell culture flasks.
2. Phosphate-buffered saline (PBS).
3. 0.05% trypsin–EDTA.

4. 10% FBS in PBS.
5. 15-mL conical tubes.
6. Trypan Blue.
7. Hemocytometer.
8. 5 mL round-bottom polystyrene tubes.
9. Antibodies (fluorescent labeled primary or unlabeled primary and fluorescent labeled secondary).
10. 10% formalin in dH_2O.

2.4.2 Cell Sorting

1. T75 cell culture flask.
2. PBS.
3. 0.05% trypsin–EDTA.
4. 10% FBS in PBS.
5. 15-mL conical tubes.
6. 1% FBS in PBS.
7. Trypan Blue.
8. Hemocytometer.
9. 5 mL round-bottom polystyrene tubes.
10. Antibodies (fluorescent labeled primary or unlabeled primary and fluorescent labeled secondary antibodies).
11. DMEM-F12 supplemented with 1% AA and 20% FBS.
12. 6-well culture plates.

2.5 Cell Doubling

1. 96-well culture plates.
2. AlamarBlue®.
3. Complete stromal medium: DMEM-F12 supplemented with 10% FBS and 1% AA.

2.6 Adipogenic Differentiation (Fig. 1)

1. 24-well cell culture plates.
2. Complete stromal medium: DMEM-F12 supplemented with 10% FBS and 1% AA.
3. PBS.
4. Adipogenic induction medium: minimum essential medium-α (MEM-α), 10% rabbit serum, 10% FBS, dexamethasone (10 nM), insulin (5 μg/mL), 5, 8, 11, 14-eicosatetraynoic acid (ETYA, 50 μM), indomethacin (100 μM).
5. 4% paraformaldehyde (PFA) in PBS.

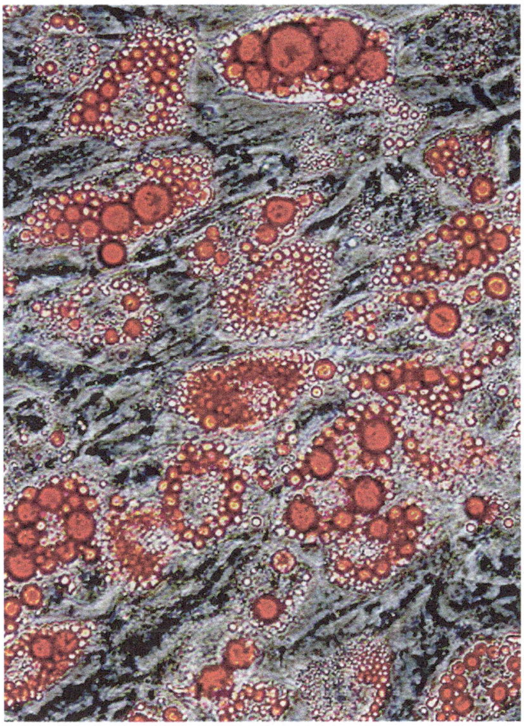

Fig. 1 Light photomicrograph demonstrating Oil Red O-stained lipid droplets following adipogenic induction of feline ASC (20x)

6. 0.3% Oil Red O in isopropanol. Filter solution with a 0.22 μm syringe filter and store at room temperature protected from light.
7. Double-distilled water.

2.7 Osteogenic Differentiation (Fig. 2)

1. 24-well cell culture plates.
2. Complete stromal medium: DMEM-F12 supplemented with 10% FBS and 1% AA.
3. PBS.
4. Osteogenic induction medium: DMEM F-12, 10% FBS, dexamethasone (100 nM), L-ascorbic acid (0.25 mM).
5. Osteogenic maintenance medium: osteogenic medium with β-glycerophosphate (10 nM).
6. 70% ethanol (EtOH) in dH_2O (ice cold).
7. 2% Alizarin Red in double-distilled H_2O (pH 4.1–4.3). Adjust pH with NaOH or HCl. Filter solution with a 0.22 μm syringe filter and store at room temperature protected from light.

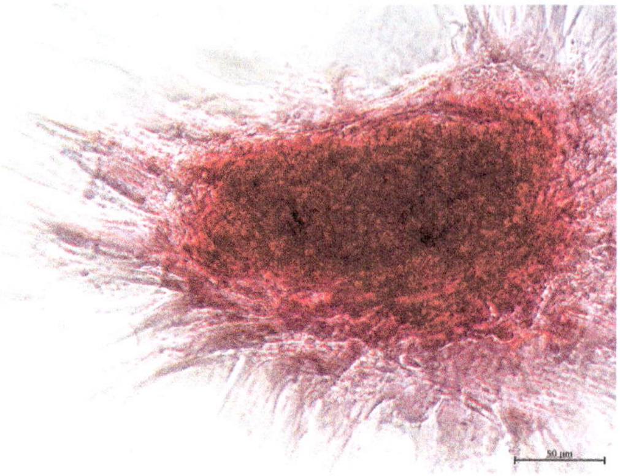

Fig. 2 Alizarin Red staining of calcium deposition a cell colony following culture of feline ASCs in osteogenic induction medium (40x)

2.8 Chondrogenic Differentiation

1. 6-well cell culture plates.
2. Complete stromal medium: DMEM-F12 supplemented with 10% FBS and 1% AA solution.
3. 15 mL centrifuge tubes. For chondrogenesis, 5×10^5 cells (1×10^6 cells/mL starting concentration) are centrifuged ($200 \times g$, 5 min) to form a cell pellet.
4. Chondrogenesis medium: low glucose (1 g/L) DMEM-F12, 3% FBS, 1% AA, ascorbate phosphate (50 μg/mL), dexamethasone (100 nM), proline (40 μg/mL), sodium pyruvate (1 mM), 1% insulin-transferrin-selenium (ITS), bone morphogenetic protein-6 (BMP-6, 10 ng/mL) and transforming growth factor-β1 (TGF-β1, 10 ng/mL).
5. 10% formalin in dH_2O.
6. 1% Alcian Blue.
7. 3% Nuclear Fast Red.

2.9 Neurogenic Differentiation (Fig. 3)

1. 24-well cell culture plates.
2. Complete stromal medium: DMEM-F12 supplemented with 10% FBS and 1% AA.
3. PBS.
4. Neurogenic induction medium: high glucose (4.5 g/L) DMEM supplemented with 20% FBS and β-mercaptoethanol (1 mM).
5. Neurogenic maintenance medium: high glucose (4.5 g/L) DMEM supplemented with 2% FBS, 2% dimethylsulfoxide and butylated hydroxyanisole (200 μM).
6. 10% formalin in dH_2O.
7. Cresyl Violet (1 mg/mL).

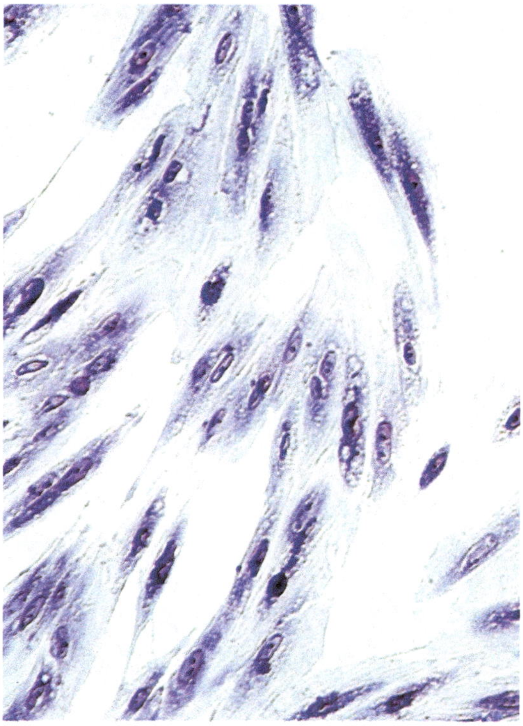

Fig. 3 Neuronal morphology of feline ASCs after culture in neurogenic induction medium (20x). Cells are stained with Cresyl Violet

2.10 Colony Forming Unit Assays

1. T75 culture flasks.
2. Stromal medium: DMEM-F12 supplemented with 1% AA and 20% FBS.
3. PBS.
4. 96-well cell culture plates.
5. Adipogenic induction medium.
6. Osteogenic induction medium.
7. Osteogenic maintenance medium.
8. 4% PFA.
9. 70% EtOH.
10. 0.3% Oil Red O.
11. 2% Alizarin Red.
12. 0.1% Toluidine Blue in dH_2O.

2.11 mRNA Isolation

1. 6-well cell culture plates.
2. RNeasy® Plus Mini Kit.
3. Nuclease-free water.
4. Polymerase chain reaction (PCR)-clean 0.5 mL tubes.
5. 70% EtOH in nuclease-free water.

3 Methods

3.1 ASC Collection and Isolation from Minimal Adipose Tissue (<0.5 g) Surrounding Excised Reproductive Organs

1. Aseptically collect tissues harvested during elective surgical sterilization procedures and place in 15 mL conical tubes with 10 mL PBS. Tissues should be maintained at room temperature and processed within a few hours of collection.
2. Rinse whole tissues with 5 mL 1% chlorhexidine in PBS 1–2 times.
3. Isolate the adipose tissue from organs using a sterile scalpel blade and forceps.
4. Place tissue in a preweighed, sterile petri dish to determine weight.
5. Mince the tissue into small pieces (1 mm) using sterile instruments.
6. Add the tissue to a sterile, glass bottle beaker with a sterile stir bar.
7. Add 33 mL of collagenase solution per 1 g adipose tissue.
8. Agitate tissue at 37 °C and a stir bar speed of 1000 rpm for 30–45 min. The solution should become homogenous.
9. Pass the mixture through a 100 μm cell strainer. If a lot of tissue pieces remain in the mixture, pass it through a 70 μm cell strainer.
10. In a 15-mL conical tube, centrifuge the digest for 5 min at 260 × *g* at room temperature.
11. Resuspend the resulting pellet in 3 mL red cell lysis buffer.
12. Centrifuge again for 5 min at 260 × *g*, remove the supernatant, resuspend in PBS and centrifuge as before.
13. Resuspend the pellet in 5 mL of stromal medium and culture in a T25 culture flask under standard culture conditions (37 °C, 5% CO_2).
14. After 24 h of culture, remove the medium, rinse the cells with PBS, and then add fresh stromal medium. Change medium every 3–4 days.

3.2 Cell Expansion

1. When cells reach about 70% confluence, remove medium from culture vessels.
2. Rinse cells with 5 mL PBS.
3. Detach cells by adding 5 mL trypsin-EDTA and incubating at 37 °C for 5 min.
4. Check cells under the microscope to ensure cell detachment. If the majority is not detached, incubate for a few more minutes (2–3).
5. Stop the reaction by adding 5 mL 10% FBS.

6. Add the mixture to 15 mL conical tubes and centrifuge (350 × *g*) for 5 min at room temperature.
7. Decant supernatant and resuspend the cells in 1 mL stromal medium (*see* **Note 1**).
8. Count cells by combining 10 μl of the cell suspension with 10 μl Trypan Blue. Add the 20 μl mixture to a hemocytometer to determine cell number.
9. Culture cells in stromal medium under standard culture conditions (37 °C, 5% CO_2), typically at a seeding density of 5×10^3 cells/cm^2 in T75 flasks.
10. Change the medium after 24 h then every 2–3 days after.

3.3 Cryopreservation

1. Remove medium from T75 flask at approximately 70% cell confluence.
2. Wash the cells with 5 mL PBS.
3. Detach cells by adding 5 mL trypsin-EDTA and incubating at 37 °C for 5 min. Incubate for a few more minutes as necessary for the majority of cells to detach.
4. Stop the reaction by adding 5 mL 10% FBS.
5. Add the mixture to 15 mL conical tubes and centrifuge (350 × *g*) for 5 min at room temperature.
6. Decant supernatant and resuspend the cells in 1 mL DMEM-F12.
7. Count cells by combining 10 μl of the cell suspension with 10 μl Trypan Blue. Add the 20 μl mixture to a hemocytometer to determine cell number.
8. Transfer 1–2 × 10^6 cells to each cryovial in 0.8–0.9 mL freezing medium.
9. Place cryovials in room temperature CoolCell freezing container.
10. Store for 12–24 h at −80 °C.
11. Remove from CoolCell freezing container and transfer tubes to liquid nitrogen storage.

3.4 Flow Cytometry

3.4.1 Immunophenotype

1. Remove medium from T75 flask at approximately 70% cell confluence.
2. Wash the cells with 5 mL PBS.
3. Detach cells by adding 5 mL trypsin-EDTA and incubating at 37 °C for 5 min. Incubate for a few more minutes as necessary for the majority of cells to detach.
4. Add 5 mL 10% FBS.
5. Add the mixture to 15 mL conical tubes and centrifuge (350 × *g*) for 5 min at room temperature.

6. Decant the supernatant and resuspend the cells in 1 mL PBS.
7. Count cells by combining 10 μl of the cell suspension with 10 μl Trypan Blue. Add the 20 μl mixture to a hemocytometer to determine cell number.
8. Transfer 1–5 × 10^6 cells to each polystyrene round bottom tube in 300 μL PBS (*see* **Note 2**).
9. Add antibody solution to each tube, typically 1–5 μL of 1 mg/mL antibody.
10. Vortex each sample to mix well, and then incubate at room temperature for 30 min. Labeled antibodies must be incubated in darkness.
11. Add 3 mL PBS, mix, and centrifuge at 300 × *g* for 5 min at room temperature. Decant all but 50–100 μL supernatant. Resuspend the cell pellet in the mixture.
12. If the primary antibody is labeled, proceed to the next step. If a labeled secondary antibody is required, vortex each sample to mix well, and then incubate at room temperature for 30 min in darkness. After incubating, add 3 mL PBS, mix, and centrifuge at 300 × *g* for 5 min at room temperature. Decant all but 50–100 μL supernatant. Resuspend the cell pellet in the mixture.
13. Fix the sample with 300 μL 10% formalin.
14. Keep the samples in darkness at 2 °C until analysis.

3.4.2 Cell Sorting

1. Remove medium from a T75 flask at approximately 70% cell confluence.
2. Wash the cells with 5 mL PBS.
3. Detach cells by adding 5 mL trypsin-EDTA and incubating at 37 °C for 5 min. Incubate for a few more minutes as necessary for the majority of cells to detach.
4. Add 5 mL 10% FBS.
5. Add the mixture to 15 mL conical tubes and centrifuge (350 × *g*) for 5 min at room temperature.
6. Decant supernatant and resuspend the cells in 1 mL 1% FBS.
7. Count cells by combining 10 μl of the cell suspension with 10 μl Trypan Blue. Add the 20 μl mixture to a hemocytometer to determine cell number.
8. Transfer 1–3 × 10^6 cells to each polystyrene round bottom tube in 300 μL 1% FBS (*see* **Note 2**).
9. Add antibody solution to each tube, typically 1–5 μL of 1 mg/mL antibody.
10. Vortex each sample to mix well, and then incubate at room temperature for 30 min. Labeled antibodies must be incubated in darkness.

11. Add 3 mL 1% FBS, mix, and centrifuge at 300 × *g* for 5 min at room temperature. Decant all but 50–100 μL supernatant. Resuspend the cell pellet in the mixture.
12. Maintain samples at 2 °C protected from light until processing.
13. Collect the cells directly into sterile round bottom polystyrene tubes containing 1 mL DMEM-F12 supplemented with 1% AA and 20% FBS. The total number of cells should be determined during the sorting process (*see* **Note 3**).
14. Add cells to 6-well culture plates at the desired seeding density (typically 5×10^3 cells/cm^2).

3.5 Cell Doubling

1. Seed cells in a 96 well plate at a density of 5000 cell/cm^2. There should be three replicates of the test sample and three stromal medium control replicates. Culture the cells under standard conditions (37 °C, 5% CO_2).
2. Allow sufficient time for the cells to attach (4–12 h), remove the medium and add 180 μL of fresh medium. The time between seeding and the first assessment must be consistent among trials. The cells should have sufficient time to adhere, but not increase in number.
3. Add 20 μL alamarBlue® to each well (final concentration 10%).
4. Culture the plate under standard conditions (37 °C, 5% CO_2) for 4 h.
5. Read the absorbance of each well at 570 and 600 nm (*see* **Note 4**).
6. Determine cell numbers from a standard curve of known cell numbers.
7. Repeat assessments at regular intervals to calculate cell doublings (CD) and doubling times (DT) according to standard formula, e.g., CD = ln(Nf/Ni)/ln(2) and DT = CT/CD (Ni: initial cell number; Nf: final cell number; CT = culture time).

3.6 Adipogenic Differentiation (Fig. 1)

1. Seed cells in a 24-well plate at a cell seeding density of 5×10^3 cells/cm^2 in stromal medium under standard conditions (37 °C, 5% CO_2). It is recommended to include three replicates of each test and control sample.
2. When cells reach 70–80% confluence, remove the medium and wash with 1 mL/well PBS.
3. Add 500 μL of adipogenic induction medium to each well.
4. Culture cells for 10 days (37 °C, 5% CO_2) with medium changes every 3–4 days.
5. At this point, cells should have visible lipid droplets in the cytoplasm. They can be harvested for genetic analysis as

described below or fixed for histochemical staining detailed in the proceeding steps.

6. Remove medium and fix cells in 500 μL/well 4% PFA for 20 min at room temperature.
7. Remove PFA and add 500 μL/well 0.3% Oil Red O and maintain at room temperature for 20 min.
8. Rinse cells three times with double-distilled water or until the rinse solution is clear.
9. Evaluate staining and colony morphology with microscopy.

3.7 Osteogenic Differentiation (Fig. 2)

1. Seed cells in a 24-well plate at a cell seeding density of 5×10^3 cells/cm^2 in stromal medium under standard conditions (37 °C, 5% CO_2). It is recommended to include three replicates of each test and stromal control sample.
2. When cells reach 70–80% confluence, remove the medium and wash with 1 mL/well PBS.
3. Add 500 μL of osteogenic induction medium to each well.
4. Culture cells for 10 days (37 °C, 5% CO_2) with medium changes every 3–4 days for 10 days.
5. Change culture medium to osteogenic maintenance medium, and culture for another 10 days with medium changes every 3–4 days.
6. Remove medium, and wash the cells with 150 mM NaCl solution.
7. At this point, mineralization should be apparent within the cell colonies. Cells can be harvested for genetic analysis as described below or fixed for histochemical staining detailed in the proceeding steps.
8. Remove NaCl and fix cells with ice-cold 70% EtOH (500 μL/well) for 1 h at room temperature.
9. Remove EtOH and add 500 μL/well 2% Alizarin Red. Maintain at room temperature for 45 min.
10. Rinse cells three times with double-distilled water or until the rinse solution is clear.
11. Evaluate staining and colony morphology with a light microscope.

3.8 Chondrogenic Differentiation

1. Seed cells in a 6-well plate at a cell seeding density of 5×10^3 cells/cm^2 in stromal medium and culture under standard conditions (37 °C, 5% CO_2). It is recommended to include three replicates of each test and control sample.
2. When cells reach 70–80% confluence, remove the medium and wash with 1–2 mL/well PBS.

3. Detach and count cells as described above (Subheading 3.2).
4. Prepare a cell suspension in stromal medium that contains about 1×10^6 cells/mL.
5. Transfer aliquots of 5×10^5 cells to 15 mL centrifuge tubes.
6. Centrifuge at $250 \times g$ for 5 min at room temperature form cell pellets.
7. Carefully remove the supernatant, add 1 mL chondrogenic differentiation medium to each tube and culture under standard culture conditions (37 °C, 5% CO_2) with media changes every 2–3 days for 21 days. Be sure to culture some pellets in complete stromal medium as no treatment controls. Tubes should remain upright with the cap loose to allow gas exchange.
8. At the end of the culture period, remove the medium and add 500 μl 10% formalin to each tube to fix the pellets. Tubes can be maintained at room temperature.
9. Pellets should be embedded in paraffin, sectioned (typically 5 μm), and stained with Alcian Blue and Nuclear Fast Red.

3.9 Neurogenic Differentiation (Fig. 3)

1. Seed cells in a 24-well plate at a cell seeding density of 5×10^3 cells/cm^2 in stromal medium under standard conditions (37 °C, 5% CO_2). It is recommended to include three replicates of each test and control sample.
2. When cells reach 60% confluence, remove the medium and wash with 1 mL/well PBS.
3. Add 500 μL of neurogenic induction medium to each well and culture under standard conditions (37 °C, 5% CO_2) for 24 h.
4. Change culture medium to neurogenic maintenance medium, and culture for another 3 days with one medium change.
5. At this point, cells should have assumed a neuronal morphology. Cells can be harvested for genetic analysis as described below or stained as detailed in the proceeding steps.
6. Fix cells in 300 μL 10% formalin for 20 min at room temperature.
7. Remove formalin and stain cells with 300 μL Cresyl Violet at room temperature for 20–45 min in darkness.
8. Remove Cresyl Violet and rinse cells three times with double-distilled water or until the rinse solution is clear.
9. Add 200 μL double-distilled water to each well to maintain moisture.
10. Evaluate staining and cell morphology with microscopy.

3.10 Colony Forming Unit Assays [1]

1. Seed cells in a T75 culture flask at a cell seeding density of 5×10^3 cells/cm^2 in stromal medium and culture under standard conditions (37 °C, 5% CO_2).
2. When cells reach 70–80% confluence, remove the medium and wash with 2–3 mL PBS.
3. Detach and count cells as described above (Subheading 3.2).
4. Prepare a cell suspension in stromal medium that contains about 1×10^6 cells/mL.
5. Prepare a series of serial dilutions (1:2) in stromal medium to give 1 mL aliquots corresponding to six cell concentrations, e.g. 5×10^3, 2.5×10^3, 1.25×10^3, 6.25×10^2, 3.12×10^2 and 1.56×10^2 cells/100 μL (*see* **Note 5**).
6. Add 100 μL of each cell concentration to eight replicate wells.
7. Culture under standard conditions (37 °C, 5% CO_2) for 10 days with medium changes every 3–4 days.
8. For colony forming unit frequency fibroblastic (CFU-F), remove stromal medium, rinse each well with 100 μL of PBS, fix the cells with 100 μL 4% PFA for 20 min at room temperature and stain with 100 μL 0.1% Toluidine Blue for 1 h at room temperature.
9. For CFU-adipocytic (CFU-Ad), culture cells for 10 days (37 °C, 5% CO_2) in adipogenic induction medium with medium changes every 3–4 days. Remove the medium and fix cells in 500 μL/well 4% PFA for 20 min at room temperature. Remove the PFA and add 500 μL/well 0.3% Oil Red O. Maintain at room temperature for 20 min. Rinse cells three times with double-distilled water or until the rinse solution is clear.
10. For CFU-Osteoblastic (CFU-Ob), add 100 μL of osteogenic induction medium to each well. Culture cells for 10 days (37 °C, 5% CO_2) with medium changes every 3–4 days. Change culture medium to osteogenic maintenance medium, and culture for another 10 days with medium changes every 3–4 days. Remove medium, and wash the cells with 150 mM NaCl solution. Fix cells with 70% EtOH (100 μL/well) for 1 h at room temperature. Remove EtOH and add 500 μL/well 2% Alizarin Red. Maintain at room temperature for 45 min. Rinse cells three times with double-distilled water or until the rinse solution is clear.
11. Count the number of stained colonies in each well. For CFU-F, CFU-Ad and CFU-Ob, wells are considered positive if there are ≥10 Toluidine Blue-stained colonies, ≥10 Oil Red O-stained colonies, or ≥1 Alizarin Red-stained colonies, respectively.

12. The CFU frequencies can be estimated with Poisson's ratio: $F = e^{-x}$ (*F*: ratio of negative to total wells within a row, *e*: natural logarithm constant 2.71, *x*: CFU per well) [8].

3.11 mRNA Isolation

1. Grow and detach cells as described above (Subheading 3.2) (*see* **Note 6**).
2. Add the mixture of cells, trypsin-EDTA and FBS to 15 mL conical tubes and centrifuge (350 × *g*) for 5 min at room temperature.
3. Remove solution and add 350 μL Buffer RLT Plus at room temperature. Vortex for 30 s or until mixture is homogenous.
4. Transfer the homogenized lysate to a gDNA eliminator spin column placed in a 2 mL collection tube.
5. Centrifuge for 30 s at a minimum of 8000 × *g* at room temperature.
6. Discard the column and save the flow-through.
7. Add 350 μL 70% ethanol and mix well by pipetting.
8. Transfer the sample to an RNeasy spin column in a 2 mL collection tube.
9. Centrifuge for 15 s at a minimum of 8000 × *g* at room temperature. Discard the flow-through.
10. Add 700 μL Buffer RW1 to the column and centrifuge for 15 s at a minimum of 8000 × *g* at room temperature. Discard the flow-through.
11. Add 500 μL Buffer RPE to the column and centrifuge for 2 min at a minimum of 8000 × *g* at room temperature.
12. Place the RNeasy spin column in a fresh 1.5 mL nuclease-free tube. Add 50 μL RNase-free water directly to the spin column membrane.
13. Centrifuge for 1 min at a minimum of 8000 × *g* at room temperature to elute the RNA.
14. Quantify the mRNA concentration spectrophotometrically. The 260/280 ratio should be around 1.90. The mRNA solution can be stored −80 °C.

4 Notes

1. Cells can be resuspended in 5 mL PBS and centrifuged again (350 × *g* for 5 min) to wash them prior to seeding if desired.
2. Make sure to include appropriate controls including cells in PBS alone and secondary antibody alone (isotype control).
3. Depending on the number of cells that are collected, a PBS wash may be added between the sorting and seeding steps by centrifuging the cells (350 × *g* for 5 min), resuspending the

pellet in 1 mL PBS, centrifuging again (350 × *g* for 5 min) and then resuspending the cells in culture medium.

4. A sample formula to calculate percent alamarBlue® reduction in each well is provided for consideration. Percent reduction of alamarBlue® = (((117,216 × test well absorbance at 570) − (80,586 × test well absorbance at 600))/((155,677 × well with media plus alamarBlue® but no cells absorbance at 600) − (14,652 × well with media plus alamarBlue® but no cells absorbance 570))) × 100.
5. Dilution range and replicate numbers can be customized for the number of available cells and anticipated CFU numbers.
6. mRNA can be isolated from cells grown in induction medium as well. Volumes should be adjusted accordingly.

Acknowledgments

The authors gratefully acknowledge funding by the Winn Feline Foundation. Many thanks to members of the Laboratory for Equine and Comparative Orthopedic Research at Louisiana State University, School of Veterinary Medicine for their assistance with the preparation of this work

References

1. Zhang N, Dietrich MA, Lopez MJ (2014) Therapeutic doses of multipotent stromal cells from minimal adipose tissue. Stem Cell Rev 10:600–611
2. Kalaszczynska I, Ruminski S, Platek AE et al (2013) Substantial differences between human and ovine mesenchymal stem cells in response to osteogenic media: how to explain and how to manage? Biores Open Access 2:356–363
3. Quimby JM, Webb TL, Habenicht LM et al (2013) Safety and efficacy of intravenous infusion of allogeneic cryopreserved mesenchymal stem cells for treatment of chronic kidney disease in cats: results of three sequential pilot studies. Stem Cell Res Ther 4:48
4. Quimby JM, Webb TL, Randall E et al (2016) Assessment of intravenous adipose-derived allogeneic mesenchymal stem cells for the treatment of feline chronic kidney disease: a randomized, placebo-controlled clinical trial in eight cats. J Feline Med Surg 18(2):165–171
5. Trzil JE, Masseau I, Webb TL et al (2014) Long-term evaluation of mesenchymal stem cell therapy in a feline model of chronic allergic asthma. Clin Exp Allergy 44:1546–1557
6. Volarevic V, Arsenijevic N, Lukic ML et al (2011) Concise review: mesenchymal stem cell treatment of the complications of diabetes mellitus. Stem Cells 29:5–10
7. Gimble JM, Nuttall ME (2011) Adipose-derived stromal/stem cells (ASC) in regenerative medicine: pharmaceutical applications. Curr Pharm Des 17:332–339
8. Spencer ND, Chun R, Vidal MA et al (2012) In vitro expansion and differentiation of fresh and revitalized adult canine bone marrow-derived and adipose tissue-derived stromal cells. Vet J 191:231–239
9. Quimby JM, Webb TL, Gibbons DS et al (2011) Evaluation of intrarenal mesenchymal stem cell injection for treatment of chronic kidney disease in cats: a pilot study. J Feline Med Surg 13:418–426

Chapter 16

Canine Adult Adipose Tissue-Derived Multipotent Stromal Cell Isolation and Characterization

Wei Duan and Mandi J. Lopez

Abstract

Adult multipotent stromal cells (MSCs) are increasingly popular for direct therapeutic applications and bioengineering. Canine patients constitute a major component of veterinary practice, and the dog is an established preclinical animal model for numerous traumatic, degenerative, and disease conditions. Current information supports the presence and relative abundance of adipose tissue-derived multipotent stromal cells (ASCs) in various canine adipose tissue depots. Refined isolation and characterization techniques contribute to collective knowledge of ASC phenotypes and subpopulations for customized, targeted applications. Continued efforts to augment understanding of canine ASCs is critical to progressive treatment advances and high-impact study outcomes. This chapter contains a description of techniques to isolate and characterize canine ASCs.

Key words Canine, Dog, Adipose, Adult multipotent stromal cell, Bioengineering

1 Introduction

Adipose tissue is an established source of canine adult multipotent stromal cells. Current information documents the expansion potential, plasticity, and cryopreservation of canine adult adipose-tissue-derived multipotent stromal cells (ASCs) [1, 2]. Comparisons to canine MSCs from bone marrow confirm similar cell expansion rates and multipotentiality [1]. Similarly, immunophenotype, growth rate, and differentiation capacity of MSCs isolated from stifle (knee) intra-articular tissues support that ASCs from infrapatellar adipose tissue have the highest in vitro expansion rates, differentiation capacity, and percentages of MSC immunophenotypes [2]. Hence, canine adipose tissue has tremendous potential as a viable and accessible tissue source of ASCs intended for both therapeutic and investigative purposes [3].

Undifferentiated cells from different tissue harvest sites vary in capacity to produce tissue-specific extracellular matrix [4]. Differences are attributed to epigenetic influences as well as cell

Bruce A. Bunnell and Jeffrey M. Gimble (eds.), *Adipose-Derived Stem Cells: Methods and Protocols*, Methods in Molecular Biology, vol. 1773, https://doi.org/10.1007/978-1-4939-7799-4_16, © Springer Science+Business Media, LLC, part of Springer Nature 2018

phenotype and maturity, among other factors [5]. It is also becoming increasingly apparent that primary tissue isolates contain heterogeneous cells that may vary widely in both in vitro and in vivo behaviors [6]. The International Society for Cellular Therapy (ISCT) recently expanded the minimum criteria for MSC status to include both protein and genetic confirmation of differentiation in addition to histochemical staining, surface marker expression, and plastic affinity [7, 8]. This reflects a growing awareness of distinctions among cells that retain the potential to assume multiple tissue lineages. Efforts to standardize isolation and identification methods are crucial to developing effective treatment strategies and bioengineering protocols. The information in this chapter includes current protocols for canine ASC isolation and evaluation.

2 Materials

2.1 Adipose Tissue Collection and ASC Isolation

1. Scalpel handles (#3, 4) and scalpel blades (#10, 20).
2. Forceps (Brown-Adson).
3. 70% ethyl alcohol (EtOH).
4. 50-mL centrifuge tubes.
5. Petri dishes.
6. Phosphate buffered saline (PBS).
7. 1% bovine serum albumin (BSA) dissolved in double-distilled water. Filter-sterilize with a 0.22 μm filter. Warm to 37 °C.
8. 0.1% Collagenase: collagenase type-1 and 1% BSA dissolved in double distilled water. Filter-sterilize with a 0.22 μm filter. Warm to 37 °C. Make a fresh solution for each use if there is over 24 hours between.
9. Red cell lysis buffer.
10. Complete stromal medium: DMEM-Ham's F12 medium supplemented with 1% antibiotic/antimycotic solution and 20% characterized fetal bovine serum (FBS; *see* **Note 1**).
11. Trypan Blue.

2.2 Cryopreservation of ASCs

1. Cryopreservation medium: 80% FBS, 10% DMEM/F-12, 10% dimethysulfoxide (DMSO).
2. Mr. Frosty™ (*see* **Note 2**) or CoolCell® freezing container.
3. Cryovials.

2.3 Cell Culture

1. Complete stromal medium: DMEM-Ham's F12 medium supplemented with 1% antibiotic/antimycotic solution and 20% characterized FBS.
2. Tissue culture flasks (T-25 or T-75).

2.4 Cell Doubling

1. PBS.
2. 0.05% trypsin (1:250): 0.05% porcine trypsin in Hank's balanced salt solution (HBSS) with 0.2 g/L ethylenediaminetetraacetic acid (EDTA). Aliquots can be stored in microtubes at −20 °C.

3. Trypan Blue.
4. Hemocytometer.
5. Complete stromal medium: DMEM-Ham's F12 medium supplemented with 1% antibiotic/antimycotic solution and 20% characterized FBS.
6. 12-well culture plates.

2.5 Adipogenic Differentiation

1. DMEM-Ham's F-12 medium.
2. FBS.
3. Antibiotic/antimycotic solution.
4. Biotin (33 μmol/L).
5. Pantothenate (17 μmol/L).
6. Dexamethasone (1 μmol/L).
7. Indomethacin (100 μmol/L).
8. Insulin (1 μmol/L).
9. Isobutylmethylxanthine (IBMX, 0.5 mmol/L).
10. Rosiglitazone (TZD, 5 μmol/L).
11. 4% formalin in PBS.
12. Oil Red-O.
13. Double-distilled water.

2.6 Osteogenic Differentiation

1. Complete stromal medium: DMEM-Ham's F12 medium supplemented with 1% antibiotic/antimycotic solution and 20% characterized FBS.
2. Dexamethasone (10 nmol/L).
3. β-glycerophosphate (10 mmol/L).
4. Sodium 2-phosphate ascorbate (50 μg/mL).
5. 70% EtOH.
6. 150 mM NaCl dissolved in double-distilled water.
7. 2% Alizarin Red solution. Dissolve Alizarin Red (2 g/100 mL) in doubled-distilled water. Adjust pH to 4.1–4.3 with dilute NaOH. Filter the solution through 0.2-μm syringe filter. The solution can be stored at room temperature covered with aluminum foil.

2.7 Chondrogenic Differentiation

1. Complete stromal medium: DMEM-Ham's F12 medium supplemented with 1% antibiotic/antimycotic solution and 20% characterized FBS.
2. Chondrogenic differentiation medium; DMEM-Ham's F12 medium supplemented with 10% FBS, ascorbate phosphate (50 μg/mL), dexamethasone (100 nmol/L), insulin–transferrin–selenium (ITS, 1%), recombinant

human transforming growth factor-β3 (rTGF-β3, 10 ng/mL), sodium pyruvate (2 mmol/L).

3. 10% formalin.
4. Paraffin.
5. 1% Alcian Blue.
6. 3% Nuclear Fast Red.

2.8 Papain Digestion

1. Papain from *Carica papaya* in double distilled water (25 mg/mL).
2. Solution A: 10 mM disodium EDTA and 0.4 M sodium acetate in double-distilled water.
3. Solution B: 200 mM L-cysteine in double-distilled water.
4. Solution C: Add 1 mL of solution B to 9 mL of solution A just before use (1,10 dilution).
5. Solution D: Add 200 μL of papain suspension to 10 mL solution C.

2.9 Hydroxyproline Assay

1. Double-distilled water.
2. Clear, flat-bottom 96-well ELISA plate.
3. 12 N hydrochloric acid (HCl). Dilute 1:2 in double distilled water for 6 N working solution.
4. 17% sodium chloride (NaCl) dissolved in double-distilled water.
5. Acetate–citrate buffer: 120 g sodium acetate trihydrate, 12 mL acetic acid, 50 g citric acid monohydrate, 34 g NaOH in 1 L double-distilled water (pH 6.0). Store at 4 °C.
6. Oxidant solution: 0.178 g chloramine-T in 15 mL of isopropanol and 10 mL of double-distilled water; add 25 mL acetate–citrate buffer. Make fresh (*see* **Note 3**).
7. Ehrlich's reagent: 1 g dimethylaminobenzaldehyde in 20 mL isopropanol; combine with 6.6 mL perchloric acid and 15.6 mL double distilled water. Make fresh (*see* **Note 3**).
8. Trans-4-hydroxy-L-proline (1 mg/mL) in double-distilled water. Store at 4 °C.

2.10 PicoGreen Double Stranded DNA Assay

1. TE buffer (20×): 200 mM Tris–HCl, 20 mM EDTA, pH 7.5; dilute 1:20 in sterile, distilled, DNase-free water for a 1× TE working solution.
2. PicoGreen® dsDNA reagent; dilute 1:200 in 1× TE. Store at 4 °C protected from light. Make fresh just before use.
3. Black, flat-bottom 96-well microtiter plate.
4. Bacteriophage lambda DNA stock (100 μg/mL in TE); dilute 1:50 in 1× TE to 2 μg/mL; Make serial dilutions from 1 ng/mL to 1 μg/mL for the standard curve. Be sure to include a blank

sample of TE buffer alone. Store at 4 °C protected from light. Make fresh just before use.

2.11 Lowry Total Protein Assay

1. Biuret reagent (0.5 mL 1% cupric sulfate, 0.5 mL 2% sodium potassium tartrate, 50 mL 2% sodium carbonate in 0.1 N NaOH).
2. 1 N Folin and Ciocalteu's reagent.
3. BSA (1 mg/mL) in distilled water.

2.12 1,9-Dimethylmethylene Blue (DMMB) Total Sulfated Proteoglycan Assay

1. 1,9-Dimethylmethylene Blue (DMMB) assay buffer: Dissolve 21 mg of DMMB in 5 mL of absolute EtOH. In a separate beaker, dissolve 2 g of sodium formate in 980 mL of distilled water. Adjust the pH to 3.5 with formic acid. Combine the solutions and bring the total volume to 1 L with double-distilled water. The DMMB assay buffer may be stored at 4 °C.
2. Shark chondroitin-6-sulfate C (1 mg/mL) in double-distilled water. Prepare standards, usually 100, 50, 40, 30, 20, 10, 5, and 0 μg/mL. Standards can be stored at −25 °C.

2.13 Transmission Electron Microscopy (TEM) Sample Preparation (Fig. 1)

1. PBS.
2. T75 culture flasks.
3. 2% paraformaldehyde.
4. Trypsin–EDTA. Store long term at −20 °C.
5. 5% sucrose in 0.1 M sodium cacodylate (CAC) buffer.

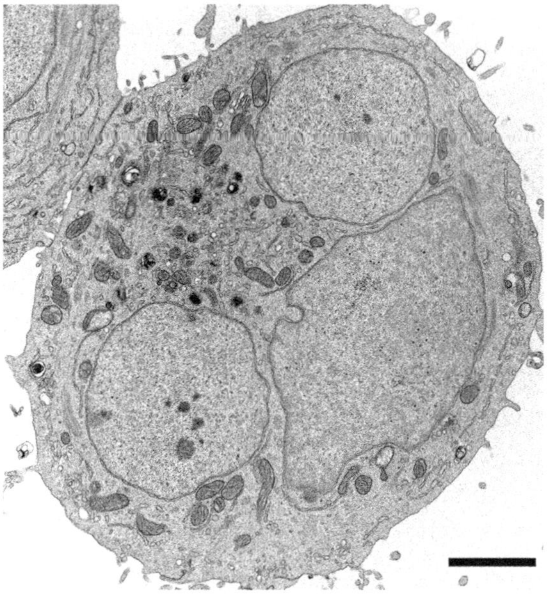

Fig. 1 Transmission electron photomicrograph of a canine adipose tissue-derived multipotent stromal cell. Scale bar = 2 μm

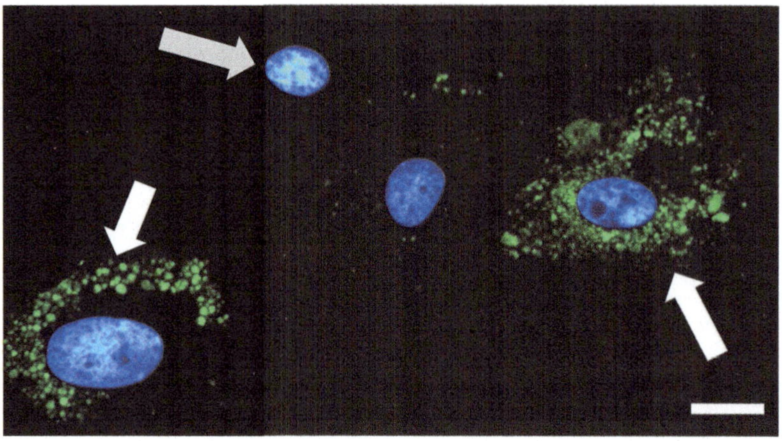

Fig. 2 Fluorescent photomicrograph demonstrating CD 29 immunocytochemical labeling of canine adipose tissue-derived multipotent stromal cells. The white arrows indicate labeled cells, and the grey arrow points to an unlabeled cell. Scale bar = 50 μm

6. 1.25% glutaraldehyde in CAC buffer (pH 7.4).
7. 1% osmium tetroxide in CAC buffer.
8. 70%, 80%, 95%, and 100% ethanol–distilled water solutions.
9. EPON™ epoxy resin.

2.14 Immunocytochemistry (Fig. 2)

1. Complete stromal medium: DMEM/F-12 supplemented with 1% antibiotic/antimycotic solution and 20% FBS.
2. 4-well chamber slides.
3. 10% formalin.
4. TBS buffer (100 mM Tris, 138 mM NaCl, 27 mM KCl, pH 7.4).
5. SDS/TBS antigen retrieval buffer (1% SDS in TBS buffer).
6. Goat serum.
7. Labeled antibodies.
8. 4′,6-Diamidino-2-phenylindole (DAPI) solution (1 μg/mL).

2.15 Protein Isolation for Western Blot

1. T75 culture flasks.
2. PBS.
3. RIPA buffer (150 mM NaCl, 1.0% Triton X-100, 0.5% sodium deoxycholate, 0.1% SDS, 50 mM Tris, 1× protease inhibitor).
4. Cell lifter.
5. Pierce™ bicinchoninic acid (BCA) protein assay kit.
6. 2 mg/mL BSA in 0.9% saline and 0.05% sodium azide.

3 Methods

Characterization of ASC isolates is critical to both clinical and investigative applications. In vitro expansion rates determine the culture time required to produce targeted cell doses. Confirmation of trilineage differentiation confirmed with histochemical staining is one of the cornerstones of multipotent stromal cell identification. Compositional analysis of tissue specific protein levels normalized to DNA content as a representation of cell number further delineates lineage differentiation capacities among cell preparations, especially for chondrogenesis. Determination of the presence and absence of cell surface marker protein expression is necessary to distinguish among MSC phenotypes. Finally, cell ultrastructure is yet another measure of cell identity. Examples of standard approaches to these contemporary measures are provided as a basis for procedures that can be customized according to specific interests.

3.1 Adipose Tissue Collection and ASC Isolation [1, 2]

1. Harvest adipose tissue from the appropriate site (subcutaneous, infrapatellar, etc.) using aseptic technique.
2. Place the tissue in a preweighed, sterile petri dish to determine tissue weight.
3. Rinse the tissue with sterile PBS.
4. Mince the tissue using a sterile scalpel and blade until the tissue is no longer fibrous.
5. Place the tissue in a 50 mL tube with an equal volume of prewarmed PBS and agitate vigorously for 30 s.
6. Allow sample to separate into phases for 3 min.
7. Remove the infranatant (lower, aqueous phase, *see* **Note 4**) and any solid pieces at the bottom of the tube.
8. Wash the remaining tissue with PBS. Specifically, suspend the tissue in PBS, agitate for 45 s, and then remove the infranatant after the adipose tissue floats to the top. Continue this process until the infranatant is clear.
9. After removing the last infranatant wash, add a volume of 0.1% collagenase type I solution equal to the adipose tissue volume, and agitate at about 75 rpm for 90 min at 37 °C or until the solution becomes homogenous. Alternatively, maintain the solution at 37 °C for 90 min, vortexing every 10–15 min.
10. Filter the mixture with a 100 μm cell strainer, and then centrifuge the solution at 260 × *g* for 5 min to form the stromal vascular fraction (SVF) pellet.

11. Carefully remove the supernatant, consisting of lipids, adipocytes, and collagenase, without disturbing the SVF pellet. It is okay to leave some fluid to avoid disrupting the pellet.
12. Resuspend the cells in an equal volume of 1% BSA (*see* **Note 5**).
13. Centrifuge solution at 260 × *g* for 5 min, and remove the supernatant. When aspirating, the tip of the pipette should remain at the top of the fluid to remove the lipid layer.
14. Resuspend the pellet in 1 mL of complete stromal medium.
15. Centrifuge at 260 × *g* for 5 min.
16. Carefully aspirate and discard the supernatant.
17. Resuspend the pellet in an equal volume of red lysis buffer, and maintain at room temperature for 5 min.
18. Centrifuge cell suspension at 260 × *g* for 5 min.
19. Resuspend the pellet in complete stromal medium
20. Remove a small aliquot (10 μL) and combine with an equal volume of Trypan Blue.
21. Determine cells/mL with a hemocytometer.
22. Plate the cells at the appropriate density in complete stromal medium.

3.2 Cryopreservation of ASCs

1. Centrifuge tubes at 260 × *g* for 5 min to pellet cells.
2. Discard supernatant and resuspend cells in freezing medium at 1–2 × 10^6 cells/mL in a cryovial.
3. Store vial in a freezing container at −80 °C for 24 h, and then transfer vials to liquid nitrogen for long-term storage.

3.3 Cell Culture

1. Remove the medium from the flasks and rinse the attached cells with PBS.
2. Add enough trypsin–EDTA to cover the surface of the flask (3–4 mL) and incubate at 37 °C for 5 min.
3. Use an inverted, light microscope to confirm that the cells are detaching. Once the cells have detached, add an equal volume of complete stromal medium to the flask. The FBS in the medium will stop the enzymatic action of the trypsin.
4. Place the cell suspension in a 15-mL tube and centrifuge for 5 min at 260 × *g*.
5. After discarding the supernatant, resuspend the cell pellet in a known volume of complete stromal medium.
6. Remove a small aliquot of the cell suspension (10 μL) and combine with an equal volume of Trypan Blue.
7. Determine cells/mL with a hemocytometer.

8. Seed the cells at the desired density, typically 3–5 × 10^3 cells/cm^2.
9. Incubate flasks at 37 °C in 5% CO_2.
10. Change the medium every 2–3 days until cells have reached 70–80% confluence, and then repeat the cell passage procedure.

3.4 Cell Doubling

1. Repeat steps for cell passage up to determining cell number with a hemocytometer.
2. Calculate cell doubling time (DT) and doubling number (CD) using the formulae:

$CD = \ln(N_f/N_i)/\ln(2)$ and $DT = CT/CD$ where CT = culture time, N_f = final cell number, and N_i = initial cell number.

3.5 Adipogenic Differentiation

1. Detach cells with trypsin–EDTA and count them.
2. Seed the cells at the desired density, typically 3–5 × 10^3 cells/cm^2, and culture in complete stromal medium under standard conditions (37 °C, 5% CO_2) for 7 days with media changes every 2–3 days until the cells are 70–80% confluent.
3. Remove the medium and add adipogenic induction medium.
4. Culture cells in adipogenic induction medium for 21 more days with media changes every 2–3 days.
5. At the end of the culture period, remove adipogenic induction medium, and then fix the cells with 4% formalin for 20 min at room temperature.
6. Remove the formalin and rinse cells three times with distilled water.
7. Remove the excess fluid, and allow the wells to air dry. Stain the cells with Oil Red O at room temperature for 20 min.
8. Rinse with distilled water three times to remove any residual dye.

3.6 Osteogenic Differentiation

1. Detach cells with trypsin–EDTA and count them.
2. Seed the cells at the desired density, typically 3–5 × 10^3 cells/cm^2, and culture in complete stromal medium under standard conditions (37 °C, 5% CO_2) for 7 days with media changes every 2–3 days until the cells are 70–80% confluent.
3. Remove the medium and culture cells in osteogenic differentiation medium for 21 more days. Change the medium every 2–3 days.
4. At the end of the culture period, remove osteogenic induction medium, wash with 150 mM NaCl solution three times, and then fix the cells with 70% EtOH for 1 h at 4 °C.

5. Prepare a 2% Alizarin Red solution in distilled water. Adjust pH to 4.1–4.3 with dilute NaOH.
6. Stain the cells with Alizarin Red at room temperature for 10 min.
7. Rinse with distilled water five times to remove residual dye.

3.7 Chondrogenic Differentiation

1. Detach cells with trypsin–EDTA and count them. Aliquots of 2.5×10^5 cells are required for chondrogenic pellet culture.
2. Centrifuge cell aliquot suspended in complete stromal medium at $250 \times g$ for 5 min at room temperature in a 1.5-mL microcentrifuge tube to form a pellet.
3. Remove the supernatant and add 1 mL chondrogenic differentiation medium to the tube and culture under standard culture conditions (37 °C, 5% CO_2) with media changes every 2–3 days.
4. The tube should remain upright and the cap should be loose enough to allow gas exchange.
5. The pellet should be ready for harvest after 14–21 days of culture.
6. After the desired culture period, remove the chondrogenic differentiation medium and fix the pellet in 10% formalin overnight at 4 °C.
7. The pellet can be embedded in paraffin for sectioning (3–5 μm) and staining with Alcian Blue and Nuclear Fast Red.

3.8 Papain Digestion

1. Add enough solution D to tissue samples to give about 10 mL of solution per 1 mg of sample.
2. Incubate at 60 °C overnight. If samples are not homogenous after vortexing, return them to the incubator. Repeat until samples are at least 70% homogenous (*see* **Note 6**).
3. Papain digested samples can be stored at −80 °C until needed.

3.9 Hydroxyproline Total Collagen Assay [9]

1. Dilute samples in an equal volume of 6 N HCl and incubate overnight or 18 h at 110 °C (*see* **Notes** 7 and **8**).
2. Prepare a series of 1:2 serial dilutions, typically 0–20 μg/mL, of trans-4-hydroxy-L-proline in distilled water for the standard curve.
3. Add 250 μL each of distilled water and 17% NaCl to each sample and mix thoroughly.
4. Add 500 μL of oxidant solution to samples and standards and incubate at room temperature for 5 min.
5. After incubation, add 500 μL of Ehrlich's reagent to each sample and standard and incubate at 60 °C for 12 min.

6. Cool samples and standards on ice for 4 min.
7. Pipet 100 μL of each sample and standard onto a microtiter plate.
8. Read absorbance at 550 nm.

3.10 PicoGreen Double Stranded DNA Assay [10]

1. Prepare bacteriophage lambda DNA standard curve samples in 1× TE buffer, typically 1 ng/mL to 1 μg/mL.
2. Samples can be diluted in 1× TE buffer so that they are within the standard curve.
3. Add 100 μL of samples and standards to a black, flat bottom, 96-well microtiter plate.
4. Add 100 μL of 1× PicoGreen® dsDNA reagent and incubate for 2–5 min in the dark.
5. The fluorescence of each well should be read at an excitation wavelength of 480 nm and an emission wavelength of 520 nm.

3.11 Lowry Total Protein Assay [11]

1. Prepare a series of serial dilutions of BSA, usually 0–40 μg/100 μL.
2. Transfer 100 μL of each sample and standard to a flat-bottom, microtiter plate.
3. Samples can be diluted with distilled water so that they are within the standard curve.
4. Add 200 μL of Biuret reagent to each well and mix thoroughly by pipetting several times.
5. Allow plate to incubate for 10–15 min at room temperature.
6. Add 20 μL/well of 1.0 N Folin and Ciocalteu's reagent. Mix thoroughly with repeated pipetting.
7. Allow color to develop for 30 min at room temperature.
8. Immediately read the plate at 650 nm absorbance.

3.12 1,9-Dimethylmethylene Blue (DMMB) Total Sulfated Glycosaminoglycan Assay [12]

1. Prepare a series of serial dilutions of chondroitin-6-sulfate in double-distilled water, usually 0–100 μg/mL.
2. Add 40 μL of each sample and standard to a flat-bottom, microtiter plate.
3. Add 250 μL of the DMMB assay buffer to each sample and standard.
4. Immediately read the plate at 530 nm immediately (*see* **Note 9**).

3.13 Transmission Electron Microscopy (TEM) Sample Preparation (Fig. 1)

1. Remove the medium from the flasks and rinse the cells (minimum of 1 × 10^6 cells, about 75% confluent) with PBS.
2. Add 3–4 mL trypsin–EDTA (enough to cover the surface of the flask) and incubate at 37 °C for 5 min.

3. Decant the solution into a 15-mL tube and centrifuge at 260 × *g* for 5 min at room temperature.
4. Resuspend the cells in 5 mL PBS and centrifuge at 260 × *g* for 5 min at room temperature in a 15-mL tube.
5. Resuspend the cells in 2 mL 2% paraformaldehyde and centrifuge at 260 × *g* for 10 min at room temperature in a 5-mL tube.
6. Decant the solution.
7. Incubate the pellet in 1.25% glutaraldehyde at room temperature for 24 h.
8. Wash the pellet with 5% sucrose in 0.1 M CAC buffer. Be careful not to disrupt pellet at this and the succeeding steps.
9. Incubate the pellet in 1% osmium tetroxide for 2 h at room temperature.
10. Dehydrate the pellet in a series of ethanol–distilled water solutions by soaking in each concentration (low to high) at room temperature for at 30 min.
11. Embed the pellet in Epon and prepare ultrathin sections for transmission electron microscopy.

3.14 Immunocy tochemistry (Fig. 2)

1. Culture the cells in complete stromal medium within 4-well chamber slides for 3 days or until about 50% confluent.
2. Remove the medium and rinse cells with PBS.
3. Fix the cells with 10% formalin (200 μL/well) for 24 h at room temperature.
4. Remove the formalin and add SDS/TBS antigen retrieval buffer (200 μL/well) for 5 min at room temperature.
5. Remove the buffer and then incubate with goat serum (1:10 in TBS buffer, 200 μL/well) overnight at 4 °C.
6. Add desired antibodies (e.g., CD34-PE 1:200 in TBS buffer) to goat serum in each well and incubate for an additional 2 h at room temperature.
7. Remove the solution from each well and wash the cells with TBS buffer.
8. Add 200 μL/well DAPI solution and incubate at room temperature for 30 min in darkness.
9. Use an inverted fluorescent microscope to view and image cells.

3.15 Protein Isolation for Western Blot

1. Remove the medium from the flasks. Cells should be about 80–90% confluent.
2. Place the flasks in ice and wash the cells with ice-cold PBS.
3. Remove the PBS, add ice-cold RIPA buffer (enough to cover the flask surface), and use the cell lifter to disrupt the cells.

4. Pour the cell suspension into a precooled 1.5 mL microfuge tube and maintain it at 4 °C for 30 min, vortexing every 5–10 min.
5. Centrifuge cell suspension at $12 \times 10^3 \times g$ for 20 min at 4 °C.
6. Aspirate the supernatant and place it into a clean 1-mL tube. Discard the pellet
7. Prepare the BCA working reagent by mixing BCA reagent A and reagent B 50:1.
8. Prepare a series of 1:2 serial BSA dilutions in double-distilled water, usually 0–2.0 mg/mL.
9. Pipette 25 μL of each standard and sample into separate wells of a 96-well plate.
10. Samples can be diluted in double-distilled water if necessary.
11. Add 200 μL of the BCA working reagent to each well and mix well by pipetting.
12. Incubate the plate at 37 °C for 30 min.
13. Allow the plate to cool to room temperature and measure the absorbance at 562 nm.
14. Protein samples can be stored at −80 °C until needed for analysis.

4 Notes

1. Complete stromal media can be made with 10% FBS instead of 20% FBS. We used 20% FBS in our initial cell culture and continued its use to remain constant throughout the experiments.
2. Mr. Frosty freezing container can be used up to 5 times before the isopropanol needs to be changed.
3. The oxidant solution and Ehrlich's reagent can be used up to 2 days after initial use if stored at 4 °C.
4. We use a small gauge, 5 in. needle to remove the infranatant after each wash step. This makes it easy to remove the liquid without losing some of the sample.
5. Transferring the solution to a new centrifuge tube helps to reduce the amount of free lipid in the sample.
6. If pieces of tissue still remain after repeated incubation at 60 °C, centrifuge the sample and collect the supernatant.
7. Use of screw cap tubes helps prevent the tubes from popping open when exposed to high temperature.
8. The sample may have a charred appearance after incubating at 110 °C. Mix the sample well to dissolve.
9. Read the plate immediately to prevent precipitation of the sample and inaccurate results.

Acknowledgments

The authors thank the members of the Laboratory for Equine and Comparative Orthopedic Research at Louisiana State University, School of Veterinary Medicine for technical assistance.

References

1. Spencer ND, Chun R, Vidal MA et al (2012) In vitro expansion and differentiation of fresh and revitalized adult canine bone marrow-derived and adipose tissue-derived stromal cells. Vet J 191:231–239
2. Zhang N, Dietrich MA, Lopez MJ (2013) Canine intra-articular multipotent stromal cells (MSC) from adipose tissue have the highest in vitro expansion rates, multipotentiality, and MSC immunophenotypes. Vet Surg 42:137–146
3. Ostrander EA, Galibert F, Patterson DF (2000) Canine genetics comes of age. Trends Genet 16:117–124
4. Xie L, Zhang N, Marsano A et al (2013) In vitro mesenchymal trilineage differentiation and extracellular matrix production by adipose and bone marrow derived adult equine multipotent stromal cells on a collagen scaffold. Stem Cell Rev 9:858–872
5. Eilertsen KJ, Floyd Z, Gimble JM (2008) The epigenetics of adult (somatic) stem cells. Crit Rev Eukaryot Gene Expr 18:189–206
6. de Mattos Carvalho A, Alves ALG, Golim MA et al (2009) Isolation and immunophenotypic characterization of mesenchymal stem cells derived from equine species adipose tissue. Vet Immunol Immunopathol 132:303–306
7. Bourin P, Bunnell BA, Casteilla L et al (2013) Stromal cells from the adipose tissue-derived stromal vascular fraction and culture expanded adipose tissue-derived stromal/stem cells: a joint statement of the International Federation for Adipose Therapeutics and Science (IFATS) and the International Society for Cellular Therapy (ISCT). Cytotherapy 15:641–648
8. Dominici M, Le BK, Mueller I et al (2006) Minimal criteria for defining multipotent mesenchymal stromal cells. The International Society for Cellular Therapy position statement. Cytotherapy 8:315–317
9. Stegemann H, Stalder K (1967) Determination of hydroxyproline. Clin Chim Acta 18:267–273
10. Labarca C, Paigen K (1980) A simple, rapid, and sensitive DNA assay procedure. Anal Biochem 102:344–352
11. Lowry OH, Rosebrough NJ, Farr AL et al (1951) Protein measurement with the Folin phenol reagent. J Biol Chem 193:265–275
12. Ratcliffe A, Billingham ME, Saed-Nejad F et al (1992) Increased release of matrix components from articular cartilage in experimental canine osteoarthritis. J Orthop Res 10:350–358

Chapter 17

Soft Tissue Reconstruction

Francesco Egro and Kacey G. Marra

Abstract

Autologous fat transplantation has revolutionized soft tissue reconstruction, but conventional methods remain unpredictable as graft resorption rates are high due to lack of vascularization. The advent of adipose-derived stem cells (ASCs) has led to improvement of fat grafting outcomes, in part to their ability to undergo facile differentiation into adipose tissue, their angiogenic properties, and their ability to express and secrete multiple growth factors. This chapter discusses the isolation and characterization of human ASCs, its expansion in vitro, and relevant in vivo models for adipose tissue engineering.

Key words Adipose tissue, Mesenchymal stem cells, Reconstruction

1 Introduction

"*We restore, repair, and make whole those parts which nature has given but fortune has taken away. Not so much that they may delight the eye but that they may buoy up the spirit and help the mind of the afflicted*". These are the words of Gaspare Tagliacozzi, who already in 1597 [1] appreciated the importance of soft tissue reconstruction, not so much for the aesthetic benefit, but for the patient's overall physical and psychological well-being.

Soft tissue reconstruction still remains one of the most significant challenges for plastic surgeons. A lack of soft tissue can occur due to trauma, congenital diseases (e.g., Romberg disease [2] or Poland syndrome [3]), or oncologic surgery. A variety of techniques have been employed to reconstruct such defects, including autologous composite tissue flap surgery where vascularized tissue is transferred from another region of the body, insertion of prosthetic implants, and transplantation of autologous adipose tissue. Each technique has its own disadvantages. Autologous composite tissue flap procedures allow a natural reconstruction of bone, adipose tissue, and skin, but are highly invasive with significant donor site morbidity, and potentially require multiple costly surgical procedures [4–7]. Implant

Bruce A. Bunnell and Jeffrey M. Gimble (eds.), *Adipose-Derived Stem Cells: Methods and Protocols*, Methods in Molecular Biology, vol. 1773, https://doi.org/10.1007/978-1-4939-7799-4_17, © Springer Science+Business Media, LLC, part of Springer Nature 2018

reconstruction has associated long-term problems of migration, leakage, infection, and capsular contracture [6].

Transplantation of autologous adipose tissue minimizes these complications, and due to its minimally invasive approach, it has increasingly gained popularity throughout the USA [8]. From the initial report by Dr. Neuber in 1893 [9], autologous fat transplantation has undergone an evolution which has aimed to reduce resorption rates and improve clinical outcomes. Various advances in techniques and instrumentation have been achieved, and the Coleman method marked a very important milestone for the use of autologous transfer of adipose tissue in soft tissue reconstruction [10]. This worldwide-used method utilizes the principles of careful lipoaspiration, centrifugation, and subsequent grafting of the concentrated fat tissue. Unfortunately, despite these advances the clinical outcome of adipose tissue transplantation remains unpredictable as there is graft resorption due primarily to lack of vascularization [11]. Adipose tissue is highly vascularized with extensive capillary networks surrounding adipocytes; therefore, it is imperative that the newly transplanted adipocytes have access to a vascular source. Furthermore, there is a constant resorption rate that can vary from 20% to 90% over time [12–16]. Immediately after harvesting of the fat graft, the suctioned aspirate can be composed of various degrees of viable fat (10–90%) [10]; and it has been hypothesized that the number of viable adipocytes at the time of transplantation correlates with the ultimate fat graft survival volume [17, 18]. The viability of adipocytes can be correlated to the extent of vascularization.

Scientists have recently considered the influence of precursor cells, such as adipose-derived stem cells (ASCs), normally found in fatty tissue as a variable that can affect graft integration [10]. ASCs are the mesenchymal stem cells that are contained within the stromal-vascular fraction (SVF) of enzymatically digested adipose tissue. Dr. Yoshimura popularized a technique called cell-assisted lipotransfer (CAL), which admixes SVF isolated from adipose tissue with fat graft material from the same patient with the aim of improving fat grafting outcomes [19].

ASCs have the ability to differentiate into many different cell types, such as osteoblasts [20, 21], chondrocytes [22, 23], myocytes [20, 24, 25], neuronal-like cells [26, 27], cardiovascular cells [28–31], hepatocytes [32–34], and adipocytes [20, 35–39]. ASCs are also capable of dividing and renewing for extended periods of time [40]. Of particular interest to reconstructive surgery is the ability of ASCs to undergo facile differentiation into the mesenchymal phenotype, particularly adipose tissue. ASCs have been shown to possess angiogenic characteristics and to differentiate into vascular endothelial cells [41, 42]. Furthermore, ASCs were shown to express and secrete multiple growth factors including IGF, HGF, TGF-b1 and VEGF, all of which are important in tissue regeneration [31, 43–46].

1.1 Animal Models

A variety of animal models have been described in the literature, but none has been identified as the gold standard. In the past decade, animal studies have shed some light on the role of ASCs in the improvement of long-term graft survival and retention. Masuda et al. showed that co-transplantation of preadipocytes with omental tissue under the skin of Wistar rats, significantly enhanced adipose tissue formation, capillary density, and VEGF production compared to transplantation of omental tissue alone [47]. In another study, Matsumoto et al. used SCID mice to show that CAL fat had a greater survival rate (35% greater mass on average) and a more prominent microvasculature than non-CAL fat [42]. Thus, showing that co-transplantation of preadipocytes with mature adipocytes may contribute to acute adipogenic differentiation of ASCs. Moseley et al. [48] showed that fat supplemented with nude mice ASCs (either freshly isolated or cultured) maintained its adipocyte-rich appearance and weighed 2.5 times more compared with grafted fat not supplemented with ASCs. Lu et al. used CAL with VEGF-transduced ASCs, and reported that fat supplemented with ASCs possessed increased transplant survival and greater capillary density in comparison with fat not supplemented with ASCs. Furthermore, histological analyses revealed that human ASC-derived endothelial cells accounted for the increased microvasculature [49]. Zhu et al. also noticed similar results in their murine fat-transplantation model, where at both 6 and 9 months after transplantation, ASCs increased graft retention by twofold and increased capillary density compared to non-ASC controls [50]. Hypoxia was found to induce cultured human ASCs to synthesize and release numerous angiogenic growth factors [46, 51, 52], and perhaps contributed to the increased microvasculature seen in the other studies [42].

The question of what concentration of SVF cells is required to enhance fat graft retention has been researched by Paik et al. [53]. The authors transferred SVF-enriched fat grafts beneath the scalps of adult Crl:NU-Foxn1nuCD-1 immunocompromised mice, and showed maximum fat graft retention, increased vascularity, and decreased inflammation at a concentration of 10,000 cells per 200 μL of fat. Interestingly, the addition of higher number of cells negatively impacted fat graft retention and led to lipodegeneration [53, 54].

Many unanswered questions still remain, including the type of animal to be used (rat versus mouse versus a larger animal), immunocompetency, follow-up duration, anatomical site, and type of injection. Further research will define the best animal model to investigate ASCs.

1.2 Role of Biomaterial Scaffolds

Tissue engineering has been employed as an attempt to overcome some of the shortfalls of current autologous adipose tissue transplantation techniques. Biomaterial scaffolds offer a 3D construct to promote ASC attachment, migration, proliferation, adipocyte differentiation, and angiogenesis. A variety of natural or synthetic materials have been utilized including poly(lactide-co-glycolide) microspheres [55–58], silk [59], gelatin [60], matrigel [61–63], alginate [64], and collagen/hyaluronic acid [65].

A variety of biomaterials have been utilized as carriers for controlled delivery of encapsulated growth factors, such as insulin, insulin-like growth factor, and basic fibroblast growth factor [55, 56, 60, 61, 65]. Choi et al. showed that injectable human extracellular matrix powders derived from human adipose tissue provided exceptional substrates for cell adhesion and growth in vitro, and supported formation of new adipose tissue in vivo [66]. Lee et al. utilized acellular allogenic dermis as a scaffold for the culturing and in vivo implantation of preadipocytes instead [67].

Despite these optimistic advances, a rich nutrient and oxygen supply by means of a sufficient and rapid neovascularization, remains the key limiting factor of any attempts to obtain large volumes of tissue engineered adipose tissue to meet the demand of large soft tissue defect. A solid organ would require a 3D capillary network connected to the host circulation for the inner cells to survive, something bioengineers and scientists have yet to discover.

A revolutionary technology, called Brava™, was devised by Dr. Khouri to bridge this gap [68–73]. This vacuum-based external breast expander allows the potential of mechanical forces to be harnessed and induce the body to generate its own 3D vascularized scaffold suitable for fat grafting. This technology has enabled successful megavolume autologous fat transfer as described below, and certainly offers great promise for successful and durative fat grafting outcomes.

1.3 Clinical Studies

Over the past decade, there has been an increase in studies looking at the long-term outcomes of fat grafting and its clinical application in soft tissue reconstruction. Coleman reported in 2006 optimal volume correction, excellent fat integration within surrounding tissue, and long-term retention (longest follow-up was ~12 months) in a case series of three patients who were treated with facial fat grafting [10]. Coleman also reported significant improvement in breast size and contouring in a case series of 17 patients who had a follow-up of up to 98 months [74].

Yoshimura et al. [75] pioneered CAL and showed in a variety of studies the clinical benefit of CAL versus conventional lipoinjection. A clinical trial of six patients suffering from facial lipoatrophy, showed that after a follow-up period of up to 13 months, patients receiving CAL had a better clinical improvement score than patients who had undergone conventional lipoinjection [75]. Optimal results were also obtained in a prospective trial of 15 patients undergoing breast augmentation following breast implant removal. After a mean follow-up of

12 months, radiological assessment confirmed a graft take of approximately 40–80% with satisfactory aesthetic results [76]. Further evidence of the benefit of ASCs was presented by a triple-blind, placebo-controlled trial conducted in 2013 by Kølle et al., which showed a statistically significant benefit in fat graft survival in patient undergoing fat grafting enriched with ex vivo expanded ASCs at a concentration 2000-times greater than what is seen in normal adipose tissue, compared to nonenriched fat grafting [77].

One of the limiting factors of the use of autologous fat transfer in soft tissue reconstruction is the inability to inject high volumes of fat due to low survival rates. The evolution of techniques and the study of adipose tissue, has led to the concept of preexpansion. Dr. Khouri's nonsurgical breast-enlargement system, Brava™ [68], utilizes the principles of mechanotransduction to convert mechanical tension (by means of low continuous vacuum distraction force to each breast) into growth-promoting signals, which creates an autologous 3D vascularized scaffold that can be used for grafting of large volumes of fat. The first case series was published in 2000, and the authors showed a stable long-term increase in breast size by 55% (range, 15–115%) with no adverse events [68]. Since then, multiple studies have proven its efficacy and safety in both the aesthetic and the reconstructive setting. Zocchi and Zuliani [78] published their 9-year experience of using Brava™ plus autologous fat grafting in 181 patients undergoing body contouring, and/or breast augmentation (+/− reduction mammoplasty or mastopexy). The authors grafted a mean volume of 375 mL per breast with a 55% retention rate at the 1-year follow-up. Khouri et al. [72] published a 9-year multicenter breast aesthetic experience of using Brava™ plus autologous fat grafting in 476 women. The augmentation group was grafted a mean volume of 367 mL per breast per operation with an 80% retention rate (mean follow-up measurements taken at 9-month follow-up). Khouri et al. [73] also published a 7-year multicenter experience of using this technology to reconstruct breasts of 488 women. The authors grafted a mean volume of 225 mL per breast per operation, leading to a mean breast mound volume of 375 mL per breast, but no data of graft retention was presented.

The retention rates presented by Khouri et al. [71, 72] (78–80%) are certainly much higher than other studies where Brava™ was utilized (55%) [78], and where CAL or autologous fat grafting alone were utilized [19, 76, 79, 80]. While all these studies lack control groups to truly assess the impact of preexpansion, the favorable outcomes and high retention rates warrant further investigation of this concept. Furthermore, randomized controlled studies are still lacking in the literature, and fundamental questions still need to be addressed.

The planning of future studies should take into account the regulatory hurdles placed by the Food and Drug Administration (FDA), that classified SVF within the drug, device, and/or biological product category, due to the use of collagenase during component separation [81]. Since the FDA regulates SVF, surgeons wishing to use it must submit an investigational new drug application and must have an approved Institutional Review Board, which is a costly and time-consuming process. Planning and consideration is warranted, and further investigation into other methods of separating cells and stroma, such as mechanical separation and rapid isolation techniques, may be needed given the implications of these regulations.

2 Materials

1. Dulbecco's Modified Eagle's Medium (DMEM/Fl2).
2. Fetal bovine serum.
3. Penicillin/Streptomycin.
4. Gentamycin.
5. Dexamethasone.
6. Collagenase solution: type II collagenase, M8B10274 (0.1%), with bovine serum albumin (BSA; 3.5%) in 1× Hanks solution.
7. Human regular medium: DMEM and DMEM/F12 at a 1:1 ratio, 10% fetal bovine serum, 0.1 mM penicillin, 0.06 mM streptomycin, 0.1 mM dexamethasone, and gentamycin sulfate (10 mg/L).
8. Erythrocyte lysis buffer.
9. Centrifuge tubes.
10. CyQUANT assay.
11. Fatty acid-free BSA.
12. PKH26.
13. Hematoxylin and Eosin.
14. PBS.
15. Formalin.

3 Methods

3.1 Adipose Tissue Removal

1. Adipose tissue (100–200 g) is harvested from the superficial abdominal depots of females or males undergoing elective abdominal reduction surgery.
2. The patient age range is 18–80 years old and all are healthy according to clinical examination and laboratory tests. The

University of Pittsburgh Institutional Review Board (IRB) approved the procedure of collecting the samples of adipose tissue.

3. Samples are not pooled, with each experiment using cells from different patients.

3.2 ASC Isolation

1. Abdominal adipose tissue is placed in sterile 50 mL conical tubes.
2. A collagenase solution is sterile filtered then added at a 3:1 ratio of solution to adipose tissue for digestion.
3. This solution is then filtered through double-layer gauze (4 × 4 in.), to remove unprocessed debris.
4. The resulting filtered solution is centrifuged at 9200 × *g* for 10 min at 20 °C. The supernatant is removed and the pellet is resuspended in 10 mL of erythrocyte lysis buffer (*see* **Note 1**).
5. Resuspended pellets are vortexed to lyse RBCs and debris, and then recentrifuged at 9200 × *g* for 10 min at 20 °C.
6. The supernatant is then removed and the pellet resuspended in 10 mL of human regular media.
7. After 6 h, the ASCs are adherent to the plates and then washed extensively with PBS to remove any contaminants.
8. Cells are passaged at confluency, and characterized as previously described by our laboratory [82, 83].

3.3 ASC Culture

1. Plating media is changed three times a week, and continued until the ASCs reach approximately 80% confluence (*see* **Note 2**).

3.4 Measurement of Proliferation

1. Approximately 2.5 × 10^3 cells from each depot are seeded in triplicate into 24 well plates.
2. After 48 and 96 h, the plates are then frozen at −80 °C for a minimum of 24 h.
3. The plates are then thawed to room temperature and a CyQUANT® cell proliferation assay kit is used to quantify cell number.
4. Fluorescence is measured at 480-nm excitation and 520-nm emission.
5. A standard curve is created and cell number obtained.

3.5 In Vivo Model

A common in vivo model for assessment of human ASCs for soft tissue reconstruction is described.

1. A dorsal subcutaneous injection is made in 6-week-old male athymic nude mice (*see* **Note 3**).
2. Each injection is 1 mL in volume and performed through a hypodermic syringe (*see* **Note 4**).

3. For histological evaluation, animals are sacrificed, and excised gel implants are rinsed in PBS and fixed in 10% buffered formalin at 37 °C.
4. The gel explants and surrounding tissues are dehydrated in a graded series of alcohols at 37 °C, and embedded in paraffin.
5. The transverse paraffin is sectioned through the center of gel implants at 5 μm thickness, and histologically processed using Hematoxylin and Eosin (H&E) stains.
6. Identification of implanted ASCs can be achieved through a variety of techniques, described in the next section.

3.6 Identification of ACSs

Labeling of ASCs prior to animal implantation can occur using:

1. Viral transfection with green fluorescent protein (GFP).
2. Cell membrane dyes such as PKH26.
3. Nuclear dyes such as CMFDAse or DiR
4. Gender mismatch (e.g., implantation of male cells into female animals and identification of the Y chromosome via FISH).

4 Notes

1. Erythrocyte lysis buffer is composed of 154 mM NH_4Cl, 10 mM $KHCO_3$, <1 mM EDTA in water, and sterile filtered.
2. It is important to avoid overconfluence as the ASCs may spontaneously differentiate into adipocytes.
3. Various delivery systems can be used, such as fibrin glue, collagen gels, or hyaluronic acid.
4. Time points can range from 1 day to 1 year. Typically, researchers use 4, 8, and 12, or 3, 6, and 12 weeks as end points, as it is important to demonstrate long-term adipose retention.

References

1. Tagliacozzi G. De curtorum chirurgia per insitionem, libri duo. In quibus ea omnia, quae ad hujus chirurgiae, narium scilicet, aurium, ac labiorum per insitionem restaurandorum cum theoricen, tum practicen pertinere videbantur, clarissima methodo cumulatissime declarantur. Additis cutis traducis instrumentorum omnium, atque deligationum iconibus, & tabulis. 1597, Venetiis: Apud Gasparem Bindonum, juniorem. 28, 94, 95, 47, 32 p
2. Finch GD, Dawe CJ (2003) Hemiatrophy. J Pediatr Orthop 23(1):99–101
3. Baudinne P, Bovy GL, Wasterlain A (1967) A case of Poland's syndrome. Acta Paediatr Belg 21(5):407–410
4. Mirzabeigi MN et al (2015) Predicting and managing donor-site wound complications in abdominally based free flap breast reconstruction: improved outcomes with early reoperative closure. Plast Reconstr Surg 135(1):14–23
5. Vega S et al (2008) 500 Consecutive patients with free TRAM flap breast reconstruction: a single surgeon's experience. Plast Reconstr Surg 122(2):329–339
6. Tsoi B et al (2014) Safety of tissue expander/implant versus autologous abdominal tissue breast reconstruction in postmastectomy breast cancer patients: a systematic review

and meta-analysis. Plast Reconstr Surg 133(2):234–249

7. Khouri RK et al (1998) A prospective study of microvascular free-flap surgery and outcome. Plast Reconstr Surg 102(3):711–721
8. Kling RE et al (2013) Trends in autologous fat grafting to the breast: a national survey of the american society of plastic surgeons. Plast Reconstr Surg 132(1):35–46
9. Neuber F (1893) Fettransplantation. Bericht uber die Verhandlungen der Deutscht Gesellsch Chir 22:66
10. Coleman SR (2006) Structural fat grafting: more than a permanent filler. Plast Reconstr Surg 118(3 Suppl):108S–120S
11. Peer L (1956) The neglected free fat graft. Plast Reconstr Surg 18:223–250
12. Nguyen A et al (1990) Comparative study of survival of autologous adipose tissue taken and transplanted by different techniques. Plast Reconstr Surg 85(3):378–386. discussion 387–389
13. Boyce RG, Nuss DW, Kluka EA (1994) The use of autogenous fat, fascia, and nonvascularized muscle grafts in the head and neck. Otolaryngol Clin North Am 27(1):39–68
14. Jackson IT et al (2001) A successful long-term method of fat grafting: recontouring of a large subcutaneous postradiation thigh defect with autologous fat transplantation. Aesthetic Plast Surg 25(3):165–169
15. Boschert MT et al (2002) Analysis of lipocyte viability after liposuction. Plast Reconstr Surg 109(2):761–765. discussion 766–767
16. Rohrich RJ, Sorokin ES, Brown SA (2004) In search of improved fat transfer viability: a quantitative analysis of the role of centrifugation and harvest site. Plast Reconstr Surg 113(1):391–395. discussion 396–397
17. Peer LA (1955) Cell survival theory versus replacement theory. Plast Reconstr Surg 16:161–168
18. Gonzalez AM et al (2007) An alternative method for harvest and processing fat grafts: an in vitro study of cell viability and survival. Plast Reconstr Surg 120(1):285–294
19. Yoshimura K et al (2008) Cell-assisted lipotransfer for cosmetic breast augmentation: supportive use of adipose-derived stem/stromal cells. Aesthetic Plast Surg 32(1):48–55. discussion 56–57
20. Zuk PA et al (2002) Human adipose tissue is a source of multipotent stem cells. Mol Biol Cell 13(12):4279–4295
21. Gabbay JS et al (2006) Osteogenic potentiation of human adipose-derived stem cells in a 3-dimensional matrix. Ann Plast Surg 57(1):89–93
22. Jin X et al (2007) Ectopic neocartilage formation from predifferentiated human adipose derived stem cells induced by adenoviral-mediated transfer of hTGF beta2. Biomaterials 28(19):2994–3003
23. Xu Y et al (2007) Analysis of the material properties of early chondrogenic differentiated adipose-derived stromal cells (ASC) using an in vitro three-dimensional micromass culture system. Biochem Biophys Res Commun 359(2):311–316
24. Rangappa S et al (2003) Cardiomyocyte-mediated contact programs human mesenchymal stem cells to express cardiogenic phenotype. J Thorac Cardiovasc Surg 126(1):124–132
25. Rodriguez LV et al (2006) Clonogenic multipotent stem cells in human adipose tissue differentiate into functional smooth muscle cells. Proc Natl Acad Sci U S A 103(32):12167–12172
26. Kokai LE, Rubin JP, Marra KG (2005) The potential of adipose-derived adult stem cells as a source of neuronal progenitor cells. Plast Reconstr Surg 116(5):1453–1460
27. Safford KM et al (2002) Neurogenic differentiation of murine and human adipose-derived stromal cells. Biochem Biophys Res Commun 294(2):371–379
28. Planat-Benard V et al (2004) Spontaneous cardiomyocyte differentiation from adipose tissue stroma cells. Circ Res 94(2):223–229
29. Casteilla L et al (2005) Plasticity of adipose tissue: a promising therapeutic avenue in the treatment of cardiovascular and blood diseases? Arch Mal Coeur Vaiss 98(9):922–926
30. Cao Y et al (2005) Human adipose tissue-derived stem cells differentiate into endothelial cells in vitro and improve postnatal neovascularization in vivo. Biochem Biophys Res Commun 332(2):370–379
31. Rehman J et al (2004) Secretion of angiogenic and antiapoptotic factors by human adipose stromal cells. Circulation 109(10):1292–1298
32. Talens-Visconti R et al (2006) Hepatogenic differentiation of human mesenchymal stem cells from adipose tissue in comparison with bone marrow mesenchymal stem cells. World J Gastroenterol 12(36):5834–5845
33. Talens-Visconti R et al (2007) Human mesenchymal stem cells from adipose tissue: differentiation into hepatic lineage. Toxicol In Vitro 21(2):324–329
34. Seo MJ et al (2005) Differentiation of human adipose stromal cells into hepatic lineage

in vitro and in vivo. Biochem Biophys Res Commun 328(1):258–264
35. Zuk PA et al (2001) Multilineage cells from human adipose tissue: implications for cell-based therapies. Tissue Eng 7(2):211–228
36. Halvorsen YD et al (2001) Thiazolidinediones and glucocorticoids synergistically induce differentiation of human adipose tissue stromal cells: biochemical, cellular, and molecular analysis. Metabolism 50(4):407–413
37. Lee JA et al (2003) Biological alchemy: engineering bone and fat from fat-derived stem cells. Ann Plast Surg 50(6):610–617
38. Lu F et al (2006) Adipose tissues differentiated by adipose-derived stem cells harvested from transgenic mice. Chin J Traumatol 9(6):359–364
39. Choi YS et al (2006) Adipogenic differentiation of adipose tissue derived adult stem cells in nude mouse. Biochem Biophys Res Commun 345(2):631–637
40. Wall ME, Bernacki SH, Loboa EG (2007) Effects of serial passaging on the adipogenic and osteogenic differentiation potential of adipose-derived human mesenchymal stem cells. Tissue Eng 13(6):1291–1298
41. Fumimoto Y et al (2009) Creation of a rich subcutaneous vascular network with implanted adipose tissue-derived stromal cells and adipose tissue enhances subcutaneous grafting of islets in diabetic mice. Tissue Eng Part C Methods 15(3):437–444
42. Matsumoto D et al (2006) Cell-assisted lipotransfer: supportive use of human adipose-derived cells for soft tissue augmentation with lipoinjection. Tissue Eng 12(12):3375–3382
43. Cai L et al (2007) Suppression of hepatocyte growth factor production impairs the ability of adipose-derived stem cells to promote ischemic tissue revascularization. Stem Cells 25(12):3234–3243
44. Yang JA et al (2010) Potential application of adipose-derived stem cells and their secretory factors to skin: discussion from both clinical and industrial viewpoints. Expert Opin Biol Ther 10(4):495–503
45. Zuk PA (2010) The adipose-derived stem cell: looking back and looking ahead. Mol Biol Cell 21(11):1783–1787
46. Lee EY et al (2009) Hypoxia-enhanced wound-healing function of adipose-derived stem cells: increase in stem cell proliferation and up-regulation of VEGF and bFGF. Wound Repair Regen 17(4):540–547
47. Masuda T, Furue M, Matsuda T (2004) Novel strategy for soft tissue augmentation based on transplantation of fragmented omentum and preadipocytes. Tissue Eng 10(11-12):1672–1683
48. Moseley TA, Zhu M, Hedrick MH (2006) Adipose-derived stem and progenitor cells as fillers in plastic and reconstructive surgery. Plast Reconstr Surg 118(3 Suppl):121S–128S
49. Lu F et al (2009) Improvement of the survival of human autologous fat transplantation by using VEGF-transfected adipose-derived stem cells. Plast Reconstr Surg 124(5):1437–1446
50. Zhu M et al (2010) Supplementation of fat grafts with adipose-derived regenerative cells improves long-term graft retention. Ann Plast Surg 64(2):222–228
51. Rubina K et al (2009) Adipose stromal cells stimulate angiogenesis via promoting progenitor cell differentiation, secretion of angiogenic factors, and enhancing vessel maturation. Tissue Eng Part A 15(8):2039–2050
52. Rasmussen JG et al (2011) Prolonged hypoxic culture and trypsinization increase the pro-angiogenic potential of human adipose tissue-derived stem cells. Cytotherapy 13(3):318–328
53. Paik KJ et al (2015) Studies in fat grafting: Part V. Cell-assisted lipotransfer to enhance fat graft retention is dose dependent. Plast Reconstr Surg 136(1):67–75
54. Bliley JM, Satish L, McLaughlin MM, Kling RE, Day JD, Grahovac TL, Kokai LE, Zhang W, Marra KG, Rubin JP (2015) Imaging the stromal vascular fraction during soft tissue reconstruction. Plast Reconstr Surg 136(6):1205–1215
55. Yuksel E et al (2000) Increased free fat-graft survival with the long-term, local delivery of insulin, insulin-like growth factor-I, and basic fibroblast growth factor by PLGA/PEG microspheres. Plast Reconstr Surg 105(5):1712–1720
56. Yuksel E et al (2000) Augmentation of adipofascial flaps using the long-term local delivery of insulin and insulin-like growth factor-1. Plast Reconstr Surg 106(2):373–382
57. Patrick CW Jr et al (2002) Long-term implantation of preadipocyte-seeded PLGA scaffolds. Tissue Eng 8(2):283–293
58. Chhaya MP et al (2015) Sustained regeneration of high-volume adipose tissue for breast reconstruction using computer aided design and biomanufacturing. Biomaterials 52:551–560
59. Mauney JR et al (2007) Engineering adipose-like tissue in vitro and in vivo utilizing human bone marrow and adipose-derived mesenchymal stem cells with silk fibroin 3D scaffolds. Biomaterials 28(35):5280–5290

60. Kimura Y et al (2003) Adipose tissue engineering based on human preadipocytes combined with gelatin microspheres containing basic fibroblast growth factor. Biomaterials 24(14):2513–2521
61. Kawaguchi N et al (1998) De novo adipogenesis in mice at the site of injection of basement membrane and basic fibroblast growth factor. Proc Natl Acad Sci U S A 95(3):1062–1066
62. Toriyama K et al (2002) Endogenous adipocyte precursor cells for regenerative soft-tissue engineering. Tissue Eng 8(1):157–165
63. Stillaert FB et al (2010) Intrinsics and dynamics of fat grafts: an in vitro study. Plast Reconstr Surg 126(4):1155–1162
64. Halberstadt C et al (2002) A hydrogel material for plastic and reconstructive applications injected into the subcutaneous space of a sheep. Tissue Eng 8(2):309–319
65. Khorsandi L et al (2015) Three-dimensional differentiation of adipose-derived mesenchymal stem cells into insulin-producing cells. Cell Tissue Res 361(3):745–753
66. Choi JS et al (2009) Human extracellular matrix (ECM) powders for injectable cell delivery and adipose tissue engineering. J Control Release 139(1):2–7
67. Lee HY et al (2014) Adipose tissue regeneration in vivo using micronized acellular allogenic dermis as an injectable scaffold. Aesthetic Plast Surg 38(5):1001–1010
68. Khouri RK et al (2000) Nonsurgical breast enlargement using an external soft-tissue expansion system. Plast Reconstr Surg 105(7):2500–2512. discussion 2513–2514
69. Smith CJ, Khouri RK, Baker TJ (2002) Initial experience with the Brava nonsurgical system of breast enhancement. Plast Reconstr Surg 110(6):1593–1595. author reply 1595–1598
70. Khouri R. Follow up presentation on fat grafting with BRAVA non surgical breast expansion, in Annual Meeting of the American Society of Aesthetic Plastic Surgery. 2009: Vancouver, BC, Canada
71. Khouri RK et al (2012) Brava and autologous fat transfer is a safe and effective breast augmentation alternative: results of a 6-year, 81-patient, prospective multicenter study. Plast Reconstr Surg 129(5):1173–1187
72. Khouri RK et al (2014) Aesthetic applications of Brava-assisted megavolume fat grafting to the breasts: a 9-year, 476-patient, multicenter experience. Plast Reconstr Surg 133(4):796–807. discussion 808–809
73. Khouri RK et al (2015) Tissue-engineered breast reconstruction with Brava-assisted fat grafting: a 7-year, 488-patient, multicenter experience. Plast Reconstr Surg 135(3):643–658
74. Coleman SR, Saboeiro AP (2007) Fat grafting to the breast revisited: safety and efficacy. Plast Reconstr Surg 119(3):775–785. discussion 786–787
75. Yoshimura K et al (2008) Cell-assisted lipotransfer for facial lipoatrophy: efficacy of clinical use of adipose-derived stem cells. Dermatol Surg 34(9):1178–1185
76. Yoshimura K et al (2010) Progenitor-enriched adipose tissue transplantation as rescue for breast implant complications. Breast J 16(2):169–175
77. Kolle SF et al (2013) Enrichment of autologous fat grafts with ex-vivo expanded adipose tissue-derived stem cells for graft survival: a randomised placebo-controlled trial. Lancet 382(9898):1113–1120
78. Zocchi ML, Zuliani F (2008) Bicompartmental breast lipostructuring. Aesthetic Plast Surg 32(2):313–328
79. Wang H et al (2008) Sonographic assessment on breast augmentation after autologous fat graft. Plast Reconstr Surg 122(1):36e–38e
80. Delay E et al (2009) Fat injection to the breast: technique, results, and indications based on 880 procedures over 10 years. Aesthet Surg J 29(5):360–376
81. Food and Drug Administration (2014) Human cells, tissues, and cellular and tissue-based products (HCT/Ps) from adipose tissue: regulatory considerations; Draft guidance
82. Schipper BM et al (2008) Regional anatomic and age effects on cell function of human adipose-derived stem cells. Ann Plast Surg 60(5):538–544
83. Aksu AE et al (2008) Role of gender and anatomical region on induction of osteogenic differentiation of human adipose-derived stem cells. Ann Plast Surg 60(3):306–322

Chapter 18

Mechanical Stimulation of Adipose-Derived Stem Cells for Functional Tissue Engineering of the Musculoskeletal System via Cyclic Hydrostatic Pressure, Simulated Microgravity, and Cyclic Tensile Strain

Rachel C. Nordberg, Josie C. Bodle, and Elizabeth G. Loboa

Abstract

It is critical that human adipose stem cell (hASC) tissue-engineering therapies possess appropriate mechanical properties in order to restore function of the load bearing tissues of the musculoskeletal system. In an effort to elucidate the hASC response to mechanical stimulation and develop mechanically robust tissue engineered constructs, recent research has utilized a variety of mechanical loading paradigms including cyclic tensile strain, cyclic hydrostatic pressure, and mechanical unloading in simulated microgravity. This chapter describes methods for applying these mechanical stimuli to hASC to direct differentiation for functional tissue engineering of the musculoskeletal system.

Key words Adipose-derived stem cells, Tissue engineering, Mechanical stimulation, Tensile strain, Hydrostatic pressure, Microgravity, Differentiation, Osteogenesis, Chondrogenesis, Adipogenesis, Protocols, Methods

1 Introduction

"Functional tissue engineering" aims to overcome the challenges of tissue-engineering translation by defining the criteria required of a therapy to restore the function of a tissue. There are a number of criteria that must be met to achieve successful tissue-engineering outcomes in vivo including biocompatibility, appropriate bioactivity, and sufficient mechanical properties [1]. In tissue engineering applications that target the load-bearing tissues of the musculoskeletal system, achieving suitable mechanical properties is of critical concern [2].

Human adipose-derived stem cells (hASC) have been studied extensively for musculoskeletal tissue engineering applications due to their relative ease of harvest, proliferative capacity, and multipotent differentiation potential [3–5]. As hASC tissue engineering therapies for the musculoskeletal system enter clinical translation, it is imperative

Bruce A. Bunnell and Jeffrey M. Gimble (eds.), *Adipose-Derived Stem Cells: Methods and Protocols*, Methods in Molecular Biology, vol. 1773, https://doi.org/10.1007/978-1-4939-7799-4_18, © Springer Science+Business Media, LLC, part of Springer Nature 2018

that the resulting tissues provide sufficient mechanical function in vivo. Traditional methods of differentiating human adipose-derived stem cells (hASC) use chemical factors to induce differentiation. However, chemically induced differentiation alone is not sufficient to achieve desired mechanical strength in tissue engineered constructs [6, 7]. Due to the wealth of evidence to support hASC responsiveness to mechanical stimuli [8], there has been an increasing interest in the concept of using mechanical stimulation to condition cells and tissue engineered constructs to promote appropriate biomechanical function prior to in vivo implantation. In addition, mechanical loading studies can help elucidate how mechanical factors influence development of the musculoskeletal system in order to better inform future hASC tissue engineering efforts.

There has been significant research on understanding how mechanical loading dictates musculoskeletal development. Theoretical models of fracture healing and distraction osteogenesis have predicted that hydrostatic tension and the resulting tensile strain favors bone formation while hydrostatic pressure favors cartilage formation [9–13]. This model has been supported by histological evidence during new bone formation [14, 15]. Studies in stem cell mechanobiology have yielded similar results. In studies using both hASC and bone-marrow derived mesenchymal stem cells (hMSC), we and others have demonstrated that specific magnitudes of cyclic tensile strain favor osteogenic differentiation [16–19], hydrostatic pressure favors chondrogenic differentiation [6, 7], and simulated microgravity or unloading favors adipogenic differentiation [20, 21] or the maintenance of an undifferentiated state in spheroid culture [22].

Tensile strain has been used as an idealized in vitro model for distraction osteogenesis. Previously, we have demonstrated that exposure to 10% cyclic tensile strain increases osteogenic differentiation in hASC and hMSC [16–19]. Similarly, research has shown that bone markers are temporally expressed in hMSC after exposure to cyclic tensile strain [23, 24]. In microarray studies, we have demonstrated that 10% cyclic tensile strain upregulates bone markers as well as genes associated with angiogenesis in both hMSC and hASC [16, 17].

Hydrostatic pressure studies have suggested that cyclic hydrostatic pressure favors cartilage formation and maintenance [12–14]. Cyclic hydrostatic pressure has been shown to stimulate cartilage extracellular matrix synthesis in chondrocytes [25], and there has been a wealth of evidence to suggest that cyclic hydrostatic pressure favors chondrogenic differentiation of hMSC [26–31]. Even in the absence of transforming growth factor beta (TGFB), chondrogenesis can be stimulated with hydrostatic pressure [7, 31]. Similar results have been documented in hASC studies. We and others have found that hydrostatic pressure alone has the potential to upregulate chondrogenic markers in hASC with [32–34] and without chondrogenic growth factors [6].

Mechanical unloading via simulated microgravity has also been studied in the context of hMSC/hASC biology and tissue engineering. In hMSC, unloading has been shown to downregulate osteogenic genes and upregulate adipogenic markers [20, 21, 35, 36]. However, conflicting evidence has been published on the effect of microgravity on chondrogenesis. It has been suggested that the p38 MAPK pathway simulates chondrogenesis in microgravity [37]. However, other studies have shown that hMSC Col2A1 production is greatly reduced in microgravity, an undesirable outcome for cartilage formation [38]. Other studies have suggested that microgravity culture could be a beneficial way to expand hMSC or hASC while maintaining differentiation capabilities for regenerative medicine applications [22, 39].

To advance hASC use for musculoskeletal applications, future studies will both: (1) elucidate the underlying biological mechanisms of hASC response to mechanical stimuli; and, (2) utilize mechanical stimulation of hASC to improve the functionality of musculoskeletal tissue engineering therapies. In an effort to facilitate such research, this chapter describes methods to stimulate hASC via cyclic tensile strain, hydrostatic pressure, and microgravity.

2 Materials

2.1 hASC Preparation Strategies

All hASC preparation strategies should be performed in a laminar flow hood with sterile dishes, pipettes, and equipment. All media and solutions that come into contact with hASC should be sterile filtered. Standard equipment is required including an autoclave, a 37 °C water bath, a 37 °C CO_2 incubator, a 4 °C refrigerator, a −25 °C freezer, and a centrifuge.

2.1.1 Media Formulations

1. Complete Growth Medium (CGM): Minimum Essential Medium Eagle, alpha-modified (α-MEM) supplemented with 10% fetal bovine serum (FBS), 2 mM L-glutamine, 100 U/ml penicillin, and 100 mg/ml streptomycin [16, 40].
2. Osteogenic Differentiation Medium (ODM): α-MEM supplemented with 10% FBS, 2 mM L-glutamine, 100 U/ml penicillin, and 100 mg/ml streptomycin, 50 mM ascorbic acid, 0.1 mM dexamethasone, and 10 mM b-glycerolphosphate [8, 22].
3. Chondrogenic Differentiation Medium: Dulbecco's Modified Eagle's media (DMEM) supplemented with 1% FBS, 100 U/ml penicillin, 100 mg/ml streptomycin, 4 mM L-glutamine, 6.25 μg/ml h-insulin, 10 ng/ml transforming growth factor-β3, 500 ng/ml BMP-6, and 50 nM ascorbate-2-phosphate [41].

2.1.2 Cytopore Bead Preparation

1. Phosphate-buffered saline (PBS).
2. Cytopore beads (GE Healthcare Life Sciences, Pittsburgh, PA).
3. Desired media formulation (Subheading 2.1.1).
4. Individually wrapped, sterile specimen cups.

2.1.3 Alginate Bead Preparation

1. 0.9% NaCl solution.
2. 0.9% NaCl, 1.2% alginate solution.
3. 102 mM $CaCl_2$ solution.
4. Individually wrapped, sterile specimen cups.
5. 1 cc syringe with 22.6 syringe needle.
6. Desired media formulation (Subheading 2.1.1) and a serum free formulation of the same media.
7. 100 ml culture dish.

2.1.4 Pellet Preparation

1. 15 ml conical tubes.
2. Desired media formulation (Subheading 2.1.1).
3. Centrifuge with 15 ml adaptors.

2.2 Hydrostatic Pressure

1. MTS Mini Bionix Load Frame (MTS, Eden Prairie, MN).
2. Hydraulic cylinder (Grainger, Lake Forest, IL).
3. Pressure vessel with gauge.
4. Stainless steel tubing.
5. Mineral oil.
6. Hot water bath.
7. Thermocouple controller.
8. Computer and with control software for load frame.
9. Polyethylene bags with round-wire closure (Fisher Scientific, Waltham, MA).
10. Desired media formulation (Subheading 2.1.1).
11. Impulse sealer (PowerSeal, Phoenixville, PA).

2.3 Simulated Microgravity

1. Rotary wall vessels (Synthecon, Huston, TX).
2. A Rotary Cell Culture System (RCCS) base (Synthecon, Huston, TX) set up within a 37 °C CO_2 incubator.
3. RCCS Power supply with tachometer (Synthecon, Huston, TX).
4. 10 cc syringes.
5. Desired media formulation (Subheading 2.1.1).

2.4 Cyclic Tensile Strain

1. Flexercell FX-4000 (or Flexcell FX-5000) strain unit (Flexcell International, Hillsborough, NC).
2. FX5K™ Tension FlexLink.
3. Vacuum Pump.
4. Computer with Flexcell control software to run tension system.
5. BioFlex® baseplate and gaskets.
6. BioFlex® Loading Stations with 25 mm diameter Loading Posts.
7. BioFlex® culture plates for 2D culture or TissueTrain® for 3D culture.
8. Vacuum pump.
9. Drying filter, water trap.
10. Vacuum tubing.
11. Grease/lubricant.
12. Collagen I 3 mg/ml solution (Advanced BioMatrix, San Diego, CA).
13. 1 N NaOH.
14. Powdered Minimum Essential Medium (MEM) with Earle's salts and L-glutamine, without sodium bicarbonate or nucleosides.
15. Sodium bicarbonate ($NaHCO_3$).

3 Methods

3.1 hASC Preparation Strategies

For hydrostatic pressure and microgravity experiments, hASC must be prepared in a way that allows them to form stable attachments with the surrounding environment while in suspension. This can be achieved by numerous methods such as using microcarriers, scaffolds, or cell aggregates. Here we describe three strategies that can be used to prepare hASC for hydrostatic loading (Subheading 3.2) or simulated microgravity (Subheading 3.3): (1) cytopore bead method, (2) alginate bead method, and (3) cell pellet method. All strategies should be performed with sterile technique in a laminar flow hood.

3.1.1 Cytopore Bead Preparation

1. Suspend the desired volume of cytopore beads in PBS (*see* **Note 1**).
2. Autoclave beads on liquid cycle.
3. Aspirate PBS and replace with desired control media (*see* **Note 2**).
4. Store at 4 °C at least over night before use (*see* **Note 3**).

5. Warm desired volume of bead suspension in 37 °C waterbath.
6. In a sterile specimen cup, seed hASC suspension onto beads and swirl (*see* **Note 4**).
7. Incubate overnight before use to allow for cell attachment.

3.1.2 Alginate Bead Preparation

1. Dissolve an hASC pellet of 500 k cells in a 500 μl of the 0.9% NaCl, 1.2% alginate solution.
2. Fill a specimen cup with 30 ml of the 102 mM $CaCl_2$ solution.
3. Transfer the alginate and cell mixture to a 1 cc syringe with a 22.6 size needle.
4. Slowly drop the alginate suspension into the $CaCl_2$ solution dropwise while swirling.
5. Incubate at room temperature for 8 min.
6. Remove supernatant and wash twice with 0.9% NaCl, 1 min per wash.
7. Wash once in serum free media for 1 min.
8. Transfer beads to 100 ml culture dish with 25 ml of media until use.

3.1.3 Pellet Preparation

1. Suspend hASC at a concentration of 500 k per ml.
2. Add 500 μl to 15 ml conical tubes (i.e., 250 k hASC per tube)
3. Centrifuge at 500 × *g* for 3 min.
4. Loosen tube caps to allow for gas exchange during incubation.
5. Incubate at 37 °C in a CO_2 incubator for 2 days to allow pellet to stiffen.

3.2 Hydrostatic Pressure

Hydrostatic pressure experiments have been carried out using a variety of frequencies, pressures, media formulations, durations, cell sources, and cell carriers. A summary of experimental conditions from select hydrostatic pressure reports is shown in Table 1. In hydrostatic pressure experiments, hASC must be prepared using a method that allows for attachments to be formed while in suspension. Refer to Subheading 3.1 or referenced studies for potential strategies.

1. Assemble hydrostatic pressure system as illustrated in Fig. 1. First, fill the hydraulic cylinder, steel tubing, and pressure vessel with mineral oil. Assemble the pressure vessel and use the steel tubing to connect the hydraulic cylinder to the pressure vessel. Secure the hydraulic cylinder into the load frame. Place the pressure vessel into the water bath (*see* **Note 5**).
2. Ensure that the entire system (hydraulic cylinder, tubing, and vessel) is filled with mineral oil and that there are no air pockets by bleeding the system through the bleed valve then test to ensure that the vessel can pressurize to the desired level (*see* **Note 6**).

Table 1
A summary of select hydrostatic pressure experiments

Author	Frequency	Peak pressure	Duration	Medium	Cell source	Cell carrier
Angele et al. [26]	1 Hz	5.03 MPa	4 weeks, 4 h/day	TGF-ß1	hMSC	Pellet culture
Carroll et al. [34]	1 Hz	10 MPa	5 weeks, 5 days/ week, 4 h/day	3 weeks of TGF-ß3	Porcine MSC and ASC	2% Agarose gels
Finger et al. [7]	1 Hz	7.5 MPa, Ramped 1 to 7.5 MPa	2 weeks, 4 h/day	No growth factors	hMSC	2% Agarose gels
Miyanishi et al. [31]	1 Hz	10 MPa	2 weeks, 4 h/day	TGF-ß3 and no growth factors	hMSC	Pellet culture
Miyanishi et al. [30]	1 Hz	0.1, 1, and 10 MPa	2 weeks, 4 h/day	TGF-ß3	hMSC	Pellet culture
Ogawa et al. [32]	0.5 Hz	0.5 MPa	1 weeks in bioprocessor	TGF-ß	hASC	Collagen scaffolds
Puetzer et al. [6]	1 Hz	7.5 MPa	3 weeks, 4 h/day	No growth factors	hASC	2% Agarose gels
Steward et al. [28]	1 Hz	10 MPa	3 weeks, 5 days/ week, 4 h/day	TGF-ß3 and no growth factors	Porcine MSC	1%, 2%, 4% Agarose gels
Steward et al. [27]	1 Hz	10 MPa	3 weeks, 5 days/ week, 4 h/day	TGF-ß3	Porcine MSC	4% Agarose gels

3. In a laminar flow hood, add prepared hASC to sealable polyethylene bags with the desired volume and formulation of media.
4. Seal the bags using an impulse sealer, taking care to minimize air bubbles within the bags (*see* **Note 7**).
5. On the pressure vessel, close off all valves and open the vessel to add the hASC bags into the pressure vessel. Make sure that the oil level is topped off before reassembling the pressure vessel and bleed the system again if necessary (*see* **Note 8**).
6. Using computer control of the load frame, lower the piston of the hydraulic cylinder to pressure the system. Pressurize the vessel to half the desired loading pressure as monitored via the pressure gauge.

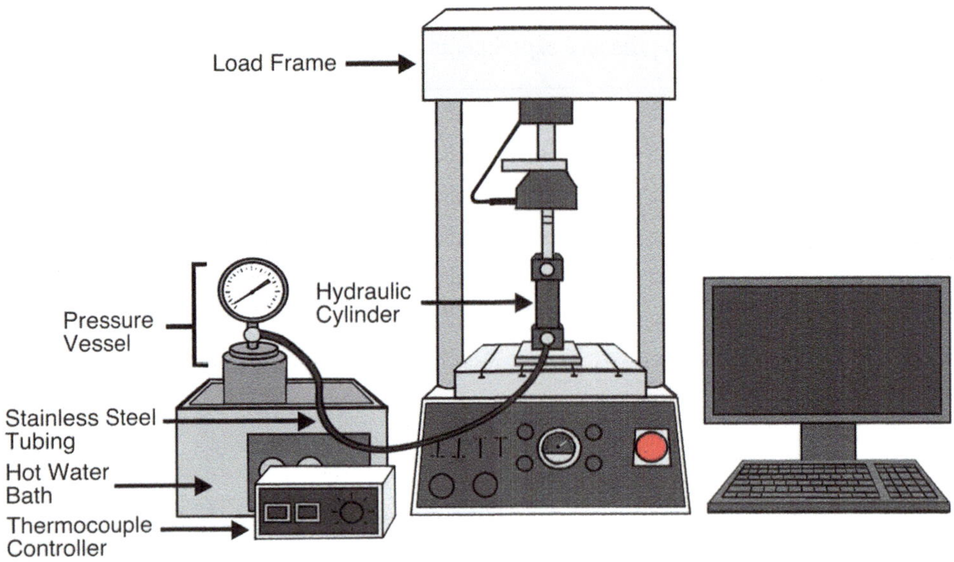

Fig. 1 Hydrostatic pressure bioreactor. The hydraulic cylinder and pressure vessel are filled with mineral oil and connected via stainless steel tubing. The hydraulic cylinder is secured to the load frame. The temperature of the pressure vessel is monitored via a thermocouple controller, and the vessel is placed in a water bath to stabilize the temperature

7. Using the computer set up, adjust the program so that the hydraulic cylinder oscillates and the pressure ranges between 0 MPa and the load pressure as monitored via the pressure gauge.
8. Monitor the vessel throughout the duration of loading to ensure pressure and temperature remain constant (*see* **Notes 9** and **10**).
9. Upon completion of loading, stop the hydraulic cylinder from cycling, disassemble vessel, and remove bags.
10. Cut bags open to remove hASC (*see* **Note 11**).

3.3 Simulated Microgravity

Microgravity can be simulated using a variety of systems including the random positioning system, the clinostat, and the rotating wall vessel bioreactor system also known as a Rotary Cell Culture System (RCCS) [42]. This chapter focuses on the commonly used RCCS system developed by NASA. The RCCS operate with the underlying principle of keeping the cells in constant free fall by applying a constant rotation to the system. As with hydrostatic pressure loading, to culture hASC in an RCCS hASC must be prepared using a method that allows the cells to attach while in suspension. Refer to Subheading 3.1 for potential strategies.

1. Sterilize rotary wall vessel by autoclaving (*see* **Note 12**).
2. Within a laminar flow hood, add the prepared hASC and media to sterile vessel through the fill port.
3. Remove any remaining air bubbles from the vessel (*see* **Note 13**).

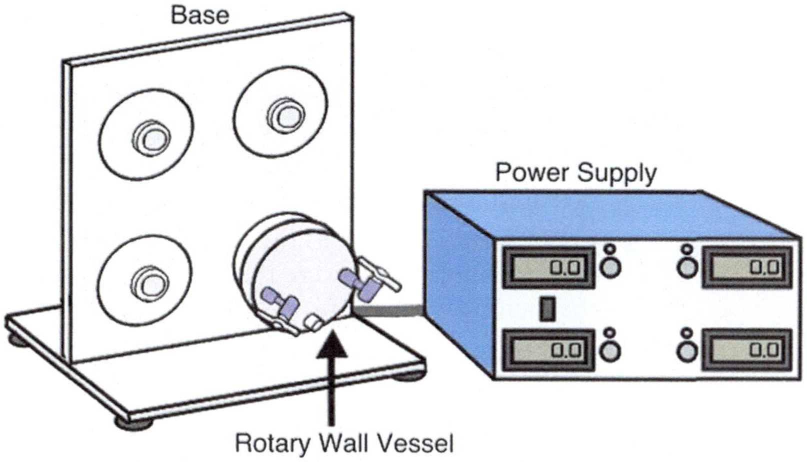

Fig. 2 Simulated microgravity system. The rotary wall vessels are screwed into the rotary base and these two components are placed within a CO_2 incubator. The base is connected to the power supply, which can control the speed of rotation from outside of the incubator

4. Screw vessels onto the RCCS base as illustrated in Fig. 2.
5. Switch the system on.
6. Adjust the speed of the tachometer using the external power supply until the hASC are suspended at the center of the rotary wall vessel (*see* **Note 14**).
7. Periodically check the vessel to ensure that no air bubbles have accumulated in the vessel. If air bubbles are present, stop the system, remove vessels, and repeat **steps 3–5**.
8. Upon completion of experiment, stop vessel, remove beads/ pellets, and sterilize vessel for future use.

3.4 Cyclic Tensile Strain

Cyclic tensile strain experiments have been conducted using a variety of strain systems, which employ different methods to apply tensile strain to cells cultured under both static and dynamic strain conditions. The most prevalent approaches use motors to stretch a flexible cell culture substrate or apply a vacuum to a stretch to the culture surface/construct over a loading post. For the purposes of this methods paper, we summarize an approach that uses a vacuum source to strain cultured cells. This approach can be applied to cells cultured in two-dimensional (2D) culture or 3-dimensional culture (3D) culture. The setup for cells in cultured in 2D strain is largely the same as typical monolayer culture on tissue culture plates, however cells cultured in 3D under cyclic tensile strain require a number of important culture considerations.

3.4.1 Cyclic Tensile Strain for Cells in 2D Culture

Cells are cultured in a standard 6-well Bioflex plate, which is composed of coated silicone membranes as a growth surface. In our applications we use Collagen I-coated plates; however, the appropriate sized plate and coating should be chosen for the specific cell type/application.

1. Seed cells at 50,000 k cells/well in 3 ml of complete growth media and allow cells to adhere and proliferate for 24–48 h. Once the cells have reached 80% confluency, the plates can be prepared for tensile loading (*see* **Note 15**).
2. Apply vacuum grease to the surface of each loading post and spacer platforms to ensure that tight suction can be applied to the membrane of the plate (*see* **Note 16**).
3. Fit the greased 25 mm "arctangle" loading posts into the wells of the Bioflex plates and fit the rubber gasket around the plate-loading post unit.
4. When the plate has been assembled with the loading posts and gasket, fit each until into the loading station baseplate, which supplies the applied vacuum from the Flexercell unit.
5. If loading fewer than the maximum four plates, ensure that the empty baseplate openings are sealed with duct or electrical tape.
6. Apply the strain regimen to test the plate setup for leaks. If vacuum leaks are evident either through a waveform of unexpected magnitude or audible leaking, make plate adjustments as necessary. Vacuum grease can be applied to rubber gaskets to improve the seal. To maintain a tight seal for the duration of an experiment, weights can be placed on the lid of the plates.
7. Maintain the system for the duration of the experiment, under the desired loading conditions. In our experience application of 10% cyclic tensile strain for 4 h/day at 1 Hz is optimal to promote osteogenic differentiation in hMSCs and hASCs.

3.4.2 Cyclic Tensile Strain for Cells in 3D Culture

Many challenges are associated with applying tensile strain to cells in 3D culture. Selecting an appropriate elastomeric culture substrate with a 3D architecture is key to the success of the strain experiment. In this section, we describe our protocol for applying cyclic tensile strain to cells embedded in 3D collagen I gels using the Flexcell TissueTrain system.

1. Assemble the TissueTrain plates onto the cell-loading toughs provided for the Flexcell TissueTrain loading system.
2. Apply vacuum grease to the surface of the linear trough loader and fit the 6-well trough loader under the 6-well TissueTrain plate with the trough aligned along the long axis of the plate and the nylon anchors.
3. Assemble the plate/trough loader unit with the rubber gasket and fit into the baseplate loading station. Ensure that your

tubing setup allows you to move your loading station between incubators and sterile biosafety cabinet.

4. Turn on the vacuum, pressure reservoir, Tension Plus, and Flexcell Computer.
5. Move the baseplate unit to a sterile hood, with the tubing running through incubator port.
6. Assign your holding regimen (we use 10–20% strain, STATIC for 90–120 min) in the Flexcell system.
7. Test the regimen on your plates and ensure that there is tight suction with no leaks. Adjust plates as necessary. Make sure that the trough is centrally aligned between the nylon anchors. Well membrane should appear completely conformed with trough shape.
8. If making fewer than four 6-well plates, use duct tape to close off the vacuum holes in the other baseplate slots.
9. Each construct will be composed of 200 μl volume of collagen I gel. Cells should be seeded at 20,0000–60,000 cells/200 μl construct in collagen suspension. And prepare to take at least 10–15% more than you need as collagen solution is viscous and volume is lost in handling (*see* **Note 17**).
10. Pellet the fraction of cells to be used for experimentation using centrifugation.
11. Prepare a small ice container to prevent the collagen solution from gelling.
12. Prepare the collagen-gel solution on ice. Make up 15–20% more than you need. When mixing solution, invert/pipette gently, avoid creating bubbles. If bubbles arise during mixing, you will experience significant volume loss in addition to the surface tension having an adverse effect of cell viability. For one TissueTrain plate with 200 μl/construct, mix the following: 840 μl Vitrogen (Collagen Solution, pH 2, 70% v/v), 240 μl 5× Growth Media (20% v/v), and 120 μl FBS (10% v/v) (*see* **Note 18**).
13. Add 1.0 N NaOH (at pH 10) to adjust pH to 7.0 to neutralize the collagen solution (about 5 μl NaOH for 1 ml of gel solution).
14. Aspirate the supernatant off of the cell pellet.
15. Start the Flexcell regimen to pull the membrane down into the troughs.
16. Resuspend the cell pellet in collagen mixture above. Pipet slowly because the gel solution will become foamy.
17. Add 200 μl of collagen–cell suspension to each trough. Ensure that the gel is covering the attachment posts at the end of each side of the well and that the posts are horizontal with the

bottom of the well. Make sure that there are no air bubbles in the gel, as this will affect gel integrity. (*see* **Note 19**)

18. Place the lid back on top of culture plate and return to incubator (37 °C).
19. Allow the collagen to polymerize and attach to nylon tabs for at least 90–120 min.
20. Stop the Flexcell regimen, gently remove plates from the baseplate, add growth media (about 3 ml/well), and return plates to incubator (*see* **Note 20**).
21. Allow the constructs to incubate for 12–24 h, then disassemble the TissueTrain plates from the Trough Loaders.
22. To apply strain to the 3D TissueTrain constructs, assemble plates on 25 mm arctangle loading posts as described in Subheading 3.4.1. Exercise care when assembling strain setup.

This chapter has described methods to mechanically stimulate hASC for functional tissue engineering of the musculoskeletal system. Specifically, in this chapter we have focused on the application of cyclic hydrostatic pressure, simulated microgravity, and cyclic tensile strain. It should be noted that additional forms of mechanical simulation, as well as other chemical and electrical stimuli should also be considered when using hASC for tissue engineering applications. However, within this chapter we have aimed to describe three fundamental methods of mechanical stimulation that are of great significance to functional tissue engineering of the musculoskeletal system. In the future, these methods will continue to both enhance our understanding of hASC mechanobiology and improve the functional properties of tissue engineered constructs for the musculoskeletal system.

4 Notes

1. The cytopore beads will settle out of solution within 1–5 min but can be resuspended via swirling. We recommend maintaining a bead to media ratio of 1:1, which can be measured when the beads are settled. Immediately after suspending beads, a quantity of beads can be pipetted assuming an even suspension (i.e., 10 ml of beads in suspension will result in 5 ml of beads once settled).
2. When changing from PBS to media, we have found it difficult to remove all PBS, even after the beads are settled. To reduce residual PBS, rinse once with the desired media before adding the final media volume.
3. The beads can be stored in media for up to 1 month at 4 °C.

4. We recommend seeding cells at a density of 100 k per 1 ml of cytopore beads.
5. We use mineral oil to pressurize the hydrostatic pressure vessel and maintain lubrication in the hydraulic cylinder. Other groups have used water to achieve an equivalent affect [27, 34], although this may reduce the life span of the hydraulic cylinder.
6. Air in the system will prevent the system from pressurizing. Air can be removed from the system by submerging the bleed hose in oil, opening the bleed valve, and pumping the hydraulic cylinder. If this is unsuccessful, air may be trapped within the hydraulic cylinder, which we have found difficult to remove via bleeding. In this case the hydraulic cylinder should be disconnected from the load frame and lowered below the elevation of the setup, which will facilitate directing the air into the main pressure vessel so that the system can be bled.
7. The impulse sealer must be set to the correct temperature to form a proper seal. Test the heat sealer with the polyethylene bags filled with water prior to using it with samples to ensure that the bags maintain seal under pressure and that heat sealer does not melt through bag.
8. We recommend submerging control bags in mineral oil throughout the duration of loading to ensure that results are due to loading and not oil submersion.
9. If the pressure begins to drift, the vessel can be reset by either readjusting the hydraulic cylinder range or recentering the midposition of the hydraulic cylinder.
10. The temperature can be monitored throughout loading using a thermocouple and stabilized by placing the vessel in a water bath set to 37 °C. In our experience, a temperature increase is evident when loading at frequencies above 3 Hz [7].
11. When cutting open the hydrostatic pressure bags, it is important to remove all oil from the outside of the bag and to use sterile scissors to cut the bags to reduce chance of contamination.
12. The rotary wall vessel should be disassembled, rinsed, and wrapped in foil for autoclaving. Alternately, vessel can be sterilized by soaking in 70% ethanol overnight and dried in a cell culture hood before use.
13. We recommend using a syringe technique to remove air bubbles from the assembled rotary wall vessel. Obtain two 10-ml slip-tip syringes and add ~5ml of media to one syringe. Attach both the empty and media-filled syringes to the syringe ports located on the rotary wall vessel. Use the empty syringe to pull out any air bubbles, which will be replaced with fresh media from media-filled syringe. Once all of the air bubbles have been removed, close off both syringe ports, remove syringes, and cap both ends.

14. Adjust the speed of the vessel so that the hASC are maintained at the center of the vessel. Optimal speed is dependent on size and shape of the carrier or pellet.
15. This strain approach can be used to apply dynamic cyclic tensile strain or static strain to cells and the appropriate ECM-coating should be selected for the particular cell type/experimental application.
16. We recommend adding vacuum grease to the loading posts every 3 days as the grease tends to dry out over time in the incubator and the loading profile will change as a result of leakiness in the system.
17. Adjust cell-seeding density according to the duration of experiment and measurement assay. Collagen constructs exhibit a high level of matrix contractility which over the course of a 14 day experiment can break, particularly when exposed to cyclic tensile strain. We recommend seeding at a lower cell density (20 k/200 μl construct) for longer duration experiments whereas seeding up to 60 k/construct may be appropriate for short duration studies.
18. To preparation the 5× MEM (Minimum Essential Medium) solution, measure out 5% less dH_2O than desired total volume (i.e., 190 ml for 200 ml total volume). Add 1 l (1 pack) of powdered MEM medium. Dilute to desired volume with dH_2O (i.e., 200 ml for 1× packet for a 1 l solution). Adjust pH to 0.2–0.3 below the desired final working pH (7.0) by slowly adding and stirring in 1 N NaOH or 1 N HCl. Sterile filter the medium (this will increase pH 0.1–0.3 units). Store the 5× media at 4 °C. Sodium bicarbonate ($NaHCO_3$) is added to the solution before use because it precipitates from 5× MEM upon storage, and hence it must be prepared separately for each experiment. Add 7.5% sodium bicarbonate to 5× MEM immediately before use.
19. When loading collagen–cell solution into the troughs, try to distribute the volume evenly throughout the length of the trough. If excess solution is loaded on one side, the structural integrity of the gel is often compromised and cannot sustain cyclic tensile loading.
20. Handle all gel constructs with care, particularly immediately following polymerization as the gels tend to be very unstable and can break easily.

Acknowledgments

This work was funded by a North Carolina Space Grant Fellowship (R.N.), NSF/CBET Grants 1133427 and 1702841 (E.L.), William R. Kenan Jr. Institute for Engineering, Technology and Science grant (E.L.), NIH CTSA grant 550KR71418 (E.L.), and NIH CTSA grant 550KR61325 (E.L.). We also thank Alison Nordberg for drawing the figures.

References

1. Awad HA, Wickham MQ, Leddy HA et al (2004) Chondrogenic differentiation of adipose-derived adult stem cells in agarose, alginate, and gelatin scaffolds. Biomaterials 25(16):3211–3222
2. Butler DL, Goldstein SA, Guilak F (2000) Functional tissue engineering: the role of biomechanics. J Biomech Eng 122(6):570–575
3. Gimble JM, Katz AJ, Bunnell BA (2007) Adipose-derived stem cells for regenerative medicine. Circ Res 100(9):1249–1260
4. Gimble J, Guilak F (2003) Adipose-derived adult stem cells: isolation, characterization, and differentiation potential. Cytotherapy 5(5):362–369
5. Mizuno H (2009) Adipose-derived stem cells for tissue repair and regeneration: ten years of research and a literature review. J Nippon Med Sch 76(2):56–66
6. Puetzer J, Williams J, Gillies A et al (2013) The effects of cyclic hydrostatic pressure on chondrogenesis and viability of human adipose- and bone marrow-derived mesenchymal stem cells in three-dimensional agarose constructs. Tissue Eng Part A 19(1–2):299–306
7. Finger AR, Sargent CY, Dulaney KO et al (2007) Differential effects on messenger ribonucleic acid expression by bone marrow-derived human mesenchymal stem cells seeded in agarose constructs due to ramped and steady applications of cyclic hydrostatic pressure. Tissue Eng 13(6):1151–1158
8. Bodle JC, Hanson AD, Loboa EG (2011) Adipose-derived stem cells in functional bone tissue engineering: lessons from bone mechanobiology. Tissue Eng Part B Rev 17(3):195–211
9. Loboa EG, Beaupre GS, Carter DR (2001) Mechanobiology of initial pseudarthrosis formation with oblique fractures. J Orthop Res 19(6):1067–1072
10. Carter DR, Beaupre GS, Giori NJ et al (1998) Mechanobiology of skeletal regeneration. Clin Orthop Relat Res (355 Suppl):S41–S55
11. Carter DR, Blenman PR, Beaupre GS (1988) Correlations between mechanical stress history and tissue differentiation in initial fracture healing. J Orthop Res 6(5):736–748
12. Loboa EG, Fang TD, Parker DW et al (2005) Mechanobiology of mandibular distraction osteogenesis: finite element analyses with a rat model. J Orthop Res 23(3):663–670
13. Carter DR, Orr TE, Fyhrie DP et al (1987) Influences of mechanical stress on prenatal and postnatal skeletal development. Clin Orthop Relat Res (219):237–250
14. Loboa EG, Fang TD, Warren SM et al (2004) Mechanobiology of mandibular distraction osteogenesis: experimental analyses with a rat model. Bone 34(2):336–343
15. Urist MR, Mazet R Jr, McLean FC (1954) The pathogenesis and treatment of delayed union and non-union; a survey of eighty-five ununited fractures of the shaft of the tibia and one hundred control cases with similar injuries. J Bone Joint Surg Am 36-A(5):931–980
16. Charoenpanich A, Wall ME, Tucker CJ et al (2014) Cyclic tensile strain enhances osteogenesis and angiogenesis in mesenchymal stem cells from osteoporotic donors. Tissue Eng Part A 20(1–2):67–78
17. Charoenpanich A, Wall ME, Tucker CJ et al (2011) Microarray analysis of human adipose-derived stem cells in three-dimensional collagen culture: osteogenesis inhibits bone morphogenic protein and Wnt signaling pathways, and cyclic tensile strain causes upregulation of proinflammatory cytokine regulators and angiogenic factors. Tissue Eng Part A 17(21–22):2615–2627
18. Hanson AD, Marvel SW, Bernacki SH et al (2009) Osteogenic effects of rest inserted and continuous cyclic tensile strain on hASC lines with disparate osteodifferentiation capabilities. Ann Biomed Eng 37(5):955–965
19. Wall ME, Rachlin A, Otey CA et al (2007) Human adipose-derived adult stem cells upregulate palladin during osteogenesis and in response to cyclic tensile strain. Am J Physiol Cell Physiol 293(5):C1532–C1538
20. Sheyn D, Pelled G, Netanely D et al (2010) The effect of simulated microgravity on human mesenchymal stem cells cultured in an osteogenic differentiation system: a bioinformatics study. Tissue Eng Part A 16(11):3403–3412
21. Zayzafoon M, Gathings WE, McDonald JM (2004) Modeled microgravity inhibits osteogenic differentiation of human mesenchymal stem cells and increases adipogenesis. Endocrinology 145(5):2421–2432

22. Zhang S, Liu P, Chen L et al (2015) The effects of spheroid formation of adipose-derived stem cells in a microgravity bioreactor on stemness properties and therapeutic potential. Biomaterials 41:15–25
23. Kearney EM, Farrell E, Prendergast PJ et al (2010) Tensile strain as a regulator of mesenchymal stem cell osteogenesis. Ann Biomed Eng 38(5):1767–1779
24. Sumanasinghe RD, Bernacki SH, Loboa EG (2006) Osteogenic differentiation of human mesenchymal stem cells in collagen matrices: effect of uniaxial cyclic tensile strain on bone morphogenetic protein (BMP-2) mRNA expression. Tissue Eng 12(12):3459–3465
25. Smith RL, Rusk SF, Ellison BE et al (1996) In vitro stimulation of articular chondrocyte mRNA and extracellular matrix synthesis by hydrostatic pressure. J Orthop Res 14(1):53–60
26. Angele P, Yoo JU, Smith C et al (2003) Cyclic hydrostatic pressure enhances the chondrogenic phenotype of human mesenchymal progenitor cells differentiated in vitro. J Orthop Res 21(3):451–457
27. Steward AJ, Kelly DJ, Wagner DR (2014) The role of calcium signalling in the chondrogenic response of mesenchymal stem cells to hydrostatic pressure. Eur Cell Mater 28:358–371
28. Steward AJ, Wagner DR, Kelly DJ (2013) The pericellular environment regulates cytoskeletal development and the differentiation of mesenchymal stem cells and determines their response to hydrostatic pressure. Eur Cell Mater 25:167–178
29. Steward AJ, Thorpe SD, Vinardell T et al (2012) Cell-matrix interactions regulate mesenchymal stem cell response to hydrostatic pressure. Acta Biomater 8(6):2153–2159
30. Miyanishi K, Trindade MC, Lindsey DP et al (2006) Dose- and time-dependent effects of cyclic hydrostatic pressure on transforming growth factor-beta3-induced chondrogenesis by adult human mesenchymal stem cells in vitro. Tissue Eng 12(8):2253–2262
31. Miyanishi K, Trindade MC, Lindsey DP et al (2006) Effects of hydrostatic pressure and transforming growth factor-beta 3 on adult human mesenchymal stem cell chondrogenesis in vitro. Tissue Eng 12(6):1419–1428
32. Ogawa R, Mizuno S, Murphy GF et al (2009) The effect of hydrostatic pressure on three-dimensional chondroinduction of human adipose-derived stem cells. Tissue Eng Part A 15(10):2937–2945
33. Ogawa R, Md P, Orgill DP, Murphy GF et al (2014) Gene expression profile on hydrostatic pressure-driven three-dimensional cartilage induction using human adipose-derived stem cells and collagen gels. Tissue Eng Part A 21(1-2):257–266
34. Carroll SF, Buckley CT, Kelly DJ (2014) Cyclic hydrostatic pressure promotes a stable cartilage phenotype and enhances the functional development of cartilaginous grafts engineered using multipotent stromal cells isolated from bone marrow and infrapatellar fat pad. J Biomech 47(9):2115–2121
35. Meyers VE, Zayzafoon M, Gonda SR et al (2004) Modeled microgravity disrupts collagen I/integrin signaling during osteoblastic differentiation of human mesenchymal stem cells. J Cell Biochem 93(4):697–707
36. Meyers VE, Zayzafoon M, Douglas JT et al (2005) RhoA and cytoskeletal disruption mediate reduced osteoblastogenesis and enhanced adipogenesis of human mesenchymal stem cells in modeled microgravity. J Bone Miner Res 20(10):1858–1866
37. Yu B, Yu D, Cao L et al (2011) Simulated microgravity using a rotary cell culture system promotes chondrogenesis of human adipose-derived mesenchymal stem cells via the p38 MAPK pathway. Biochem Biophys Res Commun 414(2):412–418
38. Mayer-Wagner S, Hammerschmid F, Redeker JI et al (2014) Simulated microgravity affects chondrogenesis and hypertrophy of human mesenchymal stem cells. Int Orthop 38(12):2615–2621
39. Yuge L, Kajiume T, Tahara H et al (2006) Microgravity potentiates stem cell proliferation while sustaining the capability of differentiation. Stem Cells Dev 15(6):921–929
40. Bodle JC, Teeter SD, Hluck BH et al (2014) Age-related effects on the potency of human adipose-derived stem cells: creation and evaluation of superlots and implications for musculoskeletal tissue engineering applications. Tissue Eng Part C Methods 20(12):972–983
41. Puetzer JL, Petitte JN, Loboa EG (2010) Comparative review of growth factors for induction of three-dimensional in vitro chondrogenesis in human mesenchymal stem cells isolated from bone marrow and adipose tissue. Tissue Eng Part B Rev 16(4):435–444
42. Ulbrich C, Wehland M, Pietsch J et al (2014) The impact of simulated and real microgravity on bone cells and mesenchymal stem cells. Biomed Res Int 2014:928507

Chapter 19

Cryopreservation Protocols for Human Adipose Tissue Derived Adult Stem Cells

Shahensha Shaik and Ram Devireddy

Abstract

The development of simple but effective storage protocols for adult stem cells will greatly enhance their use and utility in tissue-engineering applications. Cryopreservation has shown to be most promising but is a fairly complex process, necessitating the use of chemicals called cryoprotective agents (CPAs), freezing equipment, and obviously, storage in liquid nitrogen. The purpose of this chapter is to present a general overview of cryopreservation storage techniques and the optimal protocols/results obtained in our laboratory for long-term storage of adult stem cells using freezing storage.

Key words Cryopreservation, Cryoprotective agents, Dimethylsulfoxide, Polyvinylpyrrolidone, Flow cytometry, Histochemical analysis, Adipogenesis, Osteogenesis, Apoptosis

1 Introduction

Multiple studies report that primary preadipocytes derived from stromal vascular fraction (SVF) of human adipose tissue, although being able to differentiate in adipocytes under appropriate conditions, decrease their ability to differentiate after multiple passages; this coincides with a prolonged doubling time and replicative senescence [1–11]. These variable results indicating the genotype and phenotype may not remain stable during growth and passage of adult stem cells, suggest that passaged cells for higher generations may not be suitable for regenerative medical therapies. Moreover, serial passaging exposes cell cultures to the repeated risk of contamination by environmental microorganisms. These shortcomings associated with in vitro stem cell expansion can be partly alleviated by proper long-term storage (suspended animation) of stromal vascular fraction (SVF) of human adipose tissue or passage 1 (P1) adult stem cells (ASCs).

Bruce A. Bunnell and Jeffrey M. Gimble (eds.), *Adipose-Derived Stem Cells: Methods and Protocols*, Methods in Molecular Biology, vol. 1773, https://doi.org/10.1007/978-1-4939-7799-4_19, © Springer Science+Business Media, LLC, part of Springer Nature 2018

1.1 Long-Term Storage of Adipose Tissue Derived Adult Stem Cells

The three primary options of long-term storage of cells, macromolecules, proteins, and other biosystems are:

Freezing: Starting at room temperature (RT) a reduction/lowering in temperature will lead to phase change/freezing (formation of ice), and subsequent lowering of temperature below the glass transition temperature (T_g) will lead to glass formation and low-temperature stable storage. Freezing storage, typically, involves [12–17] the following:

(a) Osmotic injury or toxic injury due to the addition of chemicals (called cryoprotective agents, CPAs) that are needed to alleviate the freezing injury.

(b) Chilling injury associated with the reduction of temperature from RT to the phase change (ice formation) temperature.

(c) Cooling injury associated with the formation of extracellular and intracellular ice.

(d) Thawing injury associated with the increase in temperature from liquid nitrogen storage to RT.

Desiccation: Starting at room temperature (RT) a progressive increase in solute concentration, by the process of drying or evaporating the liquid in the system, will also lead to glass formation and shelf-temperature stable storage. Desiccation storage, typically, involves the following [18–25]:

(a) Solute effects injury or exposure to progressively increasing (and toxic) concentrations of solutes/salts.

(b) Precipitation (or crystallization) injury and the associated damage to cellular membranes and intracellular components.

(c) Long processing times, leading to hypoxia and cell death before the cells can be stabilized and properly stored.

(d) Rehydration injury associated with the reconstitution (or addition) of solvents to the desiccated extracellular and intracellular matrix.

Freeze-Drying: Starting at room temperature (RT), the temperature is first reduced leading to the formation of ice and then the solute concentration is progressively increased by sublimation, leading to glass formation and refrigeration stable storage. Freeze-drying storage, typically, involves, as described above, a combination of the damage mechanisms associated with freezing storage and desiccation storage [22, 26–33]:

(a) Osmotic or toxic injury due to the addition of CPAs.

(b) Chilling injury.

(c) Cooling injury.

(d) Precipitation (or crystallization) injury, although this is somewhat mitigated by the fact that this process occurs at subzero temperatures.

(e) Rehydration injury (and the associated thawing injury).

Note the differences between the three processes, freezing requires a substantial reduction in temperature (~150 °C), desiccation requires a substantial increase in solute concentration (~70%), while freeze-drying requires both a moderate decrease in temperature (~60 °C) and a moderate increase in solute concentration (~30%). Correspondingly, the damage mechanisms associated with the three processes are also different: And finally, the advantages/disadvantages associated with the three processes are also different.

1. Freezing storage, typically, has shown to be most promising for poststorage cell recovery and is widely used, but requires the use of costly freezing and low temperature storage equipment.
2. Desiccation storage is yet to be shown as being an efficacious technique for mammalian cells but has the advantage of ambient cell storage and ambient temperature processing.
3. Freeze-drying storage requires the use of costly and ill-developed equipment and is also not widely used due to the low recovery of cell survival poststorage.

Thus, this chapter presents the optimal protocols/results obtained in our experience for long-term storage of adult stem cells using freezing storage [34–38]. A more biophysically based approach to the development of adult stem cell cryopreservation protocols is presented elsewhere [39–41].

1.2 Freezing Storage of Adipose Tissue Derived Adult Stem Cells

The use of low temperatures to preserve (and store) biomaterials has long been known to mankind, and once frozen, these biomaterials (most commonly, fruits and vegetables) can endure storage at very low temperatures for a very long period of time. Obviously, the higher the subzero storage temperature, the shorter the time that these substances can be successfully stored. The primary reasons for the success of freezing storage are as follows:

(a) No thermally driven reactions occur in aqueous systems at the commonly used storage temperature of liquid nitrogen (~−160 °C).
(b) Liquid water does not exist below −130 °C. The crystalline and glassy states that do exist have a very high viscosity (>10^3 poise) such that diffusion is insignificant over less than geological time spans [42].

1.2.1 Choice of Cryoprotective Agents (CPAs)

Starting with a serendipitous observation by Polge and colleagues in the 1940s, the use of chemicals, denoted as cryoprotective agents (CPAs), in the freezing media has been shown to be a prerequisite for the successful outcome of a freezing-storage process [43]. Chemicals that exhibit and/or confer some form of cryoprotection range from low-molecular-weight permeating solutes like dimethylsulfoxide (DMSO), glycerol, and ethylene glycol (EG) to high-molecular-weight nonpermeating solutes like polyvinylpyrrolidone (PVP), hydroxyethylstarch (HES), and methylcellulose (MC) [44–47]. The exact nature and the extent of cryoprotection offered by these chemicals to a specific cell type are still a matter of conjecture and are ill-understood [48–51]. But the explanations range from:

(a) Enhanced colligative properties of solutions with CPAs [52–54].
(b) Modifying the cell membrane transport properties [40, 55–61].
(c) Preventing denaturing of membranes by elevated concentrations of extracellular salts at low temperatures [62–65].
(d) Forming a protective coating on sensitive plasma membranes [66–69].
(e) Prevention of seeding of the supercooled intracellular water and thus the formation of damaging intracellular ice [70–72].
(f) Ability of the CPAs to alter the physical properties of solutions during freezing rate than in any direct effects on cellular membranes [73–80].

Thus, the choice and composition of an optimal cryopreservation media is an area of intense research in the field of cryobiology and is not, as yet, completely understood. Our experiments with SVF and P1 adult stem cells were conducted with the media listed later in this chapter [36–38].

2 Materials

Where appropriate, materials should be sterile and all manipulations should be performed using a laminar flow bench (fume hood). Adequate safety equipment (gloves, face shield) should be used during all liquid nitrogen handling procedures.

1. Sterile plastic dishes/pipettes/micropipettes.
2. −80 °C freezer.
3. 37 °C CO_2 incubator and 43 °C CO_2 incubator.
4. Normal and cooling centrifuges.
5. Commercially available and programmable controlled rate freezer able to cool at rates of 1 °C/min to 40 °C/min to −80 °C (*see* **Note 1**).
6. Type-T hypodermic thermocouples and precision temperature data logger.
7. Directional solidification stage, capable of controlled cooling rates ranging from 1 °C/min to 40 °C/min to −80 °C.

8. Liquid nitrogen storage Dewars (long-term storage containers).
9. 37 °C warm water bath.
10. Digital image analyzer—typhoon laser imaging scanner.
11. Glass microslides with central grooves.
12. Warm copper block maintained at 50 °C.
13. Cryovials, such as 2-mL plastic cryotubes.
14. Diluent: DMEM, DMEM-F12, fetal calf serum, and human serum.
15. Supplies for sodium dodecyl sulfate polyacrylamide gel electrophoresis (SDS-PAGE) including distilled water, Tris (pH 8.8 and pH 6.8), acrylamide–bisacrylamide, ammonium persulfate (APS), TEMED, and EtOH. Gel casting apparatus including gel casting plates, comb, power pack, and plate holder/container.
16. Lysis buffer—RIPA buffer with protease cocktail inhibitors.
17. BCA assay kit for protein concentration.
18. Loading buffer (6×) containing Tris, glycerol, Bromophenol Blue, and SDS.
19. Western blotting antibodies (HSP-70, HSP-90, HSP-27, and β-actin).
20. Western blot chemiluminescence substrate.
21. Running buffer or 1× Tris glycine buffer.
22. Transfer buffer or 1× Tris glycine buffer and methanol.
23. Blocking buffer or 5% nonfat milk with 1× Tris buffered saline with Tween® 20 (TBST).
24. Washing buffer or 1× Tris buffered saline with Tween® 20 (TBST).
25. Cell culture medium containing 10% FBS at an equivalent of 1 mL of tissue per mL of media (*see* **Note 2**).
26. Adidogenic induction cocktail containing DMEM/F-12 Ham's with 3% FBS, 33 μM biotin, 17 μM pantothenate, 1 μM bovine insulin, 1 μM dexamethasone, 0.5 mM isobutylmethylxanthine (IBMX), 5 μM rosiglitazone, and 100 U penicillin–100 μg streptomycin–0.25 μg Fungizone.
27. Osteogenic induction cocktail containing DMEM/F-12 Ham's, 10% FBS, 10 mM β-glycerophosphate, 50 μg/mL sodium ascorbate-2-phosphate, and 100 U penicillin–100 μg streptomycin–0.25 μg Fungizone.
28. Cryoprotectants (CPAs): dimethylsulfoxide (*see* **Note 3**), methylcellulose, polyvinylpyrrolidone. All cryoprotectant solutions are prepared using DMEM supplemented with either fetal calf serum or human serum. PVP and MC are individually

autoclaved at 121 °C for 30 min before being added to DMEM (*see* **Note 4**). The DMEM-PVP solutions and DMEM-MC solutions are prepared by dissolving weighted PVP and MC in DMEM at room temperature and the solutions are then stored overnight at 4 °C to obtain a homogeneous preparation. Concentrations above 1% MC and 40% PVP have been found to be highly viscous and hard to handle and hence are not used [36–38].

29. Fluorescence-activated cell sorting (FACS) machine and associated chemicals (etoposide, hydrogen peroxide, annexin V-FITC/PI kit, Propidium Iodide). Cell Quest software to acquire the FACS data. Apoptosis control is cells incubated in fresh medium enriched with 40 mM etoposide for 24 h. Necrotic control is cells incubated in fresh medium with 5 mM hydrogen peroxide (H_2O_2) for 24 h.

3 Methods

3.1 Collection and Isolation of ASCs

The process of adult stem cell isolation (and culture/expansion/differentiation) from human liposuction aspirate (adipose tissue) is described elsewhere [81–90] and is only briefly described below.

1. Obtain liposuction aspirates from the plastic surgeon.
2. Wash tissue and collagenase-digest at 37 °C with continuous shaking for 1 h.
3. Centrifuge digest to obtain "stromal vascular fraction (SVF)" pellet (*see* **Note 5**).
4. Suspend SVF in culture media.
5. Seed flasks at a density of 0.156 mL tissue digest/cm^2 surface area as "Passage 0 (P0)." Incubate at 37 °C, 5% CO_2.
6. Wash cultures to remove nonadherent cells after 24 h of incubation.
7. Feed cultures every 2–3 days.
8. Harvest P0 by trypsin–EDTA digestion when it reaches 80–90% confluency (mean, 5.5 days).
9. Seed "Passage 1 (P1)" and subsequent passages of undifferentiated ASCs at a density of 5 × 10^3 cells/cm^2 and maintain according to **steps 6** and **7**.
10. Harvest P1 by trypsin–EDTA digestion when it reaches 80–90% confluency (mean, 5.5 days) (*see* **Note 6**).

3.1.1 Freezing Media for SVF of Adipose Tissue (Those with Asterisks Are Suboptimal)

The measured **post-freeze–thaw viability** is shown for each media. The detailed freeze–thaw procedures are described in Subheading 3.3. Additional (and corresponding) data on postthaw apoptotic and necrotic cells is presented elsewhere [37].

1.	Dulbecco's Modified Eagle's Medium (DMEM) with 80% fetal calf serum (FCS) and 10% DMSO	65.3 ± 3.3%
2.	DMEM with 80% human serum (HS) and 10% DMSO	61.9 ± 2.4%
3.	*DMEM with 0% DMSO	11.4 ± 7.4%
4.	DMEM with 2% DMSO	60.1 ± 4.4%
5.	DMEM with 4% DMSO	62.8 ± 4.5%
6.	DMEM with 6% DMSO	65.8 ± 2.4%
7.	DMEM with 8% DMSO	72.9 ± 3.3%
8.	DMEM with 10% DMSO	69.8 ± 2.5%
9.	*DMEM with 1% MC	37.3 ± 4.1%
10.	*DMEM with 1% MC and 10% FCS	22.2 ± 8.3%
11.	*DMEM with 1% MC and 10% HS	20.0 ± 6.8%
12.	DMEM with 1% MC and 10% DMSO	66.0 ± 2.1%
13.	DMEM with 10% PVP	54.6 ± 1.7%
14.	DMEM with 10% PVP and 10% FCS	52.3 ± 5.7%
15.	DMEM with 10% PVP and 40% FCS	49.7 ± 6.7%
16.	DMEM with 10% PVP and 80% FCS	58.6 ± 3.2%
17.	DMEM with 10% PVP and 10% HS	53.9 ± 5.5%

3.1.2 Freezing Media for P1 ASCs (Those with Asterisks Are Suboptimal)

The measured **post-freeze–thaw viability** is shown for each media. Additional (and corresponding) data on postthaw apoptotic and necrotic cells is presented elsewhere [36, 38].

1.	DMEM with 80% FCS and 10% DMSO	84.1 ± 7.7%
2.	DMEM with 80% HS and 10% DMSO	82.5 ± 8.3%
3.	*DMEM with 0% DMSO	25.3 ± 5.7%
4.	DMEM with 2% DMSO	83.8 ± 5.3%
5.	DMEM with 4% DMSO	80.6 ± 3.1%
6.	DMEM with 6% DMSO	86.0 ± 4.6%
7.	DMEM with 8% DMSO	85.6 ± 2.8%
8.	DMEM with 10% DMSO	87.9 ± 2.9%

9.	*DMEM with 1% MC	46.8 ± 7.6%
9.	*DMEM with 1% MC and 10% FCS	41.5 ± 7.6%
10.	*DMEM with 1% MC and 10% HS	39.8 ± 7.6%
11.	DMEM with 1% MC and 10% DMSO	79.8 ± 2.8%
12.	*DMEM with 1% PVP	5.7 ± 0.5%
13.	DMEM with 5% PVP	49.3 ± 8.0%
14.	DMEM with 10% PVP	69.7 ± 8.2%
15.	DMEM with 20% PVP	55.5 ± 11.0%
16.	*DMEM with 40% PVP	4.6 ± 1.3%
17.	DMEM with 5% PVP and 10% HS	53.2 ± 8.7%
18.	DMEM with 10% PVP and 10% HS	72.1 ± 8.1%
19.	DMEM with 20% PVP and 10% HS	61.8 ± 12.5%
20.	DMEM with 5% PVP and 10% FCS	55.4 ± 11.0%
21.	DMEM with 10% PVP and 10% FCS	69.2 ± 6.7%
22.	DMEM with 20% PVP and 10% FCS	64.3 ± 6.4%
23.	DMEM with 10% PVP and 40% FCS	70.5 ± 3.5%
24.	DMEM with 10% PVP and 80% FCS	72.5 ± 2.5%

3.2 Process of Adding CPAs and Cooling to Liquid Nitrogen Temperatures and Thawing to Room Temperature

1. Suspend the SVF or P1 ASCs at a concentration of 1.0×10^6 cells/mL in the various cryopreservation medium, as described above (*see* **Note 7**).
2. Place the cells in an ethanol-jacketed closed container overnight at −80 °C.
3. The temperature/time history experienced by the cells in the ethanol-jacketed container can be measured using a type-T hypodermic needle thermocouple (Omega Technologies, Stamford, CT, USA). Thermocouple voltages can be read by a precision temperature data logger (Veriteq Instruments Inc., Richmond, BC, Canada) and are transferred to a personal computer for further reduction and data analysis. The representative cooling rates experienced by the cells in the ethanol-jacketed container placed in a −80 °C freezer is shown in Fig. 1 (*see* **Note 8**).
4. Note that the exact magnitude of the cooling rates experienced by the cells in the ethanol-jacketed container is a function of the cryopreservation media.
5. Store the cells in liquid nitrogen vapor and/or immersed in liquid nitrogen (*see* **Notes 9–12**).
6. Rapidly thaw individual cryovials of cells in a 37 °C water bath (1–2 min of agitation (*see* **Note 13**).

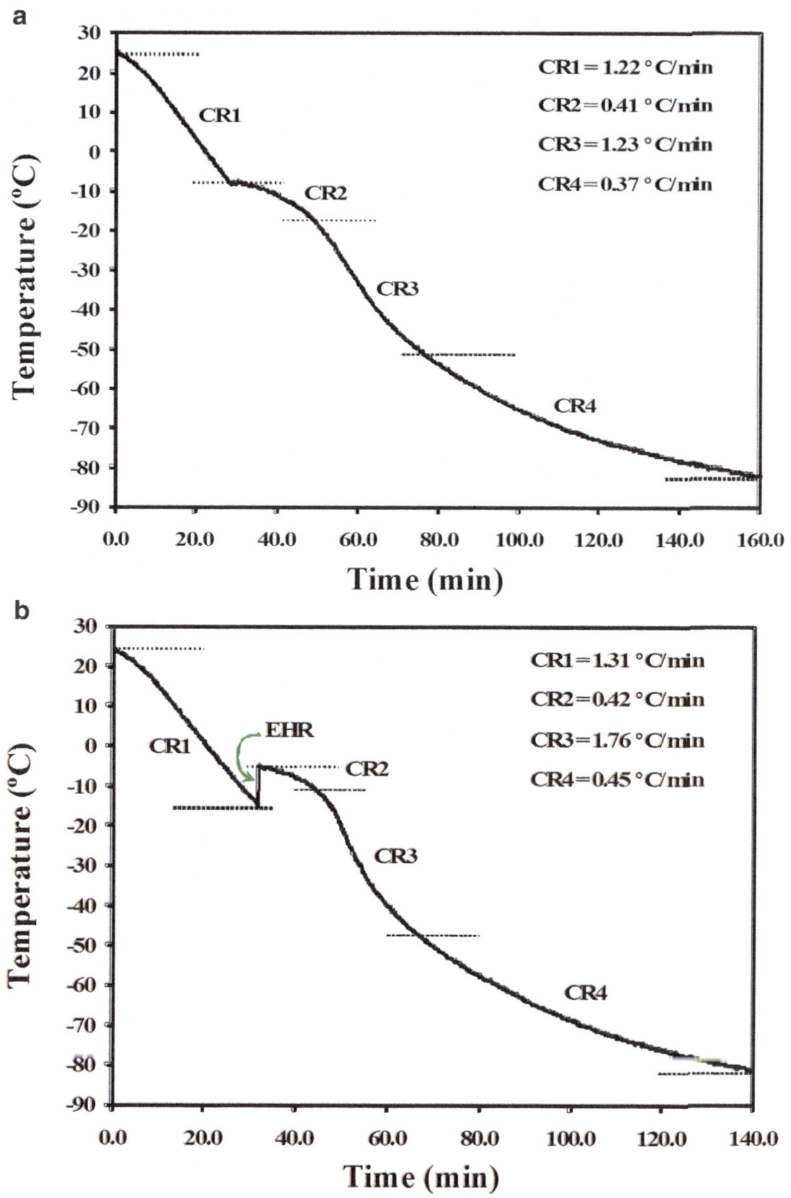

Fig. 1 Representative cooling rates experienced by the P1 ASCs in the ethanol-jacketed closed container placed in the −80 °C freezer with 80% FCS + 10% DMSO in DMEM (**a**) and 10%PVP in DMEM (**b**). The measurements are measured using type-T thermocouple coupled with a precision temperature data logger. The plot suggests that the cells are subjected to different cooling rates at different time points in different freezing media within the ethanol-jacketed container. For media with 80%FCS + 10% DMSO in DMEM, the ice nucleation is observed around −8 (±1.1) °C and subsequently, a cooling rate of ~0.4 °C/min is imposed to a temperature of ~−18 °C. The cooling rates experienced by the cells then further drops to ~1.2 °C/min until ~−50 °C and then to ~0.4 °C/min, before reaching ~−80 °C (**a**). Alternatively, for media with 10% PVP in DMEM, there is significant supercooling before any ice nucleation is observed around −17 (±2.3) °C. Due to the enthalpic heat release (EHR) the sample temperature is abruptly raised to −5 °C and a subsequent cooling rate of ~0.4 °C/min is imposed to a temperature of −10 °C. The cooling rates experienced by the cells then further drops to 1.8 °C/min until ~−50 °C and then to ~0.4 °C/min, before reaching ~−80 °C (**b**)

7. Resuspend the cells in stromal culture media, and seed into the separate wells of a 6-well plate for a 24 h incubation period at 37 °C (*see* **Note 14**).

3.3 Evaluation of Frozen–Thawed Cells

The postthaw viability and functionality of the cells is analyzed using flow cytometry [91] and histochemical staining for adipogenic and osteogenic differentiability [35, 83, 84, 86–88].

3.3.1 Cell Viability and Apoptosis/Necrosis Assessment

1. A well-established annexin V apoptosis assay is analyzed by quantitative flow cytometry.
2. Chemically induced apoptotic control is cells incubated in fresh medium enriched with 40 mM etoposide for 24 h. Apoptosis is characterized by phosphatidylserine (PS) translocation from the inner leaflet to the outer leaflet of the lipid bilayer, while the cell membrane remains intact. Annexin V-positive cells correspond to cells that have experienced PS translocation.
3. Necrotic control is cells incubated in fresh medium with 5 mM hydrogen peroxide (H_2O_2) for 24 h. PI staining of the cells indicates that the integrity of the cell membrane has been compromised and is used to distinguish living and early apoptotic cells from necrotic cells.
4. The no-treatment control is cells in fresh medium, free from inducing agents.
5. For each treatment, detached and attached cells are pooled, harvested by trypsinization (0.25% trypsin), washed with 10 mL of culture medium, and resuspended in 100 μL of 1× annexin-binding buffer (included in annexin V-FITC/PI kit).
6. Approximately 100 μL of the cell suspension is mixed with 8 μL of annexin-V-FITC and 8 μL of 100 μg/mL Propidium Iodide (PI) and incubated in the dark at room temperature for 15 min.
7. Liquid volume is removed by centrifugation and aspiration, and the cells are resuspended by gentle vortexing in 300 μL of 1× annexin-binding buffer and subsequently analyzed on the flow cytometer.
8. Apoptotic analyses for ASCs are performed on an FACS Caliber flow cytometer (BD Biosciences, San Jose, CA) utilizing 488-nm laser excitation and fluorescence emission at 530 nm (FL1) and >575 nm (FL3).
9. Forward and side scatter measurements are made using linear amplification, and all fluorescence measurements are made with logarithmic amplification. A total of 20,000 cells per sample is acquired using Cell Quest software (BD Biosciences, San Jose, CA).

10. Quadrant analysis is performed on the fluorescence dotplot to quantify the percentage of live, necrotic, and apoptotic cell populations [36–38, 91]. The fluorescent dotplots show three cell populations: live (annexin V-FITC-negative and PI-negative; annexin *V*⁻ and *PI*⁻) necrotic (annexin V-FITC-positive and PI-positive; annexin *V*⁺ and *PI*⁺), and apoptotic (annexin V-FITC-positive and PI-negative; annexin *V*⁺ and *PI*⁻). The quadrants positions are placed according to the no-treatment control and 5 mM H_2O_2 necrotic control (Fig. 2).

3.3.2 Adipogenesis

1. Confluent cultures of ASCs are induced to adipogenesis by replacing the medium with an adipocyte induction cocktail (*see* Subheading 2, **item 5**).
2. After 72 h, the adipocyte induction medium is replaced with adipocyte maintenance media, which contains the same components as the induction medium except IBMX and rosiglitazone.
3. Induced cells are maintained in culture for 9 days, with adipocyte maintenance medium replacement every 3 days. Upon the

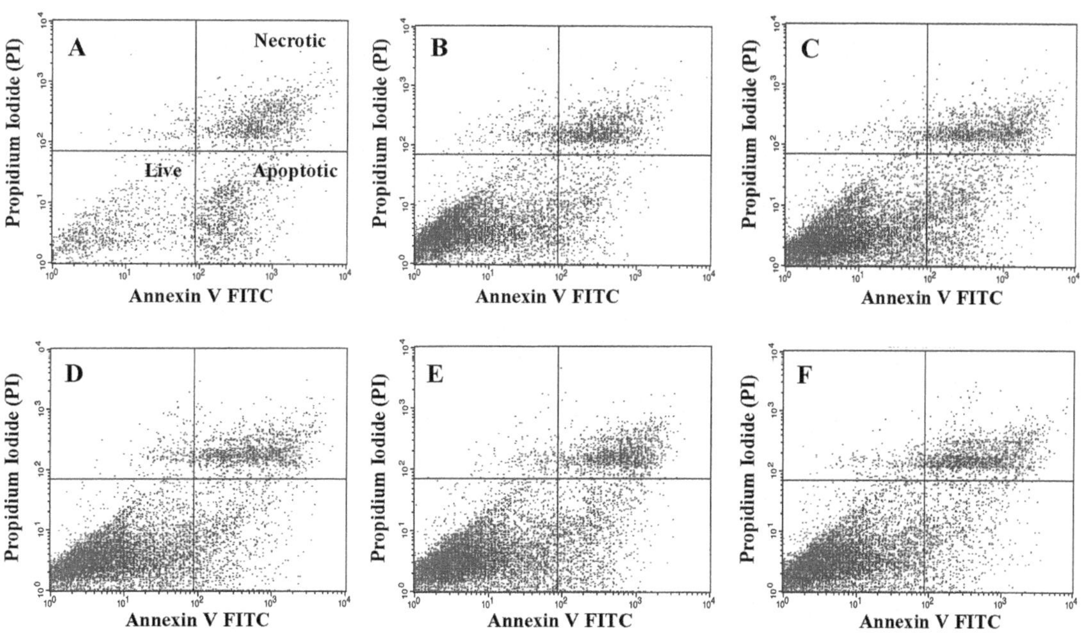

Fig. 2 Characteristic flow cytometer fluorescence dotplots showing fluorescence-activated cell sorting (FACS) analysis of SVF cells frozen/thawed in the presence various concentrations (0, 2, 4, 6, 8, or 10%) of DMSO in DMEM. Corresponding plots for other media and P1 ASCs are available in the literature [36, 38]. (**a**–**f**) represent 0%, 2%, 4%, 6%, 8% or 10% DMSO in DMEM, respectively. The fluorescent dotplots show three cell populations: live (annexin V-FITC-negative and PI-negative; annexin *V*⁻ and *PI*⁻) necrotic (annexin V-FITC-positive and PI-positive; annexin *V*⁺ and *PI*⁺), and apoptotic (annexin V-FITC-positive and PI-negative; annexin *V*⁺ and *PI*⁻). The quadrants positions were placed according to the no-treatment control, 40 μm etoposide apoptotic control, and 5 mM H_2O_2 necrotic control. Reprinted with permission from [37]

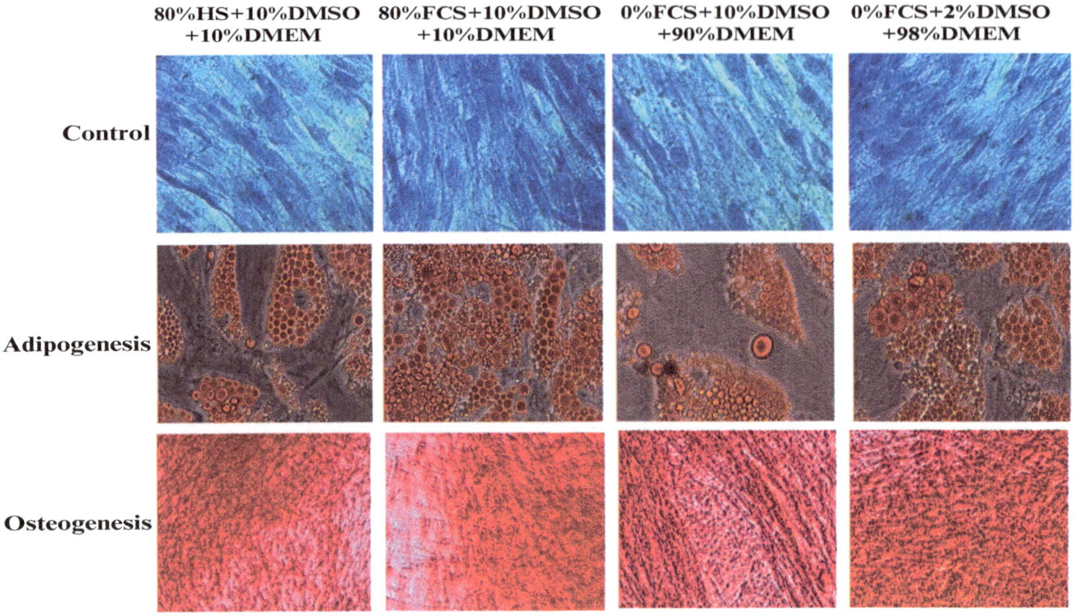

Fig. 3 Representative phase contrast photomicrographs of P1 ASCs cultured under untreated (first row; Toluidine Blue staining), adipogenic (second row; Oil Red O staining), or osteogenic (third row; Alizarin Red staining) conditions. Corresponding plots for other media and SVF cells are available in the literature [37, 38]. Adipogenic cultures were stained with Oil Red O 14 days after induction, while osteogenic cultures were stained with Alizarin Red after 21 days of culture. Images in column 1 represent cells that were cryopreserved in media containing 80% FCS with 10% DMSO in DMEM. Images in column 2 represent cells that were cryopreserved in media containing 80% HS with 10% DMSO in DMEM. Images in column 3 represent cells that were cryopreserved in media containing 0% FCS with 10% DMSO in DMEM. And finally, images in column 4 represent cells that were cryopreserved in media containing 0% FCS with 2% DMSO in DMEM. Reprinted from [36]

ninth day, the cultures are washed twice with prewarmed PBS and fixed in formalin at 4°C. Adipocyte quantification is determined by staining neutral lipids with Oil Red O (Fig. 3).

3.3.3 Osteogenesis

1. Confluent cultures of ASCs are induced to osteogenesis by replacing the medium with an osteogenic induction cocktail (*see* Subheading 2, **item 6**).
2. The induced cells are fed fresh osteogenic induction media every 3 days for 3 weeks. The cultures are then washed with 0.9% sodium chloride solution and fixed in 70% ethanol.
3. Osteoblast quantification is determined by Alizarin Red staining for calcium phosphate (Fig. 3).

3.4 Some Observations on the Effect of Various Cryoprotectants on the Post-freeze–Thaw Response of SVF of Adipose Tissue

1. The data suggests that the choice of the serum (human serum, HS; or fetal calf serum, FCS) does not significantly alter the SVF cell survival when frozen/thawed in 10% DMSO and DMEM.
2. The highest % of postthaw survival is achieved with a concentration of 8% DMSO (~73%), the postthaw survival values obtained with DMSO concentrations of 2, 4, 6, and 10% are comparable and are ~60%, ~63%, ~66%, and ~70%, respectively.

The only significant differences in the data are found when the cells are frozen with DMSO and without DMSO, i.e., even the presence of 2% DMSO is significant, and this extremely low concentration is able to cryoprotect SVF cells.

3. Replacing DMSO with MC as the CPA leads to a significant reduction (~50%) in the postthaw cell viability of SVF cells. Intriguingly, the addition of either HS or FCS with MC is extremely deleterious to postthaw cell viability.
4. DMSO is more effective in preserving cell viability than MC, i.e., the maximum % of cell survival obtained with MC (~37%) is still only half of that obtained with media containing various concentrations of DMSO (~60–70%).
5. The presence or the absence of FCS does not significantly alter the postthaw viability results for SVF cells frozen in the presence of 10% PVP.
6. The percentage of postthaw cell viability (~55%) obtained with 10% PVP and DMEM is comparable to that obtained (~64%) with control media (10% DMEM with 10% DMSO and either 80% FCS or 80% HS) and also comparable with the values obtained with DMSO and DMEM (~65%).

3.5 Some Observations on the Effect of Various Cryoprotectants on the Post-freeze–Thaw Response of P1 ASCs

1. The viability of ASCs cryopreserved in the control media containing 80% serum with 10% DMSO in DMEM is ~83%.
2. The highest % of postthaw cell survival is found for cells frozen in DMEM with 2–10% DMSO (~80% to ~88%) and the lower values are found for cells frozen in DMEM with either 40% PVP or 1% PVP (~5%); DMEM with 0% DMSO (~25%) and in DMEM with 1% MC; and either 10% HS or 10% FCS (~40%).
3. Freezing P1 ASCs in the absence of DMSO is detrimental to cell survival.
4. The addition of 1% MC significantly lowers the cell viability, i.e., the postthaw cell viability with 1% MC in DMEM (~47%) is significantly lower than that obtained with DMEM containing 1% MC and 10% DMSO (~80%).
5. The use of 1% and 40% PVP in DMEM causes a dramatic loss in cell viability (~5%), whereas the use of 10% PVP produces a maximum viability of ~65%. Although the viability of ASCs decreases when the % of PVP is increased from 10 to 20, the data analysis revealed that this increase is statistically not significant (>95% confidence level).
6. The ASCs frozen/thawed in the presence of PVP (and in the absence of both DMSO and serum) display similar morphology and growth/differentiation characteristics when compared to the ASCs cryopreserved in 80% serum with 10% DMSO in DMEM (Fig. 3).

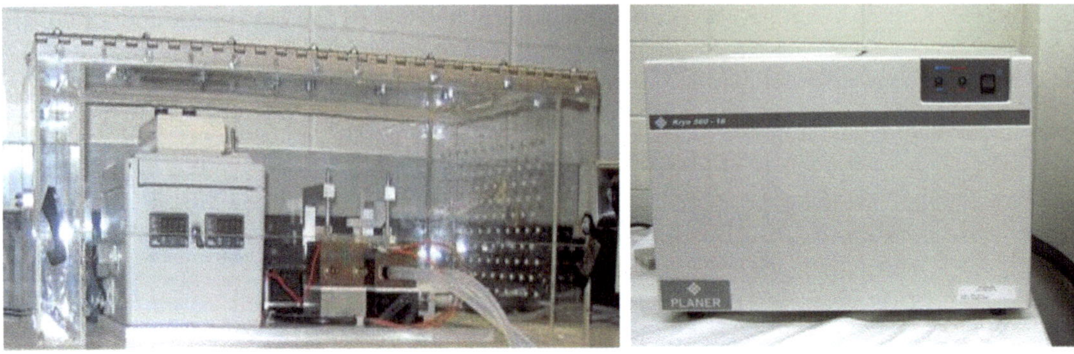

Fig. 4 Photographs of the custom-built directional solidification stage (DSS) on the left-hand side and the commercially available controlled rate freezer (CRF) on the right hand side (source: "Bioengineering Laboratory, Department of Mechanical Engineering, Louisiana State University, Baton Rouge, LA"). As described in the text, the choice of the freezing device has a significant impact on the postthaw viability of ASCs

An important (and unique) feature of our ASC freezing-storage protocols is the development of a cryopreservation media without serum (and without dimethylsulfoxide) that simplifies their use and application in in vivo tissue engineering applications.

3.6 Effect of Freezing Device on the Post-Freeze–Thaw Response of ASCs

1. The effect of directional cooling on the immediate postthaw membrane integrity of adipose tissue derived adult stem cells (ASCs) is investigated using a directional solidification stage (DSS) and a commercially available controlled rate freezer (CRF) – Fig. 4.
2. P0, P1, and P2 ASCs are cooled at either 1, 5, 20, or 40 °C/min to an end temperature of −80 °C in the presence and absence of a cryoprotective agent (dimethylsulfoxide, DMSO). For experiments in the presence of DMSO, the cell suspensions are gently mixed with 10% DMSO as a cryoprotective solution (CPA), i.e., a one-step addition of CPA. The samples are equilibrated in the cryoprotective solution for ~10 min before the DSS and CRF experiments are performed.
3. After freezing to −80 °C, the samples are thawed at 200 °C/min, and the ability of the frozen/thawed ASCs to exclude fluorescent dyes is assessed.

3.6.1 Directional Solidification Stage (DSS) Experiments

1. The directional solidification stage (DSS) built by Rubinsky and Ikeda [92] consisted of hot and cold temperature copper bases (set at T_h and T_c, respectively) separated by a distance (D_{gap}).

2. Freezing is achieved by placing the sample on a glass slide and moving the glass slide from the hot base to the cold base at a known velocity (V). The cooling rate (B) experienced by the sample is then described as follows: $B = \frac{T_h - T_c}{D_{\text{gap}}} \cdot V$
3. For the DSS freezing experiments, ~100 μL of ASC suspension is placed in a grooved microslide and traversed between the precisely controlled copper blocks at a constant velocity. Note that the first copper block is held at +4 °C and the second at −80 °C. The samples are then cooled at a nominal cooling rates of 1, 5, 20, or 40 °C/min to an end temperature of −80 °C. Due to interactions with the ambient atmosphere the actual cooling rates experienced by the cells between 0 and −40 °C are ~1.1, ~5.2, ~10.6, ~21.7, and ~43.6 °C/min, respectively [34].
4. All the glass microslides have a circular milled well, perpendicular to the direction of microslide travel, within which the cell suspension is placed. The well is of 3.2 mm width (in the direction of travel) and of 1 mm depth (into the microslide surface).
5. The thawing rate of ~200 °C/min is achieved by quickly bringing the bottom of the glass slide into contact with a warm copper block maintained at 50 °C.

3.6.2 Controlled Rate Freezer (CRF) Experiments

1. A programmable controlled rate freezing (CRF) machine (Planer Series Kryo 560-16, TS Scientific, Perkasie, PA).
2. The samples are loaded into a six-well cell culture cluster (Corning Incorporated, Corning, USA).
3. The cluster is then loaded into the cryo-machine (which is precooled to +4 °C) and kept for 1 min for equilibration. The samples are then cooled at either 1, 5, 20, or 40 °C/min to an end temperature of −80 °C.
4. The thawing rate of 200 °C/min is obtained by removing the six-well cluster from the CRF machine and quickly bringing the bottom of the cluster into contact with a warm water bath maintained at 37 °C [41].

3.7 Some Observations on the Effect of Freezing Device on the Post-Freeze–Thaw Response of ASCs

1. The viability of cell suspensions is measured using Calcein AM and Propidium Iodide (PI) [93].
2. A 10 μL sample of cell suspension is removed from the microslide and incubated with 2 μmol/L Calcein AM and 3 μmol/L PI for 15–30 min at 37 °C.
3. After incubation, cells are scored as either live or dead under a fluorescent microscope.

4. Viability measurements of cells taken directly from the stock suspension without freezing are determined before and after every freeze–thaw experiment as a control; all experimental viability values are normalized to the average control viability.
5. A comparison of the P1 ASCs postfreeze viability in the presence and absence of DMSO obtained using the DSS and the CRF is shown in Fig. 5. Corresponding data for P0 and P2 cells are available elsewhere [34].
6. The postfreeze viability of ASCs is significantly higher (confidence level > 95%) when they are frozen in the presence of DMSO, than in its absence.
7. A cooling rate of 1 °C/min results in the highest postthaw cell survival for ASCs frozen using either device.
8. The postfreeze viability of ASCs is uniformly higher (confidence level > 95%) when they are frozen using the CRF as opposed to using the DSS. This significant reduction in cell viability for the cells frozen using the DSS might be related to previously postulated mechanical/damaging interactions between the ice crystals and the cell membranes [94–97].

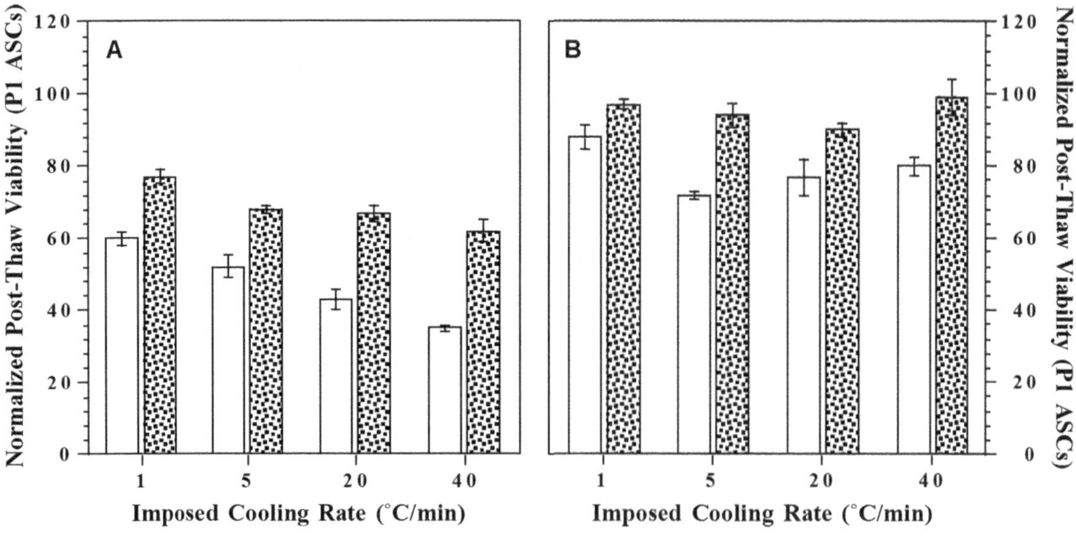

Fig. 5 Comparison of postthaw viability of P1 ASCs frozen in the absence (**a**) and presence (**b**) of 10% (v/v) DMSO is shown. In each figure, the postthaw viability obtained using the DSS (unfilled columns) and CRF (filled columns) is also shown. The error bars represent standard deviation in the data ($n = 18$). In both the figures, the immediate postthaw ASC membrane integrity is shown on the Y-axis while the cooling rate imposed on the sample (°C/min) is shown on the X-axis. Redrawn from [34]

3.8 Effect of Preheating (Inducing Heat Shock Proteins) on the Post-freeze–Thaw Response of ASCs

1. Heat shock proteins (HSPs) are known to be upregulated when the cells encounter unfavorable conditions such as elevated temperatures, hypoxia, heavy metals, ethanol, oxidative phosphorylation inhibitors, and other chemicals [98]. They function as molecular chaperones in proper folding of proteins, reduce protein aggregation, and also alleviate oxidative stress that occurs in various pathological conditions [99, 100].
2. Literature suggests that HSP70, HSP90, and HSP27 inhibit the apoptosis pathway in stressed cells and may provide cytoprotection [101–106]. For example, studies conducted on hela cells indicate that heat shock treatment significantly improves the survival rate in comparison to control hela cells [107] whereas in human fibroblasts the induction of HSPs did not provide cytoprotection [108].
3. Thus, the hypothesis of these experiments: Induction of HSPs prior to the freeze–thaw insult can improve the postthaw viability of ASCs by attenuating the effects of freezing associated stresses.

3.8.1 Collection and Isolation of ASCs

The culturing of ASCs, preparation of freezing media, and evaluation of postthaw viability are done as described earlier.

3.8.2 Heat Shock Treatment

1. The P1 ASCs after reaching 80% confluence are placed in a cell culture incubator that is adjusted to 43 (±0.2) °C for 1 h to induce the HSPs expression.
2. After 1 h of heat shock, the cells re transferred back to cell culture incubator at 37 °C and incubated for 18 h prior to freezing.

3.8.3 Freezing–Thawing Protocol

1. The heat-shocked ASCs are suspended at a density of 1.0×10^6 cells/mL with DMSO at concentrations of 10%, 1% and PVP at 10% and without any CPA in DMEM media containing 10% FBS in 1.5 mL cryovials. The ASCs without heat shock are prepared in the same way and used as controls.
2. Using a CRF the cryovials are cooled down to −35 ° C from room temperature at a cooling rate of 1 °C/min followed by plunging in liquid nitrogen. The cells are stored in liquid nitrogen for 1 day.
3. The next day the cryovials are taken out from liquid nitrogen and thawed in 37 °C water bath for 1–2 min by gentle shaking.
4. The thawed ASCs are resuspended in stromal media and cultured for 24 h before assessing the postthaw viability using Annexin V/Propidium Iodide flow cytometer based method as described earlier.

3.8.4 Casting Polyacrylamide Gels for SDS-PAGE

1. Resolving gel master mix: 400 mL distilled water, 250 mL of 1.5 M Tris buffer of pH 8.8, and 10 mL of 10% SDS.
2. Stacking gel master mix: 340 mL of distilled water, 62.5 mL of 1.0 M Tris buffer of pH 6.8, and 5 mL of 10% SDS.
3. Pouring resolving gel: To make 6 mL of resolving gel, pour 4 mL of resolving gel master mix into a 50 mL tube followed by 2 mL of 30% acrylamide and 60 μL of 10% APS. Vortex the mixture quite vigorously for 1–2 min and add 2.5 μL of TEMED. Vortex or shake the mixture vigorously again for another 1 min.
4. Immediately load gel mixture into the casing with a pipette—fill to the line on the casing. Add distilled water on top of the gel to eliminate or mitigate the presence of any air bubbles. Allow 30 min to polymerize at room temperature (25–27 °C). Remove water with a pipette or a blotting paper after the polymerization process. Use the gel immediately or store for later use at 4 °C in 1% Tris–glycine buffer.
5. Stacking gel: To make ~3 mL of stacking gel, add 2.5 mL of stacking gel master mix, 0.5 mL of 30% acrylamide, and 30 μL of 10% APS. Vortex the mixture quite vigorously for 1–2 min and add 2.5 μL of TEMED. Vortex or shake the mixture vigorously again for another 1 min.
6. Immediately or before the gel solidifies (~2 min) insert the gel comb carefully to avoid incorporating air bubbles (*see* **Note 15**).
7. Allow the gel to polymerize or solidify (~15–30 min). Carefully remove the gel comb from the top of the stacking gel (*see* **Note 16**).

3.8.5 Western Blotting

1. The ASCs recovered for 18 h post heat shock are lysed with ice-cold RIPA buffer with protease cocktail inhibitor.
2. The collected cell lysates are centrifuged at 8000 × *g* for 10 min at 4 °C using a cooling centrifuge and the supernatants are collected.
3. The total protein concentration is estimated using BCA assay.
4. Denature the proteins by adding the loading buffer. Heat the cell lysate and loading buffer for 3–5 min at 95 °C (*see* **Note 17**).
5. Let the sample cool to room temperature (25–27 °C) or ~5–10 min. Apply the appropriate power pack (160 V) until the Bromophenol Blue dye permeates to the bottom of the gel or approximately 40–45 min (*see* **Note 18**).
6. The protein samples are separated using 10% polyacrylamide gel using SDS PAGE and then transferred onto nitrocellulose membrane via the application of electrical field (70 V for 2 h) (*see* **Note 19**).

7. The membrane is then blocked using 5% nonfat dry milk prepared in 1× TBST for 1 h. Later the membrane is incubated overnight with antibodies for HSP90, HSP70, and β-actin at dilution of 1:200 and HSP27 with 1:5000 dilution at 4 °C.
8. The membrane is washed several times with 1× TBST before incubation of HRP conjugated secondary antibody at dilution of 1:2000 for 1 h.
9. One hour after incubation of secondary antibody the membrane is washed again several times and treated with chemiluminescent substrate.
10. The detection of chemiluminescence signal is done using typhoon laser imaging scanner (GE Life Sciences, Pittsburgh, PA).

3.8.6 qPCR

1. The total RNA of heat-shocked ASCs is isolated using PureLink RNA Mini Kit (Life Technologies, Grand Island, NY).
2. The isolated RNA is quantified using NanoDrop spectrophotometer and reverse transcribed using high capacity cDNA reverse transcription kit (Applied Biosystems, Grand Island, NY).
3. The synthesized cDNA is later used to carry out qPCR using SYBR select master mix (Applied Biosystems, Grand Island, NY). The qPCR data is analyzed using Δ ΔCt method.

3.8.7 Osteogenesis—Alizarin Red Staining and Quantification

To determine the effect of heat shock on osteogenesis ASCs aere given heat shock on 3rd, 9th, and 15th day of osteogenic differentiation. Osteogenic induction and heat shock are performed as described in the above sections (Subheadings 3.4 and 3.9, **step 1**).

1. Forty millimolar Alizarin Red S solution is prepared in distilled water and the pH is adjusted to 4.1 using ammonium hydroxide.
2. On the 21st day of osteogenic induction the cells are fixed with 4% paraformaldehyde for 30 min and stained with the 40 mM Alizarin Red S solution for 30 min.
3. The stained wells are washed with distilled water to remove excess unbound stain and examined with inverted microscope.
4. The Alizarin Red S stained wells are treated with 10% cetylpyiridium chloride to elute the calcium-bound stain and the absorbance of that solution is measured at 530 nm using spectrophotometer.
5. A standard plot is also generated with various concentration of Alizarin Red S stain which is used to quantify the amount of eluted dye.

3.9 Some Observations on the Effect of Inducing Heat Shock Proteins Prior to Freeze–Thaw Process

1. Western blotting expression of HSPs in ASCs is shown in Fig. 5. Heat treatment for 1 h at 41 °C *does* not significantly enhance the expression of HSPs studied, i.e., HSP27, HSP70, and HSP90. Increasing the preheat treatment temperature to 43 °C for 1 h shows an upregulation of HSP70 with no significant change in either HSP27 or HSP90.
2. The postthaw viability of heat-shocked (at 43 °C for 1 h) ASCs in comparison to controls is shown in Table 1. The lack of significant differences between the heat-treated cells and control suggests that the upregulation of HSP70 alone is *not* sufficient to provide cryoprotection in ASCs; with the caveat that the "time point" of freezing after the heat shock treatment might *not* be optimal.
3. To determine the time point where maximum expression of HSPs occur after heat shock, qPCR is performed on RNA isolated from ASCs at 3, 9, 18, 48, and 72 h after heat shock treatment and is shown in Fig. 6.
4. The data indicates that only HSP70 and HSP27 mRNA expression levels decreased with time whereas HSP90 remained virtually unchanged. Specifically, 3 h after the heat shock HSP70 and HSP27 mRNA levels reached their peak levels (Fig. 7). At 18 h after the heat shock treatment the mRNA levels of both HSP70 and HSP27 are at the basal level. However, western blotting data indicates that HSP70 protein levels are still elevated 18 h post heat shock whereas little change in HSP27 is observed.
5. The data in Fig. 7 suggests that freezing ASCs after 18 h of HSP induction is *not* optimal and that freezing ASCs 3 h after heat treatment might be more efficacious.

Table 1
Post-freeze/thaw cell viability of ASCs with (43 °C for 1 h) and without heat treatment in various cryopreservation media

	% postthaw cell viability normalized to prefreeze cells	
Cryoprotectant	**No heat shock**	**Heat shocked**
No cryoprotectant	34.7 (±7.25)	26.2(±2.56)
1% DMSO	70.9(±9.68)	72.1(±4.63)
10% DMSO	91.6(±1.67)	93.6(±1.33)
10% PVP	74.92(±3.86)	79.33(±3.65)

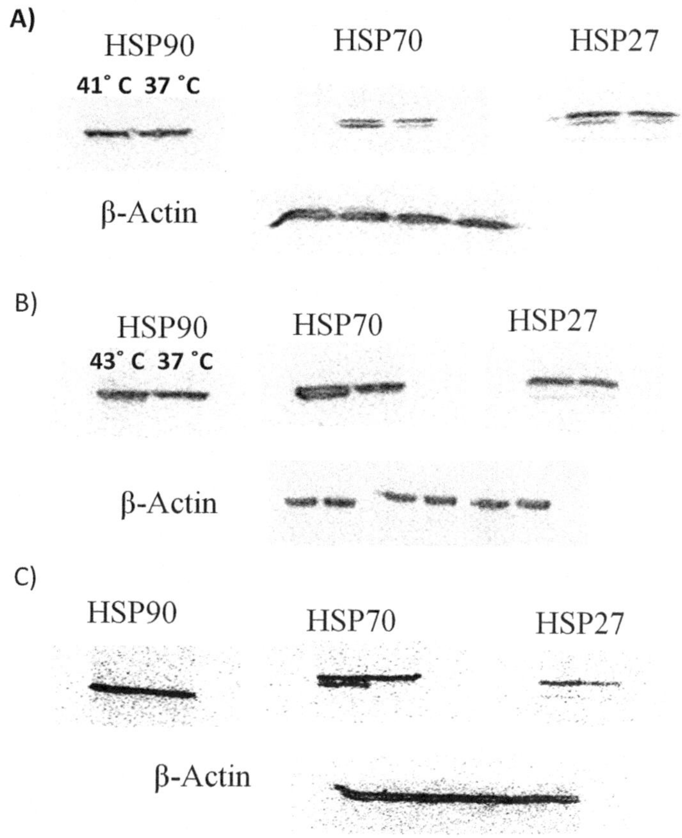

Fig. 6 Western blotting of HSP90, HSP70, and HSP27 of heat-shocked ASCs at different temperatures: 41 °C for 1 h (**a**) and 43 °C for 1 h (**b**). The elevated expression level of HSP70 is conserved in heat-shocked ASCs after cryopreservation (**c**)

6. No conclusions can be drawn, as yet, regarding the relationship between prefreeze HSP induction and the post-freeze–thaw viability of ASCs.
7. Further studies, including western blotting, are currently being pursed to assess the protein levels of HSP70 and HSP27 at various time points after the heat shock treatment (and to identify the peak time points and optimal time points for applying the freezing insult).
8. Future studies will also assess the effect of combinatorial upregulation of HSPs on the post-freeze–thaw viability of ASCs by overexpressing or knocking down certain HSPs using genetic engineering.

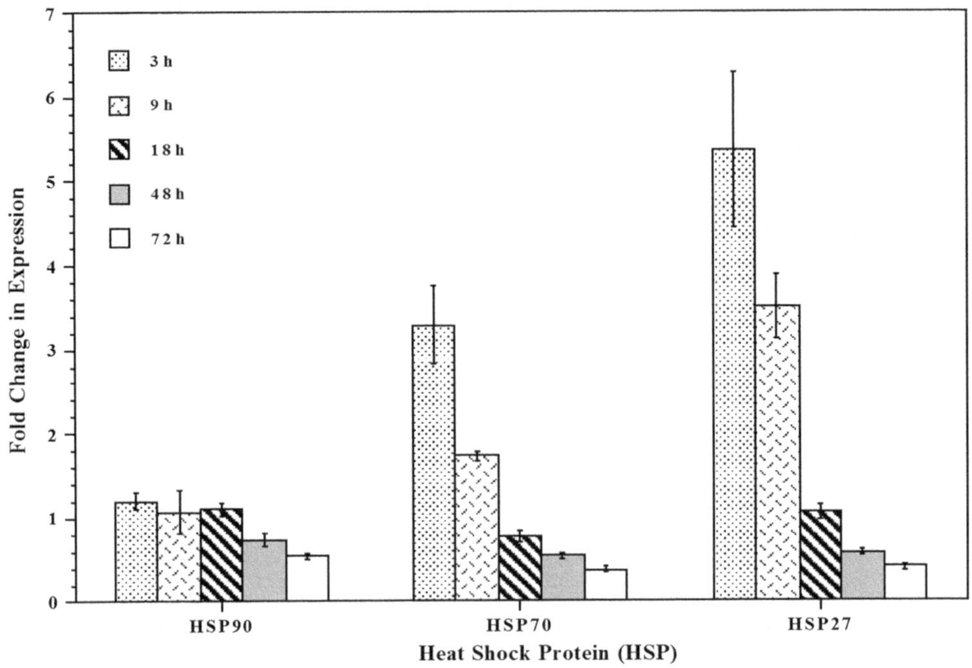

Fig. 7 The expression levels of HSP90, HSP70, and HSP27 at various time points after heat shock treatment at 43 °C for 1 h

3.10 Some Observations on the Effect of Periodic Heat Shock on ASCs During Osteogenesis

1. The hypothesis is that the periodic upregulation or elevation of HSPs might be beneficial or detrimental to ASC osteogenesis.
2. ASCs that are heat-treated (43 °C for 1 h) are periodically heat-treated (on the 3rd, 9th, and 15th day) of osteogenic differentiation. The data suggests a decrease in osteogenesis staining with periodic heat treatment and decreased the quantified Alizarin Red value from 1.35 to 1.08 mM Alizarin Red stain/9.5 cm^2 well (control value is 0.4 mM).

4 Notes

1. Commercially available programmable controlled rate freezer is costly ($20,000). Cheaper alternatives might be a −80 °C freezer.
2. Tissue culture media are highly species specific. Cryoprotectants and growth media components must be of high purity (analar grade and spectroscopic grade for DMSO).
3. DMSO freely penetrates the skin and can be irritating to the eyes and skin.
4. Where possible, cryoprotectant mixtures should be sterilized by filtration and high viscosity solutions by autoclaving.

5. Minimal *g* forces need to be applied during centrifugation.
6. The outcome of any cryopreservation procedure will depend on the quality of the prefrozen sample. Therefore, cells harvested for cryopreservation should be at their optimum viability to ensure maximal survival during freezing and after thawing.
7. The concentration, duration, and temperature of the addition of the CPAs are critical factors for the success of the whole procedure. Care should be taken to minimize the time of exposure to CPAs. Once the cells have been prepared for freezing they should be frozen as soon as possible.
8. Physical factors influencing the success of a cryopreservation procedure include the imposed cooling rate, the composition of the freezing media, the time of exposure to the cryoprotectants, the nature and sequence of the cryoprotectant addition and removal procedures, the storage temperature, and the thawing process. All of these variables need to be carefully controlled during the freezing process to ensure an optimal outcome.
9. Storage in the vapor phase of the liquid nitrogen is recommended. However, storage in the liquid phase does prevent the possibility of inadvertent warming.
10. Extreme caution should be exercised while removing the samples from the liquid phase as the penetration of liquid nitrogen into defective containers can cause excessive pressure (and possibly, explosive) build up of gas in a confined space during the thawing process. To reduce this possibility, cryovials should be allowed to equilibrate in the vapor phase before being transferred to a warm water bath for thawing. It might also be advisable to place the cryovials in a "closed metal container" (to contain any exploded material for a simpler decontamination procedure).
11. Storage temperature needs to be maintained below −139 °C.
12. Storage time does not influence the viability of the stored material. Therefore, the whole duration of low temperature storage can be as little as 1 h, greatly reducing the length of the whole procedure.
13. To retain maximum viability, cells should be thawed rapidly and uniformly, but carefully, so that the maximum temperature does not exceed normal temperature range.
14. After thawing it is necessary to remove or slowly dilute the cryoprotectant to prevent osmotic shock. DMSO will evaporate from the medium at 37 °C.
15. The presence of spurious air bubbles during the pouring of the resolving gel can be eliminated by the use of a small amount of distilled water or EtOH on the top surface of the resolving gel. However, the presence of spurious or damaging air bubbles dur-

ing the stacking gel casting can be eliminated by removing some of the stacking gel and replenishing it. To eliminate the possibility of the gel solidifying during this process, it is imperative to insert as carefully as possible within a short period of time.

16. Improper removal of the gel combs from the top of the polymerizer stacking gel will result in breakage or damage the structural integrity of the wells resulting in a catastrophic failure of the casting process and initiating the process once again from the initial pouring of the resolving gel step.
17. Heating the cell lysate (proteins) plus loading buffer for more than 5 min will result in precipitation of proteins or degraded proteins that are unsuitable for the western blot assay.
18. Gently centrifuge the sample to minimize loss of sample via condensation on the tube walls. The protein samples should be carefully loaded to avoid spillage and contamination between wells.
19. Extreme care should be taken to avoid any trapped air or air bubbles between the nitrocellulose membrane and the SDS-PAGE gel during the protein transfer process. Care should also be taken to avoid contaminating with spurious proteins onto the nitrocellulose membrane during the disassembly and subsequent handling process.

Acknowledgments

The author thanks Dr. Elizabeth Clubb and Dr. James Wade at the Pennington Biomedical Research Center (PBRC) for supplying the liposuction aspirates and their many patients for consenting to participate in this study; Marilyn Dietrick of the LSU School of Veterinary Medicine Flow Cytometry Core Facility; Prof. Jeffrey Gimble, Gang Yu, and Xiying Wu of the Stem Cell Biology Laboratory at the Pennington Biomedical Research Center (PBRC), and the clinical nutrition research unit (CNRU) Molecular Mechanism Core at PBRC for their technical assistance. Acknowledgements are also due to Dr. Gimble, a long-time colleague and friend without whose help this work would never have been initiated. In addition, acknowledgements are also due to Dr. S. Thirumala (freezing experiments) and R. Fuller (comparing the freezing devices, CRF and DSS), for the results reported in this chapter. This work was supported in part by funding from the Louisiana Board of Regents and the Department of Mechanical Engineering at the Louisiana State University (LSU). Research reported in this chapter was also partly supported by the National Institute of Diabetes and Digestive and Kidney Diseases of the National Institutes of Health under award number R21DK91852.

References

1. De Ugarte DA, Morizono K, Elbarbary A, Alfonso Z, Zuk PA, Zhu M, Dragoo JL, Ashjian P, Thomas B, Benhaim P, Chen I, Fraser J, Hedrick MH (2003) Comparison of multi-lineage cells from human adipose tissue and bone marrow. Cells Tissues Organs 174:101–109
2. Kang SK, Putnam L, Dufour J, Ylostalo J, Jung JS, Bunnell BA (2004) Expression of telomerase extends the lifespan and enhances osteogenic differentiation of adipose tissue-derived stromal cells. Stem Cells 22:1356–1372
3. Mitchell JB, McIntosh K, Zvonic S, Garrett S, Floyd ZE, Kloster A, Di Halvorsen Y, Storms RW, Goh B, Kilroy G, Wu X, Gimble JM (2006) Immunophenotype of human adipose-derived cells: temporal changes in stromal-associated and stem cell-associated markers. Stem Cells 24:376–385
4. Rubio D, Garcia-Castro J, Martin MC, de la Fuente R, Cigudosa JC, Lloyd AC, Bernad A (2005) Spontaneous human adult stem cell transformation. Cancer Res 65:3035–3039
5. Wall ME, Bernacki SH, Loboa EG (2007) Effects of serial passaging on the adipogenic and osteogenic differentiation potential of adipose-derived human mesenchymal stem cells. Tissue Eng 13:1291–1298
6. Gimble JM, Katz AJ, Bunnell BA (2007) Adipose-derived stem cells for regenerative medicine. Circ Res 100:1249–1260
7. Zhang FB, Li L, Fang B, Zhu DL, Yang HT, Gao PJ (2005) Passage-restricted differentiation potential of mesenchymal stem cells into cardiomyocyte-like cells. Biochem Biophys Res Commun 336:784–792
8. Tsutsumi S, Shimazu A, Miyazaki K, Pan H, Koike C, Yoshida E, Takagishi K, Kato Y (2001) Retention of multilineage differentiation potential of mesenchymal cells during proliferation in response to FGF. Biochem Biophys Res Commun 288:413–419
9. Zeng X, Rao MS (2007) Human embryonic stem cells: long term stability, absence of senescence and a potential cell source for neural replacement. Neuroscience 145: 1348–1358
10. Zimmermann S, Martens UM (2008) Telomeres, senescence, and hematopoietic stem cells. Cell Tissue Res 331:79–90
11. Piacibello W, Gammaitoni L, Pignochino Y (2005) Proliferative senescence in hematopoietic stem cells during ex-vivo expansion. Folia Histochem Cytobiol 43:197–202
12. Pegg DE (1987) Mechanisms of freezing damage. Symp Soc Exp Biol 41:363–378
13. Day JG (2007) Cryopreservation of microalgae and cyanobacteria. Methods Mol Biol 368:141–151
14. Pegg DE (2007) Principles of cryopreservation. Methods Mol Biol 368:39–57
15. Bischof JC, Rubinsky B (1993) Large ice crystals in the nucleus of rapidly frozen liver cells. Cryobiology 30:597–603
16. Toner M, Cravalho EG, Stachecki J, Fitzgerald T, Tompkins RG, Yarmush ML, Armant DR (1993) Nonequilibrium freezing of one-cell mouse embryos. Membrane integrity and developmental potential. Biophys J 64:1908–1921
17. Venkatasubramanian RT, Grassl ED, Barocas VH, Lafontaine D, Bischof JC (2006) Effects of freezing and cryopreservation on the mechanical properties of arteries. Ann Biomed Eng 34:823–832
18. Schill RO, Mali B, Dandekar T, Schnolzer M, Reuter D, Frohme M (2009) Molecular mechanisms of tolerance in tardigrades: new perspectives for preservation and stabilization of biological material. Biotechnol Adv 27:348–352
19. Franca MB, Panek AD, Eleutherio EC (2007) Oxidative stress and its effects during dehydration. Comp Biochem Physiol A Mol Integr Physiol 146:621–631
20. Liang Y, Sun WQ (2002) Rate of dehydration and cumulative desiccation stress interacted to modulate desiccation tolerance of recalcitrant cocoa and ginkgo embryonic tissues. Plant Physiol 128:1323–1331
21. Hoekstra FA, Golovina EA, Buitink J (2001) Mechanisms of plant desiccation tolerance. Trends Plant Sci 6:431–438
22. Allison SD, Randolph TW, Manning MC, Middleton K, Davis A, Carpenter JF (1998) Effects of drying methods and additives on structure and function of actin: mechanisms of dehydration-induced damage and its inhibition. Arch Biochem Biophys 358:171–181
23. Setlow P (1995) Mechanisms for the prevention of damage to DNA in spores of Bacillus species. Annu Rev Microbiol 49:29–54
24. Crowe LM, Crowe JH (1992) Anhydrobiosis: a strategy for survival. Adv Space Res 12:239–247
25. Clegg JS, Seitz P, Seitz W, Hazlewood CF (1982) Cellular responses to extreme water loss: the water-replacement hypothesis. Cryobiology 19:306–316

26. Yu J, Anchordoquy TJ (2009) Synergistic effects of surfactants and sugars on lipoplex stability during freeze-drying and rehydration. J Pharm Sci 98:3319–3328
27. Jiang W, Schwendeman SP (2008) Stabilization of tetanus toxoid encapsulated in PLGA microspheres. Mol Pharm 5:808–817
28. Santivarangkna C, Kulozik U, Foerst P (2008) Inactivation mechanisms of lactic acid starter cultures preserved by drying processes. J Appl Microbiol 105:1–13
29. Stoner GD, Chen T, Kresty LA, Aziz RM, Reinemann T, Nines R (2006) Protection against esophageal cancer in rodents with lyophilized berries: potential mechanisms. Nutr Cancer 54:33–46
30. Hansen RK, Zhai S, Skepper JN, Johnston MD, Alpar HO, Slater NK (2005) Mechanisms of inactivation of HSV-2 during storage in frozen and lyophilized forms. Biotechnol Prog 21:911–917
31. Arakawa T, Prestrelski SJ, Kenney WC, Carpenter JF (2001) Factors affecting short-term and long-term stabilities of proteins. Adv Drug Deliv Rev 46:307–326
32. Milano MT, Bernhard WA (1999) The influence of packing on free radical yields in solid-state DNA: film compared to lyophilized frozen solution. Radiat Res 152:196–201
33. Kreilgaard L, Frokjaer S, Flink JM, Randolph TW, Carpenter JF (1998) Effects of additives on the stability of recombinant human factor XIII during freeze-drying and storage in the dried solid. Arch Biochem Biophys 360:121–134
34. Fuller R, Devireddy RV (2008) The effect of two different freezing methods on the immediate post-thaw membrane integrity of adipose tissue derived stem. Int J Heat Mass Transf 51:5650–5654
35. Goh BC, Thirumala S, Kilroy G, Devireddy RV, Gimble JM (2007) Cryopreservation characteristics of adipose-derived stem cells: maintenance of differentiation potential and viability. J Tissue Eng Regen Med 1:322–324
36. Thirumala S, Gimble J, Devireddy RV (2009) Evaluation of methylcellulose and dimethylsulfoxide as the cryoprotectants in a serum free freezing media for cryopreservation of adipose derived adult stem cells. Stem Cells Dev 19(4):513–522
37. Thirumala S, Gimble J, Devireddy RV (2009) Cryopreservation of stromal vascular fraction of adipose tissue in a serum free freezing media. J Tissue Eng Regen Med 4(3):224–232
38. Thirumala S, Wu X, Gimble J, Devireddy RV (2009) Evaluation of polyvinylpyrollidone (PVP) as a cryoprotectant for adipose derived adult stem cells (ASCs). Tissue Eng Part C Methods 16(4):783–792
39. Devireddy RV, Thirumala S, Gimble JM (2005) Cellular response of adipose derived passage-4 adult stem cells to freezing stress. J Biomech Eng 127:1081–1086
40. Thirumala S, Gimble JM, Devireddy RV (2005) Transport phenomena during freezing of adipose tissue derived adult stem cells. Biotechnol Bioeng 92:372–383
41. Thirumala S, Zvonic S, Floyd E, Gimble JM, Devireddy RV (2005) Effect of various freezing parameters on the immediate post-thaw membrane integrity of adipose tissue derived adult stem cells. Biotechnol Prog 21:1511–1524
42. Mazur P (1984) Freezing of living cells: mechanisms and implications. Am J Phys 247:C125–C142
43. Polge C, Smith AU, Parkes AS (1949) Revival of spermatozoa after vitrification and dehydration at low temperatures. Nature 164:666
44. McGrath JJ (1997) Quantitative measurement of cell membrane transport: technology and applications. Cryobiology 34:315–334
45. Anchordoguy TJ, Cecchini CA, Crowe JH, Crowe LM (1991) Insights into the cryoprotective mechanism of dimethyl sulfoxide for phospholipid bilayers. Cryobiology 28:467–473
46. Anchordoguy T, Carpenter JF, Loomis SH, Crowe JH (1988) Mechanisms of interaction of amino acids with phospholipid bilayers during freezing. Biochim Biophys Acta 946:299–306
47. Yu ZW, Quinn PJ (1994) Dimethyl sulphoxide: a review of its applications in cell biology. Biosci Rep 14:259–281
48. Fahy GM, Lilley TH, Linsdell H, Douglas MS, Meryman HT (1990) Cryoprotectant toxicity and cryoprotectant toxicity reduction: in search of molecular mechanisms. Cryobiology 27:247–268
49. Fahy GM (2010) Cryoprotectant toxicity neutralization. Cryobiology 60(3 Suppl):S45–S53
50. Fahy GM, Wowk B, Wu J, Paynter S (2004) Improved vitrification solutions based on the predictability of vitrification solution toxicity. Cryobiology 48:22–35
51. Fahy GM (1986) The relevance of cryoprotectant "toxicity" to cryobiology. Cryobiology 23:1–13
52. Farrant J (1969) Is there a common mechanism of protection of living cells by polyvinyl-

pyrrolidone and glycerol ding freezing? Nature 222:1175–1176

53. Hey JM, MacFarlane DR (1998) Crystallization of ice in aqueous solutions of glycerol and dimethyl sulfoxide 2: ice crystal growth kinetics. Cryobiology 37:119–130
54. Shaw JM, Kuleshova LL, MacFarlane DR, Trounson AO (1997) Vitrification properties of solutions of ethylene glycol in saline containing PVP, Ficoll, or dextran. Cryobiology 35:219–229
55. Alapati R, Stout M, Saenz J, Gentry GT Jr, Godke RA, Devireddy RV (2009) Comparison of the permeability properties and post-thaw motility of ejaculated and epididymal bovine spermatozoa. Cryobiology 59(2):164–170
56. Hagiwara M, Choi JH, Devireddy RV, Roberts KP, Wolkers WF, Makhlouf A, Bischof JC (2009) Cellular biophysics during freezing of rat and mouse sperm predicts postthaw motility. Biol Reprod 81(4):700–706
57. Alapati R, Goff K, Kubisch HM, Devireddy RV (2008) Water transport in epididymal and ejaculated rhesus monkey (Macaca mulatta) sperm during freezing. Cryobiology 57:182–185
58. Kardak A, Leibo SP, Devireddy R (2007) Membrane transport properties of equine and macaque ovarian tissues frozen in mixtures of dimethylsulfoxide and ethylene glycol. J Biomech Eng 129:688–694
59. Li G, Saenz J, Godke RA, Devireddy RV (2006) Effect of glycerol and cholesterol-loaded cyclodextrin on freezing-induced water loss in bovine spermatozoa. Reproduction 131:875–886
60. Pinisetty D, Huang C, Dong Q, Tiersch TR, Devireddy RV (2005) Subzero water permeability parameters and optimal freezing rates for sperm cells of the southern platyfish, Xiphophorus maculatus. Cryobiology 50:250–263
61. Thirumala S, Huang C, Dong Q, Tiersch TR, Devireddy RV (2005) A theoretically estimated optimal cooling rate for the cryopreservation of sperm cells from a live-bearing fish, the green swordtail Xiphophorus helleri. Theriogenology 63:2395–2415
62. Meryman HT (1974) Freezing injury and its prevention in living cells. Annu Rev Biophys Bioeng 3:341–363
63. Meryman HT (1956) Mechanics of freezing in living cells and tissues. Science 124:515–521
64. Meryman HT (2007) Cryopreservation of living cells: principles and practice. Transfusion 47:935–945
65. Connor W, Ashwood-Smith MJ (1973) Cryoprotection of mammalian cells in tissue culture with polymers; possible mechanisms. Cryobiology 10:488–496
66. Holt WV, North RD (1994) Effects of temperature and restoration of osmotic equilibrium during thawing on the induction of plasma membrane damage in cryopreserved ram spermatozoa. Biol Reprod 51:414–424
67. Barrios B, Perez-Pe R, Gallego M, Tato A, Osada J, Muino-Blanco T, Cebrian-Perez JA (2000) Seminal plasma proteins revert the cold-shock damage on ram sperm membrane. Biol Reprod 63:1531–1537
68. Rodgers FG, Davey MR (1982) Ultrastructure of the cell envelope layers and surface details of Legionella pneumophila. J Gen Microbiol 128:1547–1557
69. De Leeuw FE, De Leeuw AM, Den Daas JH, Colenbrander B, Verkleij AJ (1993) Effects of various cryoprotective agents and membrane-stabilizing compounds on bull sperm membrane integrity after cooling and freezing. Cryobiology 30:32–44
70. Barnard T (1987) Rapid freezing techniques and cryoprotection of biomedical specimens. Scanning Microsc 1:1217–1224
71. Hirsh AG, Williams RJ, Meryman HT (1985) A novel method of natural cryoprotection : intracellular glass formation in deeply frozen Populus. Plant Physiol 79:41–56
72. Rall WF, Mazur P, McGrath JJ (1983) Depression of the ice-nucleation temperature of rapidly cooled mouse embryos by glycerol and dimethyl sulfoxide. Biophys J 41:1–12
73. Elliott R, Szleifer I, Schick M (2007) A microscopic model calculation of the phase diagram of ternary mixtures of cholesterol and saturated and unsaturated phospholipids. Methods Mol Biol 398:303–317
74. Fahy GM (1980) Analysis of "solution effects" injury. Equations for calculating phase diagram information for the ternary systems NaCl-dimethylsulfoxide-water and NaCl-glycerol-water. Biophys J 32:837–850
75. Guignon B, Aparicio C, Otero L, Sanz PD (2009) Prediction of ice content in biological model solutions when frozen under high pressure. Biotechnol Prog 25:454–460
76. Kleinhans FW, Mazur P (2007) Comparison of actual vs. synthesized ternary phase diagrams for solutes of cryobiological interest. Cryobiology 54:212–222
77. Morris GJ, Goodrich M, Acton E, Fonseca F (2006) The high viscosity encountered during freezing in glycerol solutions: effects on cryopreservation. Cryobiology 52:323–334

78. Watson PF, Duncan AE (1988) Effect of salt concentration and unfrozen water fraction on the viability of slowly frozen ram spermatozoa. Cryobiology 25:131–142
79. Rall WF, Mazur P, Souzu H (1978) Physical-chemical basis of the protection of slowly frozen human erythrocytes by glycerol. Biophys J 23:101–120
80. Shepard ML, Goldston CS, Cocks FH (1976) The H_2O-NaCl-glycerol phase diagram and its application in cryobiology. Cryobiology 13:9–23
81. Guilak F, Lott KE, Awad HA, Cao Q, Hicok KC, Fermor B, Gimble JM (2006) Clonal analysis of the differentiation potential of human adipose-derived adult stem cells. J Cell Physiol 206:229–237
82. DeLany JP, Floyd ZE, Zvonic S, Smith A, Gravois A, Reiners E, Wu X, Kilroy G, Lefevre M, Gimble JM (2005) Proteomic analysis of primary cultures of human adipose-derived stem cells: modulation by adipogenesis. Mol Cell Proteomics 4:731–740
83. Hicok KC, Du Laney TV, Zhou YS, Halvorsen YD, Hitt DC, Cooper LF, Gimble JM (2004) Human adipose-derived adult stem cells produce osteoid in vivo. Tissue Eng 10:371–380
84. Safford KM, Safford SD, Gimble JM, Shetty AK, Rice HE (2004) Characterization of neuronal/glial differentiation of murine adipose-derived adult stromal cells. Exp Neurol 187:319–328
85. Aust L, Devlin B, Foster SJ, Halvorsen YD, Hicok K, du Laney T, Sen A, Willingmyre GD, Gimble JM (2004) Yield of human adipose-derived adult stem cells from liposuction aspirates. Cytotherapy 6:7–14
86. Awad HA, Wickham MQ, Leddy HA, Gimble JM, Guilak F (2004) Chondrogenic differentiation of adipose-derived adult stem cells in agarose, alginate, and gelatin scaffolds. Biomaterials 25:3211–3222
87. Gimble JM, Guilak F (2003) Differentiation potential of adipose derived adult stem (ADAS) cells. Curr Top Dev Biol 58:137–160
88. Gimble J, Guilak F (2003) Adipose-derived adult stem cells: isolation, characterization, and differentiation potential. Cytotherapy 5:362–369
89. Safford KM, Hicok KC, Safford SD, Halvorsen YD, Wilkison WO, Gimble JM, Rice HE (2002) Neurogenic differentiation of murine and human adipose-derived stromal cells. Biochem Biophys Res Commun 294:371–379
90. Gronthos S, Franklin DM, Leddy HA, Robey PG, Storms RW, Gimble JM (2001) Surface protein characterization of human adipose tissue-derived stromal cells. J Cell Physiol 189:54–63
91. Thirumala S, Forman JM, Monroe WT, Devireddy R (2007) Freezing and post-thaw apoptotic behaviour of cells in the presence of palmitoyl nanogold particles. Nanotech 18:195104
92. Rubinsky B, Ikeda M (1985) A cryomicroscope using directional solidification for the controlled freezing of biological mat. Cryobiology 22:55–68
93. Garner DL, Johnson LA (1995) Viability assessment of mammalian sperm using SYBR-14 and propidium iodide. Biol Reprod 53:276–284
94. Takamatsu H, Rubinsky B (1999) Viability of deformed cells. Cryobiology 39:243–251
95. Takamatsu H, Takeya R, Naito S, Sumimoto H (2005) On the mechanism of cell lysis by deformation. J Biomech 38:117–124
96. Wolfe J, Bryant G (1999) Freezing, drying, and/or vitrification of membrane- solute-water systems. Cryobiology 39:103–129
97. Rubinsky B (2000) Cryosurgery. Annu Rev Biomed Eng 2:157–187
98. Lindquist S (1986) The heat-shock response. Annu Rev Biochem 55:1151–1191
99. Young JC, Agashe VR, Siegers K, Hartl FU (2004) Pathways of chaperone-mediated protein folding in the cytosol. Nat Rev Mol Cell Biol 5:781–791
100. Kalmar B, Greensmith L (2009) Induction of heat shock proteins for protection against oxidative stress. Adv Drug Deliv Rev 61:310–318
101. Beere HM, Wolf BB, Cain K, Mosser DD, Mahboubi A, Kuwana T, Tailor P, Morimoto RI, Cohen GM, Green DR (2000) Heat-shock protein 70 inhibits apoptosis by preventing recruitment of procaspase-9 to the Apaf-1 apoptosome. Nat Cell Biol 2:469–475
102. Saleh A, Srinivasula SM, Balkir L, Robbins PD, Alnemri ES (2000) Negative regulation of the Apaf-1 apoptosome by Hsp70. Nat Cell Biol 2:476–483
103. Li CY, Lee JS, Ko YG, Kim JI, Seo JS (2000) Heat shock protein 70 inhibits apoptosis downstream of cytochrome c release and upstream of caspase-3 activation. J Biol Chem 275:25665–25671
104. Pandey P, Saleh A, Nakazawa A, Kumar S, Srinivasula SM, Kumar V, Weichselbaum R, Nalin C, Alnemri ES, Kufe D, Kharbanda S (2000) Negative regulation of cytochrome c-mediated oligomerization of Apaf-1 and

activation of procaspase-9 by heat shock protein 90. EMBO J 19:4310–4322

105. Gorman AM, Szegezdi E, Quigney DJ, Samali A (2005) Hsp27 inhibits 6-hydroxydopamine-induced cytochrome c release and apoptosis in PC12 cells. Biochem Biophys Res Commun 327:801–810

106. Schepers H, Geugien M, van der Toorn M, Bryantsev AL, Kampinga HH, Eggen BJ, Vellenga E (2005) HSP27 protects AML cells against VP-16-induced apoptosis through modulation of p38 and c-Jun. Exp Hematol 33:660–670

107. Wang P, Shu Z, He L, Cui X, Wang Y, Gao D (2005) The pertinence of expression of heat shock proteins (HSPs) to the efficacy of cryopreservation in HELAs. Cryo Letters 26:7–16

108. Park SJ, Choi HR, Nam KM, Na JI, Huh CH, Park KC (2013) Immediate induction of heat shock proteins is not protective against cryopreservation in normal human fibroblasts. Cryo Letters 34:239–247

Chapter 20

Bone Regeneration with a Combination of Adipose-Derived Stem Cells and Platelet-Rich Plasma

Satoshi Tajima, Morikuni Tobita, and Hiroshi Mizuno

Abstract

Mesenchymal stem cells (MSCs) have the potential to directly differentiate into osteogenic cells and efficiently regenerate bone tissue. Adipose-derived stem cells (ASCs) have the potential to differentiate into an osteogenic lineage, too. In addition, ASCs can be readily harvested in large numbers with low donor-site morbidity. Meanwhile, recent reports have demonstrated that platelet-rich plasma (PRP) contains a variety of growth factors and may be a powerful biological autologous cocktail of growth factors for tissue engineering.

We have shown that ASC/PRP admixture had dramatic effects on bone regeneration in a rat calvarial defect model, not only through the osteogenic potential of ASCs, but also through the release of cytokines by platelets in PRP, which, in turn, support ASCs.

In this chapter, we introduce the bone regeneration using a combination of ASCs and PRP in a rat calvarial defect model.

Key words Bone regeneration, Adipose-derived stem cells, Platelet-rich plasma, Growth factors, Rat model

1 Introduction

Bone regeneration is a complex physiological process of bone formation and is involved in continuous remodeling throughout adult life. However, there are complex clinical conditions in which bone regeneration is required in large quantity, for example infection, tumor resection, and skeletal abnormalities [1]. Several techniques have been developed to reconstruct bony defects, including autografts [2], allografts [3], xenografts [4], and alloplasts [5]. However, these procedures have each limitation.

In recent years, adipose-derived stem cells (ASCs) can be differentiated toward the osteogenic lineage [6, 7]. Several studies have reported that ASCs have great potential for bone regeneration in vivo [8–10]. In addition, ASCs were found to secrete many kinds of growth factors for tissue regeneration [11, 12]. Meanwhile,

Bruce A. Bunnell and Jeffrey M. Gimble (eds.), *Adipose-Derived Stem Cells: Methods and Protocols*, Methods in Molecular Biology, vol. 1773, https://doi.org/10.1007/978-1-4939-7799-4_20, © Springer Science+Business Media, LLC, part of Springer Nature 2018

platelet-rich plasma (PRP) contains several growth factors, stimulates the mesenchymal progenitor cells, and supports bone regeneration with stem cells [13].

In this chapter, we describe the adjustment of ASC/PRP admixture and bone regeneration experiment in a rat calvarial defect model.

2 Materials

2.1 Harvesting ASCs from Rat Adipose Tissue

1. Inbred male F344 rat.
2. 2% isoflurane.
3. Surgical blade and scissors.
4. 0.12% Collagenase dissolved in Phosphate buffered saline (PBS), freshly prepared prior to use.
5. Dulbecco's modified Eagle medium (DMEM) supplemented with 10% fetal bovine serum (FBS) and 1% antibiotic–antimycotic as a control medium.
6. 100 mm culture dish.
7. 50 mL centrifuge tube.
8. Cell filter
9. Centrifuge
10. Trypan Blue.

2.2 ASCs Culture

1. DMEM supplemented with 10% FBS and 1% antibiotic–antimycotic as a control medium.
2. Trypsin (0.25%) and ethylenediaminetetraacetic acid (EDTA) (1 mM).
3. Centrifuge

2.3 Generation of Calvarial Bone Defect

1. 2% isoflurane.
2. Xylocaine (0.5 mL of 1%).
3. Surgical blade.
4. Micro motor (Rotex 780).
5. Dental round bar (ISO standard 005).

2.4 PRP Preparation

1. 2% isoflurane.
2. 100 mm culture dish and 15-mL centrifuge tube.
3. Blood collection tubes (5 mL EDTA-2Na blood collection tube).
4. Scissors, 18 gauge long cannula, and 1 mL syringe.
5. Centrifuge

2.5 Enzyme-Linked Immunosorbent Assay (TGF-β1, VEGF, IGF-1, HGF, and PDGF-AB): PRP and ASCs Culture Supernatant

1. Rat Quantikine ELISA Kit (TGF-β1, VEGF, IGF-1, HGF, and PDGF-AB).
2. Microplate Reader.

2.6 Transplantation of ASCs/PRP Admixture

1. 18 gauge long cannula and 1 mL syringe.
2. 10% calcium chloride solution.
3. 6-0 suture.

2.7 Scanning Electron Microscopy (SEM): PRP with ASCs

1. Phosphate-buffered glutaraldehyde fixative.
2. 2% osmium tetroxide.
3. Aluminum slabs with carbon paste.
4. Scanning electron microscope.

2.8 Calvarial Bone Dissection

1. Dental burs (Straight Fissure Crosscut, #557, Shank type; HP).
2. Micro motor (Rotex 780).
3. Surgical blade.

2.9 Tissue Preparation and Micro-Computed Tomography (CT) Analysis

1. Micro-CT.
2. Analysis software.

2.10 Fixation and Decalcification and Hematoxylin and Eosin Staining

1. 4% Paraformaldehyde PBS.
2. Decalcification solution.
3. 10% formaldehyde.
4. Hematoxylin.
5. Eosin.

2.11 Immuno-histochemistry for Osteocalcin and Osteopontin

1. Anti-osteocalcin monoclonal antibody (1:500 dilution).
2. Anti-osteopontin polyclonal antibody (1:1000 dilution).
3. Tris/HCL buffer.
4. Biotin-conjugated anti-mouse IgG secondary antibody.
5. Biotin-conjugated anti-rabbit IgG secondary antibody.

2.12 Transplantation of ASCs Derived from a Green Fluorescent Protein (GFP) Transgenic Mouse into Nude Rats

1. GFP transgenic mouse (C57BL/6-Tg [CAG-EGFP]).
2. Nude rats (F344/NJcl-rnu/rnu).
3. Multi-purpose cryosection preparation kit (Kawamoto method).
4. Tungsten carbide knife.

5. Cryostat chamber.
6. Mouse monoclonal anti-osteocalcin antibody (1:200 dilution).
7. Rabbit polyclonal anti-osteocalcin antibody (1:200 dilution).
8. Rabbit polyclonal anti-GFP antibody (1:1000 dilution).
9. Alexa Fluor 488-conjugated goat anti-rabbit IgG.
10. Alexa Fluor 594-conjugated goat anti-mouse IgG.

3 Methods

3.1 Harvesting ASCs from Rat Adipose Tissue

1. Anesthetize the F344 rats with 2% isoflurane.
2. After the skin incision of the inguinal region, harvest the inguinal fat pads, place them in the hood, and wash extensively with PBS.
3. Mince the fat pad for 10 min with fine scissors in culture dishes and transfer the material into a 50 mL centrifuge tube. Add 0.12% collagenase and enzymatically digest at 37 °C for 30 min.
4. Add an equal volume of control medium to neutralize the collagenase.
5. Centrifuge the cell suspension at 260 × *g* for 5 min to obtain a high density ASC pellet.
6. Aspirate the supernatant, being careful not to disturb the cell pellets. Resuspend the cell pellets in control medium and incubate at room temperature for 10 min.
7. Filter the cell suspension through a 40 μm nylon mesh to remove cellular debris.
8. Count the cells using Trypan Blue and plate at a concentration of 5×10^5 cells/100 mm tissue culture dish.

3.2 ASCs Culture

1. Maintain the primary ASCs in control medium at 37 °C in 5% carbon dioxide. Change the culture medium every 3 days.
2. Once adherent cells become subconfluent, aspirate the culture medium and wash the cells with 5 mL of PBS three times. Add 3 mL of trypsin–EDTA at 37 °C for 5 min to detach the cells.
3. Resuspend the ASCs with an equal volume of control medium.
4. Centrifuge the cell suspension again and resuspend in a small volume of control medium.
5. Calculate the cell concentration and adjust the volume to obtain 1.5×10^7 cells/mL for subsequent transplantation.

3.3 Generation of Calvarial Bone Defect

1. Anesthetize the F344 rats with 2% isoflurane.
2. Place the rat in the prone position and inject the Xylocaine (0.5 mL of 1%) intradermally in the middle of the calvarium.

3. Make a midline calvarial incision and elevate the cranial skin flaps. Divide the subcutaneous fascia and reflect periosteal flaps bilaterally.
4. Generate a 5 mm bone defect in the left calvarial bone with a dental round bar (*see* **Note 1**).

3.4 PRP Preparation

1. Anesthetize the F344 rats with 2% isoflurane.
2. Draw 10 mL of whole blood preoperatively via cardiac puncture with an 18 gauge needle into the blood collection tubes.
3. Centrifuge the blood in a standard laboratory centrifuge for 10 min at 450 × *g*.
4. Collect the supernatant plasma and the buffy coat, which contains platelets and leukocytes, using 18 gauge long cannula, and transfer into a 15 mL centrifuge tube.
5. Centrifuge at 850 × *g* for 15 min to concentrate the platelets.
6. Guide the top of an 18 gauge long cannula to the bottom of the plasma and gently aspirate 1 mL of the infranatant plasma containing the buffy coat. Discard the surpernatant plasma.

3.5 Enzyme-Linked Immunosorbent Assay (TGF-β1, VEGF, IGF-1, HGF, and PDGF-AB): PRP and ASCs Culture Supernatant

1. Fifty microliters of sample is added to 96-well microplates that are coated with a monoclonal antibody against the factor of interest and incubated for 2 h.
2. After washing with washing buffer, a horseradish peroxidase-conjugated cytokine is added to each well, after which the wells are incubated for 1 or 2 h and then washed.
3. The substrate solution is added and incubated for 30 min, and the reaction is terminated by addition of the stop solution.
4. Growth factor concentrations are determined by measuring the optical density at 450 nm using a microplate reader.

3.6 Transplantation of ASCs/PRP Admixture

1. After ASCs are expanded to passage 3, a pellet of 1×10^7 ASCs is mixed with 1 mL of PRP using 18 gauge long cannula and 1 mL syringe (Fig. 1).
2. ASCs–PRP admixture is activated with a 10% calcium chloride solution for transplantation (Fig. 2).
3. After 10 min, ASCs–PRP admixture (2×10^5/20 μL) is transplanted into the calvarial defect with tweezers (Fig. 3).
4. The periosteum and skin flaps are placed back using 6-0 sutures.

3.7 Scanning Electron Microscopy (SEM): PRP with ASCs

1. ASCs–PRP admixture is immediately immersed after activation in a phosphate-buffered glutaraldehyde fixative for 24 h at 4 °C and post-fixed in 2% osmium tetroxide for 2 h.
2. Samples are dehydrated in a graded series of ethanol, substituted with *t*-butyl alcohol, and dried at −20 °C in a vacuum.

Fig. 1 ASCs pellet. After centrifuging the cell suspension, the surpernatant was discarded and the pellet mixed with PRP. Arrow: ASCs pellet

Fig. 2 Activated ASCs–PRP admixture. ASCs–PRP admixture was activated with a 10% calcium chloride solution. Gel formation takes place in 10 min

3. Samples are made electrically conductive by mounting on aluminum slabs with a carbon paste, followed by coating in an osmium plasma coater to a thickness of approximately 10 nm.
4. Specimens are attached to an acrylic plate with glue tape. Subsequently, half of the plate bottom is cut with a diamond disc and vertically divided into two pieces with a chisel and hammer.
5. The cut face of each specimen is examined with a scanning electron microscope at 3 kV.

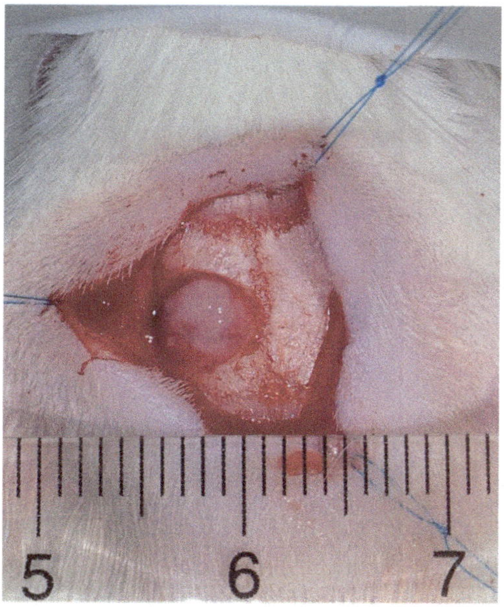

Fig. 3 Transplantation of ASCs–PRP admixture. ASCs–PRP admixture was transplanted into the calvarial defect. Then the periosteum and skin flaps were placed back using 6-0 sutures

3.8 Calvarial Bone Dissection

1. Euthanize rat with an overdose of intraperitoneal injection of pentobarbital sodium.
2. Place the rat in the prone position and inject the xylocaine (0.5 mL of 1%) intradermally in the middle of the calvarium.
3. Make a midline calvarial incision and elevate the cranial skin flaps. Divide the subcutaneous fascia and remove the calvarial bone using a straight fissure bur with a micromotor.

3.9 Tissue Preparation and Micro-Computed Tomography (CT) Analysis

1. Specimens are fixed in 4% paraformaldehyde (PFA), after which Micro-CT was performed.
2. The surface area and volume of newly formed bone in the defect site are calculated using Analysis software (Fig. 4, *see* **Note 2**). A representative image of Micro-CT is shown in Fig. 5.

3.10 Fixation, Decalcification, and Histological Analysis

1. Fix in 4% paraformaldehyde solution for 7 days at room temperature.
2. Decalcify the specimens using decalcification solution for 2–3 days (*see* **Note 3**).
3. Cut the specimens at the center of calvarial bone using surgical blade.
4. Wash the specimens with water and then embed them in paraffin.

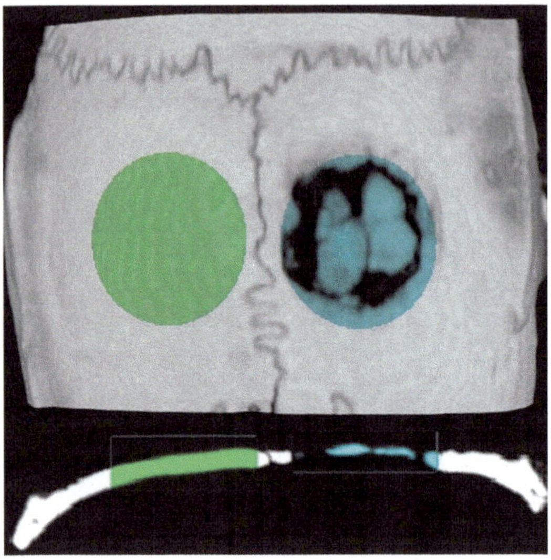

Fig. 4 Micro-CT analysis. The surface area and volume of newly formed bone in the defect site was calculated based on the healthy bone in the opposite side

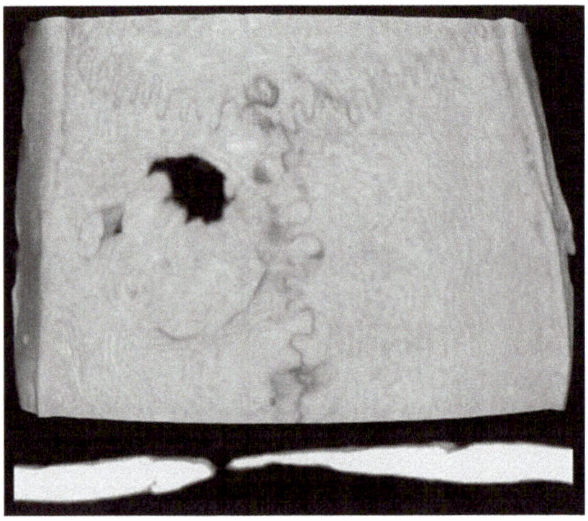

Fig. 5 Micro-CT image after the surgery of 8 weeks. Most of the defect was covered with newly formed bone

5. Section the tissue at 5 μm increments using microtome and place on the glass slides.
6. Immerse the slides in xylene I for at least 10 min.
7. Bathe the slides in xylene II and then in xylene III for 2–3 min each.
8. Immerse the slides in 100% ethanol three times for 2–3 min each, followed by immersion in 90%, 80%, 70%, and 50% ethanol for 2–3 min each.
9. Wash the slides in tap water.

10. Immerse the slides in Carazzi's Hematoxylin for 20 min (*see* **Note 4**).
11. Wash the slides two times in tap water for 2–3 min.
12. Immerse the slides in Eosin solution for 15 min.
13. Dehydrate by immersing in a graded ethanol series three times (70%, 90%, and 100%) for 2–3 min each, followed by immersion in xylenes I–III three times for 2–3 min each.
14. Mount the slides using mounting medium.
15. Observe the stained slides under a light microscope.

3.11 Immunohistochemistry for Osteocalcin and Osteopontin

1. Section the tissue specimens in 5 μm increments using microtome and place on the glass slides.
2. Immerse the slides in a series of xylene, ethanol, and water as described above in Subheading 3.7, **steps 6–9**.
3. Wipe the area around the sections to carefully to remove excess solution.
4. Immerse the sections in 3% hydrogen peroxide to block endogenous peroxidase activity for 10 min at 37 °C.
5. Wash the slides in Tris–HCl buffer for 5 min.
6. Immerse the sections in normal goat serum for 30 min.
7. Remove the slides from normal goat serum and wipe gently around each section.
8. Cover tissue with primary antibody (osteocalcin monoclonal antibody or osteopontin polyclonal antibody) overnight at 4 °C.
9. Wash the slides in Tris–HCl buffer for 5 min.
10. Wipe gently around each section and cover with a biotin-conjugated anti-mouse or anti-rabbit IgG secondary antibody at room temperature for 30 min.
11. Wash the slides in Tris–HCl buffer for 5 min.
12. Immerse the slides in Carazzi's Hematoxylin for 5 s.
13. Wash the slides in tap water.
14. Dehydrate by immersing in a graded ethanol series three times (70%, 90%, and 100%) for 2–3 min each, followed by immersion in xylenes I–III three times for 2–3 min each.
15. Mount slides using mounting medium.
16. Observe stained slides under a light microscope.

3.12 Transplantation of ASCs Derived from a Green Fluorescent Protein (GFP) Transgenic Mouse into Nude Rats

1. To evaluate the origin of newly-formed bone, ASCs are harvested from a 7-week-old inbred male GFP-transgenic mouse and cultured until passage 3.
2. A 5 mm bone defect is generated at the left calvarial bone under general anesthesia by inhalation of 2% isoflurane.

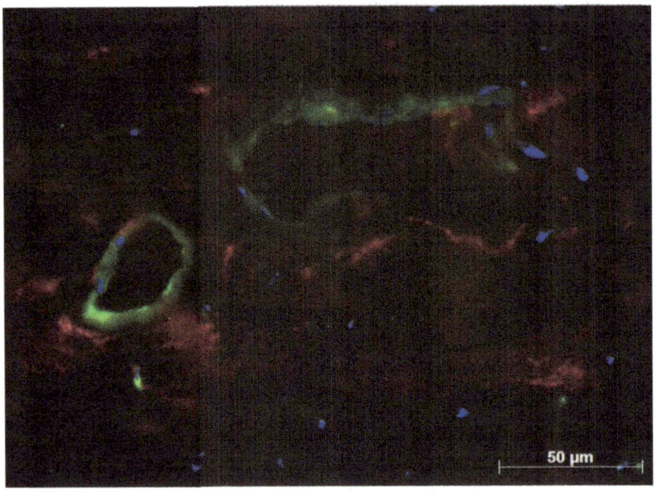

Fig. 6 Fluorescence immunohistochemical staining using anti-rat osteocalcin and anti-mouse osteocalcin antibodies. Representative image of the ASC–PRP admixture at 4 weeks after transplantation. Newly formed bone was composed of rat-derived osteocalcin (red) and mouse-derived osteocalcin (green)

3. The GFP-positive ASC–PRP admixture (2×10^5/20 μL) is transplanted into the calvarial defect of 11-week-old male nude rats. Samples are collected 4 weeks after transplantation.
4. Fresh-frozen sections of these samples are made according to the Kawamoto method using a Multi-Purpose Cryosection Preparation Kit.
5. Cryofilm type 2C is applied to the cutting surface of the completely frozen block, and the block is cut with a tungsten carbide knife at −25 °C in a cryostat chamber.
6. The sections are fixed with 4% PFA for 3 min and then washed with PBS for 3 min.
7. A mouse monoclonal anti-osteocalcin antibody, a rabbit polyclonal anti-osteocalcin antibody, and a rabbit polyclonal anti-GFP antibody are used as primary antibodies.
8. An Alexa Fluor 488-conjugated goat anti-rabbit IgG and an Alexa Fluor 594-conjugated goat anti-mouse IgG are used as secondary antibodies.
9. The sections are mounted using mounting medium containing DAPI.
10. After fixation, specimens are observed by fluorescence microscopy. A representative image is shown in Fig. 6.

4 Notes

1. Avoid the superior sagittal sinus when generating the bone defect with a dental round bar.
2. Calculate the surface area and volume of the newly formed bone in the defect site based on the healthy bone in the opposite side.
3. The harvested samples in decalcification solution are kept in 4 °C refrigerator for 2–3 days. Decalcification solution is changed every day.
4. Application time is subject to change depending on the room temperature.

Acknowledgments

We would like to thank Ryuji Tsukada and Toyoko Hidano (Division of Experimental Surgery and Biomedical Resources, Juntendo University Graduate School of Medicine) for their expert assistance with animal experiments; Takako Ikegami and Tomomi Ikeda (Laboratory of Molecular and Biochemical Research, Research Support Center, Juntendo University Graduate School of Medicine) for technical assistance; Noriyoshi Sueyoshi, Shinji Nakamura, Yuko Kojima, Yasuko Toi, and Katsumi Miyahara (Laboratory of Biomedical Imaging Research, Biomedical Research Center, Juntendo University Graduate School of Medicine) for their expert assistance with histological analyses; and Junichi Nakamoto (Laboratory of Ultrastructural Research, Juntendo University Graduate School of Medicine) for his expert assistance with SEM analyses. This work was supported in part by the Japanese Mnistry of Education, Sports, and Culture (Grant 26861514).

References

1. Marquez L, de Abreu FA, Ferreira CL, Alves GD, Miziara MN, Alves JB (2013) Enhanced bone healing of rat tooth sockets after administration of epidermal growth factor (EGF) carried by liposome. Injury 44:558–564
2. Oikarinen J, Korhonen LK (1979) The bone inductive capacity of various bone transplanting materials used for treatment of experimental bone defects. Clin Orthop Relat Res (140):208-215
3. Salgado AJ, Coutinho OP, Reis RL (2004) Bone tissue engineering: state of the art and future trends. Macromol Biosci 4:743–765
4. Goldstein SA (2002) Tissue engineering: functional assessment and clinical outcome. Ann N Y Acad Sci 961:183–192
5. Eppley BL, Sadove AM (2000) Management of alveolar cleft bone grafting--state of the art. Cleft Palate Craniofac J 37:229–233
6. Pittenger MF, Mackay AM, Beck SC, Jaiswal RK, Douglas R, Mosca JD et al (1999) Multilineage potential of adult human mesenchymal stem cells. Science 284:143–147
7. Dimitriou R, Jones E, McGonagle D, Giannoudis PV (2011) Bone regeneration: current concepts and future directions. BMC Med 9:66

8. Parrilla C, Saulnier N, Bernardini C, Patti R, Tartaglione T, Fetoni AR et al (2011) Undifferentiated human adipose tissue-derived stromal cells induce mandibular bone healing in rats. Arch Otolaryngol Head Neck Surg 137:463–470
9. Tajima S, Tobita M, Orbay H, Hyakusoku H, Mizuno H (2015) Direct and indirect effects of a combination of adipose-derived stem cells and platelet-rich plasma on bone regeneration. Tissue Eng Part A 21:895–905
10. Mizuno H, Tobita M, Uysal AC (2012) Concise review: adipose-derived stem cells as a novel tool for future regenerative medicine. Stem Cells 30:804–810
11. Rehman J, Traktuev D, Li J, Merfeld-Clauss S, Temm-Grove CJ, Bovenkerk JE et al (2004) Secretion of angiogenic and antiapoptotic factors by human adipose stromal cells. Circulation 109:1292–1298
12. Blaber SP, Webster RA, Hill CJ, Breen EJ, Kuah D, Vesey G et al (2012) Analysis of in vitro secretion profiles from adipose-derived cell populations. J Transl Med 10:172
13. Marx RE, Carlson ER, Eichstaedt RM, Schimmele SR, Strauss JE, Georgeff KR (1998) Platelet-rich plasma: Growth factor enhancement for bone grafts. Oral Surg Oral Med Oral Pathol Oral Radiol Endod 85:638–646

Chapter 21

Isolation and Differentiation of Murine Primary Brown/Beige Preadipocytes

Hui Yu, Margo Emont, Heejin Jun, and Jun Wu

Abstract

Studies on thermogenic adipose cells (brown and beige fat) hold promise to treat obesity and its associated metabolic disorders due to the ability of these cells to dissipate energy in the form of heat. Although immortalized adipose cell lines have been widely used to investigate cell autonomous regulations, many physiological functions of thermogenic fat need to be studied with primary fat cells. Here, we present a detailed protocol of the isolation, culture, and differentiation of primary brown and beige preadipocytes from the stromal vascular fraction of murine interscapular and inguinal adipose depots.

Key words Beige fat cells, Brown fat cells, Stromal vascular fraction, Preadipocytes

1 Introduction

Obesity has become an epidemic that increases the risk for cardiovascular disease, type II diabetes, hypertension, and many types of cancer [1, 2]. Obesity results from an energy imbalance, in which energy intake exceeds expenditure. The energy surplus is stored as lipids primarily in the white adipocytes. Thermogenic fat cells dissipate energy as heat, through a process mediated by mitochondrial uncoupling protein-1 (UCP1). There are two types of thermogenic adipocytes identified to date, brown and beige adipocytes. Classical brown fat cells arise from a shared lineage with skeletal muscle [3, 4], and majority of these cells in rodents can be found in the interscapular depot between the shoulder blades. Recently identified beige adipocytes intersperse in the white adipose tissue and arise from a different lineage [3, 5]. Beige fat has also been referred to as "inducible" brown fat cells, since they store energy like white adipocytes in the basal state, but induce thermogenesis upon environmental stimulation such as cold. This new type of thermogenic fat cell is most prominent in the subcutaneous depots.

Bruce A. Bunnell and Jeffrey M. Gimble (eds.), *Adipose-Derived Stem Cells: Methods and Protocols*, Methods in Molecular Biology, vol. 1773, https://doi.org/10.1007/978-1-4939-7799-4_21, © Springer Science+Business Media, LLC, part of Springer Nature 2018

In humans, it had long been assumed that only infants have thermogenic fat. However, in 2009, positron emission tomography imaging using fluorodeoxyglucose (^{18}F) identified UCP1 positive cells in the neck and upper-chest regions of human adults [6–8]. The developmental lineage(s) and molecular identities of these UCP1 positive fat cells in humans are still being investigated [5, 9–13]. Increasing evidence has revealed that these thermogenic fat cells play an important role in regulating not only energy homeostasis but also overall metabolic fitness in humans [14–17].

Several immortalized preadipose cell lines have been established for in vitro studies, such as 3T3-L1, 3T3-F442A, or HIB1 [18–20]. These cell lines are easy to access and culture, and have provided many insights we currently have in regard to adipogenesis and fat cell regulation. However, to achieve a detailed understanding of physiological function of different types of adipocytes, thermogenic fat cells in particular, primary adipocytes present a more suitable system. Primary adipocytes can be obtained through isolation and differentiation of committed fat cell precursors, or preadipocytes. These are a subset of fibroblastic cells that can be found within the stromal vascular fraction (SVF) of the fat tissues. We present here a detailed method for the isolation, culture, and differentiation of primary brown preadipocytes from the interscapular fat depot, and a significant percentage of beige preadipocytes together with white preadipocytes from the inguinal subcutaneous depot. These two fat depots are of substantial size and yield a significant number of preadipocytes, which differentiate well in culture and respond robustly to thermogenic stimuli, therefore constituting a vigorous experimental system to study physiological function of thermogenic fat cells.

2 Materials

2.1 Stromal Vascular Fraction Isolation and Culture Materials

1. Mouse dissection tools—forceps and scissors (Roboz Surgical Instrument Company, Gaithersburg, MD, USA) for dissection, and spring scissors (Fine Science Tools, Foster City, CA, USA) for mincing tissues.
2. Spray bottles.
3. 70% ethanol (Thermo Fisher Scientific, Waltham, MA, USA).
4. Digestion solutions.
   (a) Digestion solution for isolation of SVF from inguinal adipose tissue.
   (b) Collagenase D (final conc., 1.5 U/mL) and dispase II (final conc., 2.4 U/mL) mixture in phosphate-buffered saline (PBS) supplemented with 10 mM $CaCl_2$. Enzyme can be prepared fresh before each experiment or prepared in batches and stored at −80 °C as ready-to-use aliquots. For example, dissolve 2.5 g collagenase D (Roche,

Indianapolis, IN, USA, 0.21 U/mg, *see* **Note 1**) and 0.8 g dispase II (Roche, 1.05 U/mg) in 300 mL PBS (pH 7.4, Life Technologies, Carlsbad, CA, USA). Add 3.5 mL of 1 M $CaCl_2$ and additional PBS to a total volume of 350 mL. Fully dissolved solution is then stored at −80 °C as 3–6 mL aliquots.

(c) Digestion solution for isolation of SVF from brown adipose tissue.

(d) Collagenase B (final conc., 1.5 U/mL) and dispase II (final conc., 2.4 U/mL) mixture in PBS supplemented with 10 mM $CaCl_2$. Similar to digestion solution for inguinal adipose tissue, this can be prepared either fresh or as premade aliquots.

5. Sterile syringe filters, (0.2-μm, VWR International, Radnor, PA, USA).
6. Cell strainers, 100 and 40 μm (Thermo Fisher Scientific).
7. Conical sterile centrifuge tubes (50 mL, Thermo Fisher Scientific).
8. Parafilm wrap (Sigma-Aldrich, St. Louis, MO, USA).
9. *Washing medium*—Dulbecco's Modified Eagle Medium/Nutrient Mixture F-12 GlutaMAX (DMEM/F-12, Life Technologies), supplemented with 8% fetal bovine serum (FBS, Sigma-Aldrich), and penicillin–streptomycin (Life Technologies. final conc., 50 U/mL of penicillin and 50 μg/mL streptomycin).
10. *Growth medium*—DMEM/F-12 (Life Technologies) supplemented with 15% FBS (Sigma-Aldrich), and penicillin–streptomycin (Life Technologies. final conc., 50 U/mL of penicillin and 50 μg/mL streptomycin).
11. Collagen coated plates. Dilute collagen I (Corning Life Sciences, Tewksbury, MA, USA, from rat tail, 3 mg/mL) in sterile 0.02 N acetic acid to a final concentration of 50 μg/mL. Coat cell culture plates with 5 μg/cm^2. For example, add 8 mL diluted solution to coat a 100 mm dish. Incubate at room temperature for an hour with periodical agitation. Wash the dishes twice with PBS to remove acid. Air-dry the dishes in the cell culture hood. The coated dishes can be stored at 4 °C, dry environment under sterile conditions for up to 2–3 weeks.
12. Trypsin–EDTA (0.05%) mixture (Life Technologies).
13. Hemocytometer.

2.2 Differentiation Materials

1. *Differentiation components* (*see* **Note 2**).

(a) Insulin (Sigma-Aldrich) stock. Dissolve insulin in 0.02 M HCl to a final concentration of 5 mg/mL and sterilize with 0.2-μm filters. Aliquot in 1.5 mL centrifuge tubes and store at −20 °C. This is a 10,000-fold concentrated stock solution.

(b) Isobutylmethylxanthine (IBMX) stock. Dissolve IBMX (Sigma-Aldrich) in dimethyl sulfoxide (DMSO, Sigma-Aldrich) to a final concentration of 0.25 M and aliquot in 1.5 mL centrifuge tubes. This is a 500-fold concentrated stock solution and can be stored at −20 °C.

(c) Dexamethasone (Dex) stock. Dissolve Dex (Sigma-Aldrich) in ethanol to a final concentration of 5 mM. Aliquot and store at −20 °C. This is a 1000-fold concentrated stock solution.

(d) Rosiglitazone. Dissolve rosiglitazone (Cayman Chemical, Ann Arbor, MI, USA) in DMSO to a final concentration of 10 mM. This is a 10,000-fold concentrated solution. Aliquot and store at −20 °C.

2. *Base differentiation medium*—DMEM/F-12 (Life Technologies) supplemented with 10% FBS (Sigma-Aldrich), and penicillin–streptomycin (Life Technologies, final conc., 50 U/mL of penicillin and 50 μg/mL streptomycin).
3. *Differentiation induction medium—Base differentiation medium* supplemented with 0.5 μg/mL insulin, 5 μM dexamethasone, 1 μM rosiglitazone, and 0.5 mM IBMX. Prepare fresh before use.
4. *Maintenance medium—Base differentiation medium* supplemented with 0.5 μg/mL insulin. Prepare fresh before use.

2.3 Equipment

1. 37 °C shaking water bath.
2. Centrifuge (can hold 50 mL tubes, and go up to 500 × *g*).
3. Biosafety cabinet (for dissecting mice).
4. Sterile tissue culture hood.
5. CO_2 cell-culture incubator.
6. Inverted microscope.

3 Methods

3.1 Isolation of Adipose Stromal Vascular Fractions

Carry out **steps 1–11** in a biosafety cabinet and **steps 12–17** in a sterile tissue culture hood. *See* Fig. 1 for the flow diagram of this procedure.

1. Freshly prepare or thaw aliquot(s) of digestion solution (*see* **Note 3**). Keep the digestion solution on ice once it is prepared, or thawed, to avoid the decrease of enzymatic activity.
2. Label the petri dish with mice genotype and depot information, add PBS to the dish (approximate 6–8 mL/dish) and keep the dish on ice.

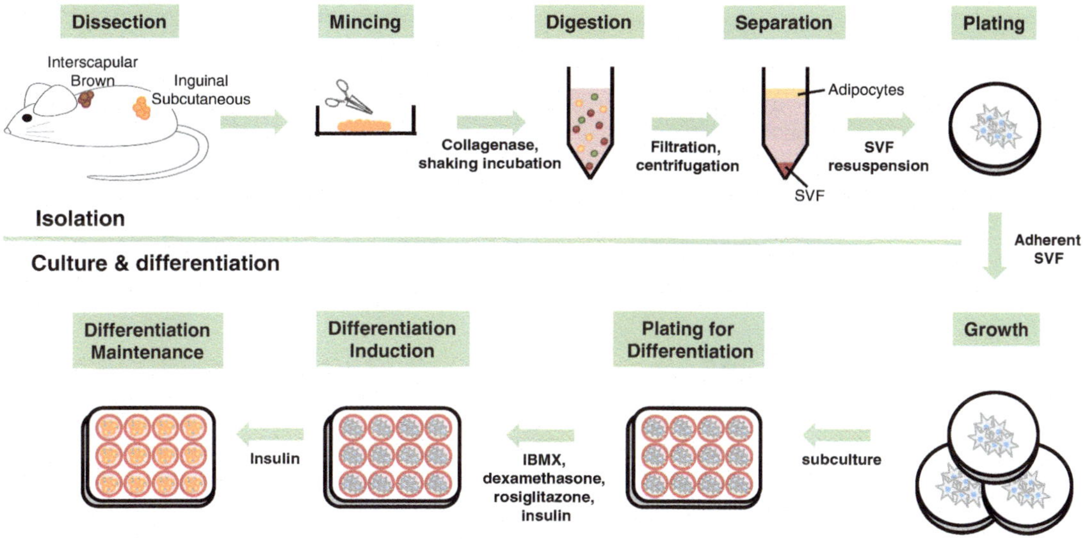

Fig. 1 Flow scheme for the isolation, culture, and differentiation of primary SVF. The interscapular brown and subcutaneous inguinal fat depots are dissected from mice, minced, and digested with collagenase–dispase solutions. After removal of floating, mature adipocytes, the pellet that contains the stromal vascular fraction is cultured and stimulated to undergo adipogenesis. Six to seven days after adipogenic induction, more than 95% of the precursors will differentiate into mature fat cultures that respond to thermogenic stimulations

3. Check and confirm that the shaking water bath is set at 37 °C before starting.
4. Sterilize the biosafety cabinet with 70% ethanol.
5. Euthanize mice through carbon dioxide asphyxiation. Sterilize mouse by spraying thoroughly with 70% ethanol. Carefully dissect out the inguinal fat (Fig. 2). The dissection can be initiated from either dorsal or ventral position. Remove any attached muscle, skin, and connective tissue. Isolated fat tissue consists of dorsolumbar, more stringently defined "inguinal" and gluteal fat depots (*see* **Notes 4** and **5**). Place the tissue in the petri dish with PBS.
6. Dissect the brown adipose tissue (BAT) from the interscapular area (Fig. 2). Lift the fat between the shoulder blades to reveal a connected fat depot with a distinguishing butterfly- or kite-like shape. Remove this depot and trim off the surrounding white adipose tissue. This should result in two much smaller, separate pieces of BAT. Place the tissue in the petri dish with PBS.
7. Repeat the dissection steps and pool inguinal and interscapular fat tissues separately from multiple mice (*see* **Note 6**).
8. Quickly wash the dissected adipose tissues in PBS and transfer to another clean petri dish with forceps. The carryover PBS helps to keep the tissue stay moist through the mincing process.

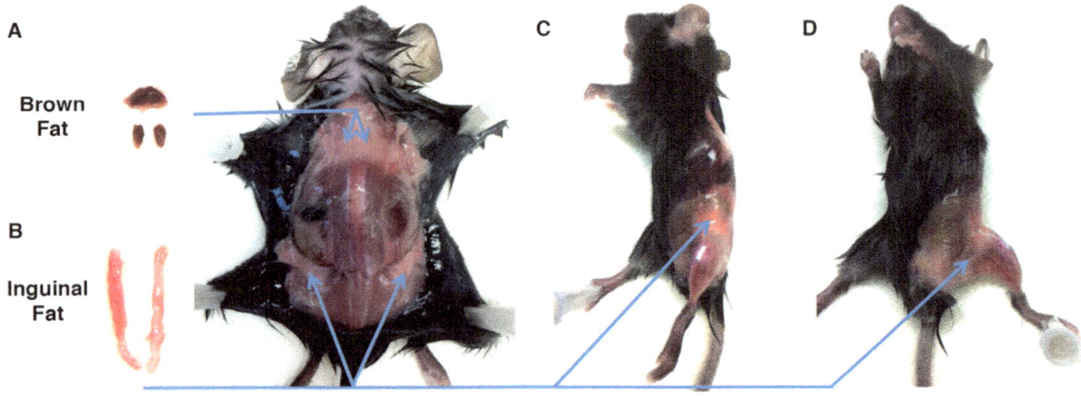

Fig. 2 Gross anatomical locations of the interscapular brown fat and subcutaneous inguinal fat depots in mice. (**a**) Brown adipose tissue in the interscapular depot between the shoulder blades in dorsal view. After removal of the surrounding white adipose tissue, two lobes of brown fat tissue will be ready for isolation of SVF. (**b–d**) Commonly referred to as "inguinal fat," the fat tissue used for SVF isolation actually consists of three continuous subcutaneous depots: dorsolumbar depot (**b**, shown in dorsal position), inguinal depot (**c**, seen in side view), and gluteal depot (**d**, in ventral position)

9. Mince the tissue with spring scissors until it is smooth with minimal visible chunks (*see* **Note 7**).
10. Add the appropriate digestion solution—collagenase D and dispase II mixture for inguinal fat, and collagenase B and dispase II mixture for brown fat—to the minced tissue. Transfer the minced tissue mixed with digestion solution into a 50 mL conical tube (*see* **Note 8**). Seal the tube cap with Parafilm to prevent potential contamination during the incubation. Place the tube in the 37 °C shaking water bath. Set the shaking speed at 150–180 strokes/min. Check the progress of digestion every 5 min. Stop digestion when majority of the tissue chunks have disappeared and the solution appears cloudy. Usually this step takes around 20 min.
11. When digestion is completed, wipe off the water on the outside of the tube, sterilize the tube with 70% ethanol, and bring it into the sterile tissue culture hood.
12. Add 10 mL of *washing medium* to stop the enzymatic activity in the digestion mixture (*see* **Note 9**). Pipet up and down to mix and pass this mixture through the 100 μm cell strainer into a new 50 mL conical tube. Collect any remaining cells with another 10–15 mL of *washing medium* and filter through the 100 μm cell strainer. Centrifuge the filtered suspension at 300–500 × *g* for 5 min at room temperature.
13. Carefully decant the supernatant into an extra tube and leave a small amount of liquid (approximately 0.5 mL) to help the dissociation of the pellet (*see* **Note 10**). When there are no significant chunks of the pellet remaining, add 10–15 mL of *washing*

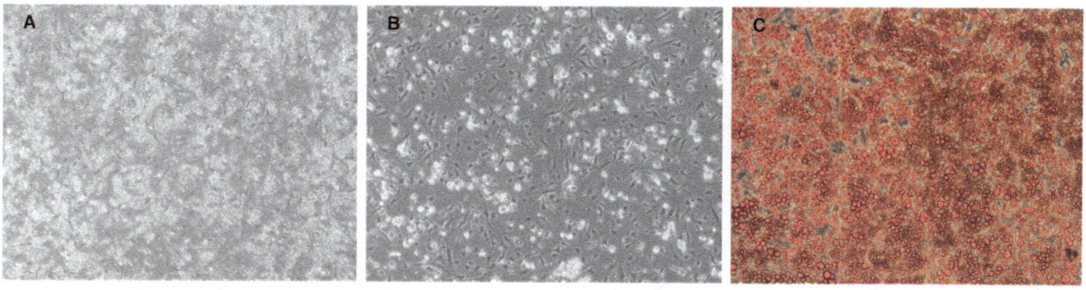

Fig. 3 Preadipocytes from inguinal depot at either passage 0 (1 day after isolation) or after adipogenic differentiation. (**a**) Preadipocytes are fibroblast-like cells that adhere to the plates, whereas red blood cells and other immune cells remain suspended in the medium. (**b**) Cells have been washed with PBS and refed with fresh growth medium to remove contaminating cell types from the culture. (**c**) Oil Red O staining of accumulated lipids in the cells 7 days after the induction of adipogenic differentiation

medium to resuspend the disrupted pellet and pass through a 40 μm cell strainer into a 50 mL conical tube. Collect and filter any remaining cells with another 10–15 mL of *washing medium*. Centrifuge at 300–500 × *g* for 5 min at room temperature.

14. Decant the supernatant and break the pellet as described in **step 13**. Add *growth medium* to resuspend and evenly spread the cells into collagen-coated 100 mm plates (one plate for brown fat and two plates for inguinal fat of 5–6 mice).
15. Culture cells in a humidified 37 °C/8% CO_2 cell culture incubator.
16. Refresh *growth medium* after overnight incubation (Fig. 3a, b). Contaminating cell types in suspension can be washed away with prewarmed PBS.
17. Refresh *growth medium* every other day until the cells approach 70–80% confluency. Subculture and seed these precursors with >40% culture confluency at any given point (*see* **Notes 11** and **12**).

3.2 Differentiation of Primary Preadipocytes from Stromal Vascular Fraction

1. Trypsinize and reseed the SVF (within 2–4 passages) into 12-well collagen-coated plates at a high density (around 5–6 × 10^5 cells/well) (*see* **Note 13**). This should result in over “100% confluency,” and the cells should be tightly packed after overnight culture.
2. Induce differentiation the next day by switching the *growth medium* to the *differentiation induction medium*. This is differentiation day 0 (D0).
3. Forty-eight hours later, on D2, switch the *induction medium* to the *maintenance medium*. Cells can be cultured in the *maintenance medium* for 4–6 more days until they reach full differentiation. Gently change the medium every other day.

4. Further studies, such as how brown and beige adipocytes respond to various thermogenic stimuli, can be performed when over 90% of the cells have differentiated into mature adipocytes (Fig. 3c).

4 Notes

1. Enzyme digestive units (given as units per milligram) of collagenase B/D and dispase II can vary significantly from lot to lot, and therefore each preparation should be calculated using the specific enzyme digestive units of the particular lot provided by the manufacturer.
2. Make small volume aliquots for differentiation reagents to avoid repeated freeze–thaw cycles, particularly for insulin and rosiglitazone, since the stocks for these two are 10,000× concentrated.
3. 2–2.5 mL digestion solution per gram adipose tissue can be used. For example, 5 mL of digestion solution is enough for SVF isolation of inguinal fat of five mice (6–8 weeks old, each inguinal fat pad weighs around 0.2 g) and 1–2 mL of digestion solution for brown fat of the same mice (each lobe of brown fat weighs around 0.05 g).
4. The subcutaneous white fat tissue used for SVF isolation is commonly referred to as inguinal fat. This posterior subcutaneous depot located around the inguinal fold actually consists of three continuous subcutaneous depots: dorsolumbar depot, stringently defined "inguinal" depot on the side, and gluteal depot. Beige adipocytes and beige preadipocytes are interspersed within the subcutaneous white fat depot at variable densities at different locations. Dissecting the whole fat pad instead of arbitrarily choosing parts of it will help to minimize variations between isolations.
5. A few lymph nodes can be observed in each inguinal fat depot, slightly darker in color and generally no larger than 1–2 mm in diameter. It is not necessary to remove these lymph nodes during the SVF isolation procedure, since the contaminating cell types (e.g., immune cells) will be removed through subsequent culture. Removal of lymph node prolongs the whole dissection process, which may reduce the viability of the precursors. Furthermore, increasing evidence suggests close interactions between adipocytes and the lymphatic system [21, 22], and therefore, experimental reproducibility may be compromised due to variable loss of surrounding fat tissue during lymph node removal.
6. Adipose tissues dissected from five to six mice (6–8 weeks of age) should provide enough preadipocytes for many functional assays after 2–4 rounds of subcultures. Procedures with much bigger cohorts of mice (e.g., >10) prolong the dissection process and may lead to decreased viability of isolated precursors.

7. The mincing process usually takes 5–10 min. Insufficient mincing will result in larger tissue chunks, extended digestion, and ultimately reduced yield of precursors; prolonged mincing will cause unnecessary mechanical stress and may lower viability and/or adipogenic potential of isolated precursors. Preliminary experiments are recommended to optimize these conditions.
8. When transferring minced tissue from the petri dish to the 50 mL tube, an eppendorf pipet (P-1000) with wide-orifice pipet tips can be used to avoid tissue loss sticking to the inside wall of serological pipets.
9. After digestion, ammonium chloride and potassium carbonate containing erythrocyte lysis buffer can be used to eliminate the red blood cells isolated together with the preadipocytes. However, this additional step is not necessary since these contaminating cell types will be removed through subculture.
10. Decanting supernatant instead of using vacuum aspiration helps to avoid loss of the pellet. Fatty acids released from mature adipocytes broken during the digestion often facilitate the formation and reformation of cell clumps. Always break up the pellet by tapping the bottom of the tube against the table before adding additional medium.
11. Higher concentration of trypsin–EDTA (e.g., 0.25%) can be used if 0.05% trypsin–EDTA is ineffective to detach cells, particularly during the first round of subculture.
12. Mitotic and adipogenic potentials of isolated precursors decrease in vitro. Two to four rounds of subculture will ensure the elimination of contaminated cell types from the culture, and the preadipocytes can still be readily differentiated into fat cells.
13. RNA isolated from cells within each well of 12-well plates is sufficient for qPCR gene expression analysis.

References

1. Freedman DS (2011) Obesity - United States, 1988-2008. MMWR Surveill Summ 60(Suppl):73–77. su6001a15 [pii]
2. Ogden CL, Carroll MD, Kit BK, Flegal KM (2014) Prevalence of childhood and adult obesity in the United States, 2011-2012. JAMA 311(8):806–814. https://doi.org/10.1001/jama.2014.732
3. Seale P, Bjork B, Yang W, Kajimura S, Chin S, Kuang S, Scime A, Devarakonda S, Conroe HM, Erdjument-Bromage H, Tempst P, Rudnicki MA, Beier DR, Spiegelman BM (2008) PRDM16 controls a brown fat/skeletal muscle switch. Nature 454(7207):961–967. https://doi.org/10.1038/nature07182. nature07182 [pii]
4. Lepper C, Fan CM (2010) Inducible lineage tracing of Pax7-descendant cells reveals embryonic origin of adult satellite cells. Genesis 48(7): 424–436. https://doi.org/10.1002/dvg.20630
5. Wu J, Bostrom P, Sparks LM, Ye L, Choi JH, Giang AH, Khandekar M, Virtanen KA, Nuutila P, Schaart G, Huang K, Tu H, van Marken Lichtenbelt WD, Hoeks J, Enerback S, Schrauwen P, Spiegelman BM (2012) Beige adipocytes are a distinct type of thermogenic fat cell in mouse and human. Cell 150(2):366–376. https://doi.org/10.1016/j.cell.2012.05.016
6. van Marken Lichtenbelt WD, Vanhommerig JW, Smulders NM, Drossaerts JM, Kemerink GJ, Bouvy ND, Schrauwen P, Teule GJ (2009)

Cold-activated brown adipose tissue in healthy men. N Engl J Med 360(15):1500–1508. https://doi.org/10.1056/NEJMoa0808718. 360/15/1500 [pii]
7. Cypess AM, Lehman S, Williams G, Tal I, Rodman D, Goldfine AB, Kuo FC, Palmer EL, Tseng YH, Doria A, Kolodny GM, Kahn CR (2009) Identification and importance of brown adipose tissue in adult humans. N Engl J Med 360(15):1509–1517. https://doi.org/10.1056/NEJMoa0810780. 360/15/1509 [pii]
8. Virtanen KA, Lidell ME, Orava J, Heglind M, Westergren R, Niemi T, Taittonen M, Laine J, Savisto NJ, Enerback S, Nuutila P (2009) Functional brown adipose tissue in healthy adults. N Engl J Med 360(15):1518–1525. https://doi.org/10.1056/NEJMoa0808949. 360/15/1518 [pii]
9. Sharp LZ, Shinoda K, Ohno H, Scheel DW, Tomoda E, Ruiz L, Hu H, Wang L, Pavlova Z, Gilsanz V, Kajimura S (2012) Human BAT possesses molecular signatures that resemble beige/brite cells. PLoS One 7(11):e49452. https://doi.org/10.1371/journal.pone.0049452
10. Lidell ME, Betz MJ, Dahlqvist Leinhard O, Heglind M, Elander L, Slawik M, Mussack T, Nilsson D, Romu T, Nuutila P, Virtanen KA, Beuschlein F, Persson A, Borga M, Enerback S (2013) Evidence for two types of brown adipose tissue in humans. Nat Med 19(5):631–634. https://doi.org/10.1038/nm.3017
11. Jespersen NZ, Larsen TJ, Peijs L, Daugaard S, Homoe P, Loft A, de Jong J, Mathur N, Cannon B, Nedergaard J, Pedersen BK, Moller K, Scheele C (2013) A classical brown adipose tissue mRNA signature partly overlaps with brite in the supraclavicular region of adult humans. Cell Metab 17(5):798–805. https://doi.org/10.1016/j.cmet.2013.04.011
12. Cypess AM, White AP, Vernochet C, Schulz TJ, Xue R, Sass CA, Huang TL, Roberts-Toler C, Weiner LS, Sze C, Chacko AT, Deschamps LN, Herder LM, Truchan N, Glasgow AL, Holman AR, Gavrila A, Hasselgren PO, Mori MA, Molla M, Tseng YH (2013) Anatomical localization, gene expression profiling and functional characterization of adult human neck brown fat. Nat Med 19(5):635–639. https://doi.org/10.1038/nm.3112
13. Shinoda K, Luijten IH, Hasegawa Y, Hong H, Sonne SB, Kim M, Xue R, Chondronikola M, Cypess AM, Tseng YH, Nedergaard J, Sidossis LS, Kajimura S (2015) Genetic and functional characterization of clonally derived adult human brown adipocytes. Nat Med 21(4):389–394. https://doi.org/10.1038/nm.3819
14. Yoneshiro T, Aita S, Matsushita M, Kayahara T, Kameya T, Kawai Y, Iwanaga T, Saito M (2013) Recruited brown adipose tissue as an antiobesity agent in humans. J Clin Invest 123(8):3404–3408. https://doi.org/10.1172/JCI67803
15. van der Lans AA, Hoeks J, Brans B, Vijgen GH, Visser MG, Vosselman MJ, Hansen J, Jorgensen JA, Wu J, Mottaghy FM, Schrauwen P, van Marken Lichtenbelt WD (2013) Cold acclimation recruits human brown fat and increases nonshivering thermogenesis. J Clin Invest 123(8):3395–3403. https://doi.org/10.1172/JCI68993
16. Lee P, Smith S, Linderman J, Courville AB, Brychta RJ, Dieckmann W, Werner CD, Chen KY, Celi FS (2014) Temperature-acclimated brown adipose tissue modulates insulin sensitivity in humans. Diabetes 63(11):3686–3698. https://doi.org/10.2337/db14-0513
17. Chondronikola M, Volpi E, Borsheim E, Porter C, Annamalai P, Enerback S, Lidell ME, Saraf MK, Labbe SM, Hurren NM, Yfanti C, Chao T, Andersen CR, Cesani F, Hawkins H, Sidossis LS (2014) Brown adipose tissue improves whole-body glucose homeostasis and insulin sensitivity in humans. Diabetes 63(12):4089–4099. https://doi.org/10.2337/db14-0746
18. Green H, Kehinde O (1974) Sublines of mouse 3t3 cells that accumulate lipid. Cell 1(3):113–116. https://doi.org/10.1016/0092-8674(74)90126-3
19. Green H, Kehinde O (1976) Spontaneous heritable changes leading to increased adipose conversion in 3T3 cells. Cell 7(1):105–113
20. Ross SR, Choy L, Graves RA, Fox N, Solevjeva V, Klaus S, Ricquier D, Spiegelman BM (1992) Hibernoma formation in transgenic mice and isolation of a brown adipocyte cell line expressing the uncoupling protein gene. Proc Natl Acad Sci U S A 89(16):7561–7565
21. Pond CM, Mattacks CA (2002) The activation of the adipose tissue associated with lymph nodes during the early stages of an immune response. Cytokine 17(3):131–139. https://doi.org/10.1006/cyto.2001.0999
22. Benezech C, Mader E, Desanti G, Khan M, Nakamura K, White A, Ware CF, Anderson G, Caamano JH (2012) Lymphotoxin-beta receptor signaling through NF-kappaB2-RelB pathway reprograms adipocyte precursors as lymph node stromal cells. Immunity 37(4):721–734. https://doi.org/10.1016/j.immuni.2012.06.010

Index

Bruce A. Bunnell and Jeffrey M. Gimble (eds.), *Adipose-Derived Stem Cells: Methods and Protocols*, Methods in Molecular Biology, vol. 1773, https://doi.org/10.1007/978-1-4939-7799-4, © Springer Science+Business Media, LLC, part of Springer Nature 2018

INTERNATIONAL ASTRONOMICAL UNION
UNION ASTRONOMIQUE INTERNATIONALE

International Astronomical Union

IAU

STAR CLUSTERS: BASIC GALACTIC BUILDING BLOCKS THROUGHOUT TIME AND SPACE

PROCEEDINGS OF THE 266th SYMPOSIUM OF
THE INTERNATIONAL ASTRONOMICAL UNION
HELD IN RIO DE JANEIRO, BRAZIL
10–14 AUGUST 2009

Edited by

RICHARD DE GRIJS
Kavli Institute for Astronomy and Astrophysics, Peking University, China
The University of Sheffield, United Kingdom

and

JACQUES R. D. LÉPINE
Universidade de São Paulo, Brazil

CAMBRIDGE
UNIVERSITY PRESS